# Network Innovation through OpenFlow and SDN

## Principles and Design

# OTHER COMMUNICATIONS BOOKS FROM AUERBACH

# Network Innovation through OpenFlow and SDN

## Principles and Design

Edited by
## FEI HU

CRC Press
Taylor & Francis Group
Boca Raton   London   New York

CRC Press is an imprint of the
Taylor & Francis Group, an **informa** business

CRC Press
Taylor & Francis Group
6000 Broken Sound Parkway NW, Suite 300
Boca Raton, FL 33487-2742

© 2014 by Taylor & Francis Group, LLC
CRC Press is an imprint of Taylor & Francis Group, an Informa business

No claim to original U.S. Government works

Version Date: 20131227

International Standard Book Number-13: 978-1-4665-7209-6 (Hardback)

**Visit the Taylor & Francis Web site at**
**http://www.taylorandfrancis.com**

**and the CRC Press Web site at**
**http://www.crcpress.com**

To Fang, Gloria, Edward and Edwin (twins)
I always love you all

# Contents

# Part IV Advanced Topics

# Preface

Software-defined network (SDN)/OpenFlow research and development have attracted the attention of many researchers and companies. The key idea of an SDN is to split the network forwarding function, performed by the data plane, from the network control function, performed by the control plane. This allows a simpler and more flexible network control and management, and also network virtualization. OpenFlow is the main SDN implementation. The network controller communicates with OpenFlow switches using the OpenFlow protocol through a secure channel. Using this connection, the controller is able to configure the forwarding tables of the switch. The figure (in the right) shows a simple OpenFlow architecture.

OpenFlow-based SDN technologies enable us to address the high-bandwidth, dynamic nature of computer networks; adapt the network functions to different business needs easily; and reduce network operations and management complexity significantly. Many large companies (such as Cisco, Microsoft, Google, etc.) have produced OpenFlow-supported products (such as switches).

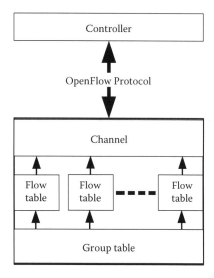

Open flow Switch

Needless to say, SDN/OpenFlow will become one of the most important trends for the future Internet and regional networks. In the future, you do not need to program, configure, or debug each network device individually. Instead, you can just sit in a centralized management office and customize your network based on the different needs in scientific experiments, business management, home network control, and local community communications. Those centralized control commands can be run in any vendor's switches and can be executed by a standard networking operating system. Isn't that awesome?

**Features of the Book**

Compared with other similar books, this book emphasizes both OpenFlow engineering design and basic principles. This is perhaps the first book that systematically discusses the different design aspects in an SDN/OpenFlow written by experts worldwide. It is different from some similar books that mainly introduce the basic principles. This book covers the entire system architecture, language and programming issues, switches, multimedia support, quality of service (QoS), network operating system, how to smoothly transfer from conventional networks to SDN/OpenFlow, security, OpenFlow for optical networks, and others.

**Targeted Audiences**

This book is suitable to the following types of readers:

(1) Industry/engineers: We have provided a detailed SDN/OpenFlow design process. Thus, company engineers could use those principles in their product design.
(2) College students: This book can serve as a textbook or reference book for college courses on advanced networking, especially on OpenFlow. Such courses could be offered in computer science, electrical and computer engineering, information technology and science, or other departments.
(3) Researchers: Because each chapter is written by top experts, the contents are very useful to researchers (such as graduate students and professors) who are interested in this field.

**Book Architecture**

This book uses four parts to cover the most important aspects in SDN/OpenFlow. Those four parts include basic concepts, engineering design, QoS, and advanced topics. The following flowchart shows the book organization.

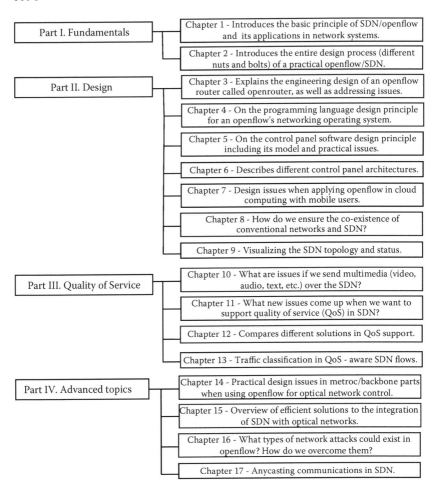

Part I. Fundamentals
- Chapter 1 - Introduces the basic principle of SDN/openflow and its applications in network systems.
- Chapter 2 - Introduces the entire design process (different nuts and bolts) of a practical openflow/SDN.

Part II. Design
- Chapter 3 - Explains the engineering design of an openflow router called openrouter, as well as addressing issues.
- Chapter 4 - On the programming language design principle for an openflow's networking operating system.
- Chapter 5 - On the control panel software design principle including its model and practical issues.
- Chapter 6 - Describes different control panel architectures.
- Chapter 7 - Design issues when applying openflow in cloud computing with mobile users.
- Chapter 8 - How do we ensure the co-existence of conventional networks and SDN?
- Chapter 9 - Visualizing the SDN topology and status.

Part III. Quality of Service
- Chapter 10 - What are issues if we send multimedia (video, audio, text, etc.) over the SDN?
- Chapter 11 - What new issues come up when we want to support quality of service (QoS) in SDN?
- Chapter 12 - Compares different solutions in QoS support.
- Chapter 13 - Traffic classification in QoS - aware SDN flows.

Part IV. Advanced topics
- Chapter 14 - Practical design issues in metroc/backbone parts when using openflow for optical network control.
- Chapter 15 - Overview of efficient solutions to the integration of SDN with optical networks.
- Chapter 16 - What types of network attacks could exist in openflow? How do we overcome them?
- Chapter 17 - Anycasting communications in SDN.

## Disclaimer

We have tried our best to provide credits to all cited publications in this book. Because of the time limit, this book could have some errors or missing contents. Moreover, we sincerely thank all the authors who have published materials and who directly/indirectly contributed to this book through our citations. If you have questions on the contents of this book, please contact the publisher and we will correct the errors and thus improve this book.

# Editor

 **Dr. Fei Hu** is an associate professor in the Department of Electrical and Computer Engineering at the University of Alabama (main campus), Tuscaloosa, Alabama. He obtained his PhD at Tongji University (Shanghai, People's Republic of China) in the field of signal processing (in 1999) and at Clarkson University (New York) in the field of electrical and computer engineering (in 2002). He has published more than 150 journal/conference articles and book chapters.

Dr. Hu's research has been supported by the U.S. National Science Foundation (NSF), Department of Defense (DoD), Cisco, Sprint, and other sources. His research expertise can be summarized as 3S: Security, Signals, Sensors. (1) Security: This is about how to overcome different cyber attacks in a complex wireless or wired network. Recently, he focused on cyber-physical system security and medical security issues. (2) Signals: This mainly refers to intelligent signal processing, that is, using machine learning algorithms to process sensing signals in a smart way to extract patterns (i.e., achieve pattern recognition). (3) Sensors: This includes microsensor design and wireless sensor networking issues.

# Contributors

**Antônio Abelém**
Federal University of Pará
Pará, Brazil

**Dharma P. Agrawal**
School of Computing Sciences
    and Informatics
University of Cincinnati
Cincinnati, Ohio

**Pedro A. Aranda Gutiérrez**
Telefónica, Investigación y
    Desarrollo
GCTO Unit Network
    Virtualisation
Madrid, Spain

**Jun Bi**
Institute for Network Sciences
    and Cyberspace
Tsinghua University
Beijing, China

**Eduardo Cerqueira**
Federal University of Pará
Pará, Brazil

**Carlos N. A. Corrêa**
Federal University of the State of
    Rio de Janeiro
Rio de Janeiro, Brazil

**Colby Dickerson**
Department of Electrical and
    Computer Engineering
The University of Alabama at
    Tuscaloosa
Tuscaloosa, Alabama

**Chuan Du**
Computer Network Information
    Center
Chinese Academy of Sciences
Beijing, China

**Yuepeng E**
Computer Network Information
  Center
Chinese Academy of Sciences
Beijing, China

**Fernando Farias**
Federal University of Pará
Pará, Brazil

**Muhammad Farooq**
Department of Electrical and
  Computer Engineering
The University of Alabama at
  Tuscaloosa
Tuscaloosa, Alabama

**Natalia Castro Fernandes**
Laboratório Mídiacom
Departamento de Engenharia de
  Telecomunicações
Universidade Federal
  Fluminense
Niterói, Rio de Janeiro, Brazil

**Xingang Fu**
Department of Electrical and
  Computer Engineering
The University of Alabama at
  Tuscaloosa
Tuscaloosa, Alabama

**Jingguo Ge**
Computer Network Information
  Center
Chinese Academy of Sciences
Beijing, China

**Ashley Gerrity**
Department of Electrical and
  Computer Engineering
The University of Alabama at
  Tuscaloosa
Tuscaloosa, Alabama

**Hongxiang Guo**
Beijing University of Posts and
  Telecommunications
Beijing, China

**Kuheli L. Haldar**
School of Computing Sciences
  and Informatics
University of Cincinnati
Cincinnati, Ohio

**Nagaraj Hegde**
Department of Electrical and
  Computer Engineering
The University of Alabama at
  Tuscaloosa
Tuscaloosa, Alabama

**Fei Hu**
Department of Electrical and
  Computer Engineering
The University of Alabama at
  Tuscaloosa
Tuscaloosa, Alabama

**Jay Iyer**
Cisco Systems, Inc.
San Jose, California

**Raj Jain**
Washington University in Saint
    Louis
St. Louis, Missouri

**Sunil Kumar**
Department of Electrical and
    Computer Engineering
San Diego State University
San Diego, California

**Sanping Li**
Department of Electrical and
    Computer Engineering
University of Massachusetts
    Lowell
Lowell, Massachusetts

**Lei Liu**
Department of Electrical and
    Computer Engineering
University of California Davis
Davis, California

**Diego R. Lopez**
Telefonica I+D
Madrid, Spain

**Sidney Lucena**
Federal University of the State of
    Rio de Janeiro
Rio de Janeiro, Brazil

**Yan Luo**
Department of Electrical and
    Computer Engineering
University of Massachusetts
    Lowell
Lowell, Massachusetts

**Rui Ma**
Department of Electrical and
    Computer Engineering
The University of Alabama at
    Tuscaloosa
Tuscaloosa, Alabama

**Luiz Claudio Schara Magalhães**
Laboratório Mídiacom
Departamento de Engenharia de
    Telecomunicações
Universidade Federal Fluminense
Niterói, Rio de Janeiro, Brazil

**Eric Murray**
Department of Electrical and
    Computer Engineering
University of Massachusetts
    Lowell
Lowell, Massachusetts

**Lyndon Y. Ong**
Ciena Corporation
Cupertino, California

**Dave Oran**
Cisco Systems, Inc.
San Jose, California

**Subharthi Paul**
Washington University in Saint
    Louis
St. Louis, Missouri

**Christian Esteve Rothenberg**
Faculty of Electrical and
    Computer Engineering
University of Campinas
Sao Paulo, Brazil

**Marcos Rogerio Salvador**
Telecomm. Research and
    Development Center
Sao Paulo, Brazil

**João Salvatti**
Federal University of Pará
Pará, Brazil

**Takehiro Tsuritani**
KDDI R&D Laboratories Inc.
Ohara Fujimino-shi, Saitama,
    Japan

**Allan Vidal**
Telecomm. Research and
    Development Center
Sao Paulo, Brazil

**Yulei Wu**
Computer Network Information
    Center
Chinese Academy of Sciences
Beijing, China

**Junling You**
Computer Network Information
    Center
Chinese Academy of Sciences
Beijing, China

**Ting Zhang**
Department of Electrical and
    Computer Engineering
The University of Alabama at
    Tuscaloosa
Tuscaloosa, Alabama

# PART I

# FUNDAMENTALS

# 1

# SDN/OpenFlow: Concepts and Applications

## ASHLEY GERRITY AND FEI HU

**Contents**

## Introduction

The Internet has become an important part of everyday life. From searching via Google to buying from Amazon to keeping up with friends on Facebook, it is a part of people's daily lives. Google has even become prolific enough to warrant becoming a dictionary term. Behind each of these multibillion dollar companies are networks that have to adapt to ever-changing needs. Because the Internet becomes more invaluable than ever, it is important for it to be more accessible and easier to maintain.

The networks, which make up the backbone of the Internet, need to adapt to changes, without being hugely labor intensive in hardware or software changes. By updating switches so that the data plane and the control plane can be separated, a centralized controller can create more optimized routes to appropriately forward traffic. These forwarding tables can use rule input to the controller to optimize routing and make networks more efficient.

With new protocols, such as OpenFlow, which is becoming more standard in the industry, software-defined network (SDN) is becoming easier to implement. SDN decouples the control plane from the data plane, thus allowing switches to compartmentalize their two main functions. The control plane generates the routing table, whereas the data plane, using the control plane tables, determines where the packets should be sent to [1]. This separation allows the network to be abstracted further, which simplifies networking analysis. Many companies use OpenFlow protocols within their data center networks to simplify operations. OpenFlow and SDN allow data centers and researchers to innovate their networks with new ways because it is easier to abstract the network.

**SDN Application 1: Internet Research**

The Internet was originally built for research. As such, during its origin, its architects never conceived of the vast network that it has now become. Security and mobility were not considered during its formation because computers were not mobile, and researchers wanted an open environment for ideas to spread. With the ever-increasing user base, the Internet is not as idyllic as when it was originally envisioned. Most of its main concepts have not changed much since its founding. There are ways to make it better and allow it to work with future hardware developments that will be developed according to Moore's law. As technology evolves, the Internet is not able to meet the new requirements as well as it could if it had been designed to do so from the start [2].

Updating the Internet brings many challenges because it is constantly being used; it is difficult to test new ideas and strategies to solve the problems found in an existing network. SDN technologies provide a means of testing ideas for a future Internet without changing the current network. In Ref. [2], it is pointed out that SDN "allows new networking protocols to be defined and readily experimented in real conditions over production networks." Because SDN allows control and data traffic to be separated with an OpenFlow switch, it is easier to separate the hardware from the software. This separation allows experimenting with new addressing schemes so that new Internet architecture schemes can be tested.

Generally, it is difficult to experiment with new types of networks. Because new types of networks often use different addressing schemes and include other nonstandard aspects, it is difficult to incorporate these changes into existing networks. OpenFlow allows routers, switches, and access points from many different companies to use the separation of the control and data planes. The devices forward data packets that were received based on defined rules from the controller. If the devices do not have a rule for the data packet that has arrived, the devices forward the packet to the controller for review. The controller determines what to do with the packet and, if necessary, sends a new rule to the device so that it can handle future data packets in the same manner.

OpenFlow has gone through much iteration, and it is currently on version 1.3; however, only version 1.0 is available for the hardware and the software. OpenFlow uses both hardware and software switches. Based on the headers in these fields, the controller can decide on what to do with the packet. The second and subsequent versions of OpenFlow changed the match structures so that the number and bit count of each header field could be specified. Thus, it would be easier to implement new protocols. A special controller is used to separate control bits from data bits, which allows the network infrastructure to be shared more easily [2]. A server is often used for the controller portion of OpenFlow systems.

Currently, there are several projects that use OpenFlow, which includes Europe and Brazil. In Europe, eight islands are currently interconnected using OpenFlow, whereas in Brazil, there are plans to create a network that will work with that in Europe to create a more widespread testbed. The project in Brazil is particularly important because replacing the Internet is a global endeavor, and a network that will only work with landmasses clustered together is not a viable solution [2].

Several other projects are currently being developed, including MobilityFirst, eXpresssive Internet Architecture, content centric internetworking architecture (CONET), and Entity Title Architecture. MobilityFirst determined mobility as the most important goal. It uses a globally unique identifier (GUID) mapped to a network address and a generalized storage-aware routing to handle data transfer at different quality links. Basically, if a link is good, then it immediately sends; otherwise, data are stored until it can be sent out. eXpressive uses a directed acyclic graph addressing scheme. This architecture is in a prototype stage using Linux, and it plans to move toward an OpenFlow-based approach. CONET, as the name suggests, is a content-centric approach. It uses two approaches: one of a clean slate packet and another that uses a traditional IPv4 or IPv6 packet. In this approach, an end node requests content, and the request is forwarded along until it reaches a node with the content in its cache or until it reaches a serving node for the content. The content is then taken in reverse back to the original node. Finally, entity title architecture uses an open form of communication to make the physical layer as efficient as possible. It is similar to user datagram protocol (UDP), where delivery, ordering, and other guarantees are only made if required. For example, if secrecy is required, then the means to set that up are provided, but extra steps have to be set up to ensure these aspects [2].

Each of these projects is important in creating a better Internet for current and, hopefully, future needs. The Internet has brought about many wonderful advantages that most people could not conceive of being without, but it has also brought about new threats, such as domain naming service (DNS) attacks, personal identification theft, and other security issues that were previously far less widespread or nonexistent. Perhaps one of these projects, or a future project, will help bring a less inherently flawed Internet, which will lead to even more innovations in the future.

**SDN Application 2: Rural Connectivity**

SDN has not just led to innovations in researching new Internet protocols; it also has many other applications, such as making the Internet more widespread than it already is. If a person lives away from vastly populated areas, it is almost impossible to use the Internet. Rural areas are often forgotten or ignored by large companies because the profit margins are small, and it is difficult to update the required network hardware [3].

SDN simplifies complex data center and enterprise networks; it can further be used to simplify rural Wi-Fi networks. The main issues with rural environments include sparse populations, small profit margins and resource constraints, and others. However, there have been many recent innovations that help alleviate these issues. These innovations include long-distance Wi-Fi, a larger unlicensed frequency band (e.g., adding the microwave spectrum), and SDN itself [3].

SDN is beneficial because it separates network construction and network configuration by placing the control and management functionality into the central controller. This separation allows companies to decrease startup costs in rural environments, thereby allowing them to gain more profit. As rural networks become more profitable, more companies will be willing to give access to more and more rural areas [3].

Separating the hardware and the software aspects of networking allows for flexible network control. SDN allows for virtualization of networks so that one can look at the network more abstractly and focus on the big picture rather than on all the gritty details. Furthermore, SDN would allow for network management to be conducted off-site by a separate company. Then, a local company needs only to focus on the physical construction and maintenance of the components needed for the network [3].

Another benefit of SDN would be that all of the rural networks would then be viewing the network from a global perspective, and the companies could then share software solutions with others, rather than rely on ad-hoc methods that are only applicable to a specific situation or environment [3].

SDN would allow rural networking companies to work with established companies rather than to compete. The rural companies could each work on the maintenance and construction side, whereas the more established telecommunication companies could work on the management of the rural networks. The idea of working together is foreign to the mindsets of many large corporations. Generally, companies are striving against one another, but this solution would allow both companies to mutually benefit in separate areas of expertise (hardware and software).

There are several difficulties involved in setting up this rural network solution. These difficulties include a network such as this that has never been set up in practice anywhere, and it is more difficult to set up controls in rural environments than in data centers, where it is easier to separate control traffic from data traffic [3]. A true test environment in a rural area would need to be conducted to find the feasibility of this solution. Companies would also have to learn accepting the idea that cooperation would be beneficial to all those involved.

### SDN Application 3: Updating Networks by Changing Silicon

SDN is limited by the hardware that can use it [4]. If one were to change the silicon from a common merchant silicon to one that was designed with SDN in mind, SDN could even be more beneficial.

The basic hardware limitations found with merchant silicon include small-size tables to implement SDN data planes, lack of support for flexible actions, slow bus speeds between pipeline and on-chip CPU, and slow on-chip CPUs. In Ref. [4], by using a combination of hardware and software, they implemented a solution that uses merchant hardware and an network processing unit (NPU)-drive software subsystem to keep costs down. They benchmarked common merchant silicon switch chips and found the limitations previously mentioned [4].

They were able to make their version of switches capable of P2P detection and prevention, detect intrusion, and encrypt on-demand flows. This solution was able to work well with the existing software implementation of SDN. Because it requires a hardware change, it could have potential issues in upgrading. Moreover, it cannot efficiently support networks with tight delay requirements. It also requires the two planes (data and control) to have consistent state information, which involves communication between the controller and the switch.

Upgrade issues as well as the sharing of current state information may be too difficult to overcome. Further testing would need to be implemented to see if this solution is viable. Different switch and hardware companies would each have to use the new silicon solution to create a widespread adoption of it. If all of the major companies do not upgrade their hardware, then the solution will never move past small trials.

### SDN Application 4: Updating Data Center Networks

It is currently difficult to connect different data center networks. Often, data center networks use proprietary architectures and topologies, which create issues when merging different networks; however, there is often a need to merge two divergent networks. For example, when a new company is acquired, the original network must be expanded to incorporate the new network, but with the aforementioned problems, this merge is often time consuming.

SDN brings a solution to this networking issue. In Ref. [5], they proposed to use an OpenFlow-based network infrastructure service software to connect data center networks. They further stated that these interconnected data center networks could solve problems with latency by moving the workload to underused networks. If a network is busy at a certain time of the day, the workload might be completed sooner in a network of a different time zone or in a network that is more energy efficient.

They also tested their solution by using two data center network topologies. The established infrastructure service software application sets OpenFlow rules for network connectivity under different operations. Generally, it would take a long time for the switches to discover all of the rules, so they created a resource description that contained all of the available data center resources. Two different types of rules that were used are global and specific. Global rules are based on the data center topology, whereas specific rules determine how the operations are handled by the hardware.

They found some limitations in their solution. It was found that it is difficult to scale because it is trying to match a large number of packet fields of multiple protocols at different layers [5]. Furthermore, in severe cases, the lookup times can be slow because of collisions.

Such an approach requires that a number of rules be established. This number of rules is proportional to the number of switches and servers in the network. As more rules are introduced, it is more difficult to ensure that every rule is valid, and none of the rules conflicts with a previously defined rule. They, however, created a generic set of configuration rules for data centers, which allows for data centers using OpenFlow to be interconnected easily. The original setup time of the OpenFlow rules is in the 20- to 40-ms time frame, depending on whether the network is a standard network or a virtual machine, respectively.

In the future, we should test this approach in a real-world situation or, at least, create a virtual testbed with more than two data centers. Google has at least 12 data centers; although many companies have fewer data centers, it would be a good idea to at least test this software on more than two data centers.

**SDN Application 5: Cloud Data Scalability**

As previously mentioned, data centers are an integral part of many companies. For example, Google has a large number of data centers, so they can quickly and accurately provide data when requested. Similarly, many other companies use data centers to provide data to clients in a quick and efficient manner, but data centers are expensive to maintain. OpenFlow allows companies to save money on setting up and configuring networks because it allows switches to be managed from a central location [1].

OpenFlow allows testing of new routing techniques and changes in protocols without affecting anything in a production environment. This ability to test new ideas allows researchers to invent new networking protocols that can work more efficiently for many applications [1,2]. In Ref. [1], it would be interesting to see if cloud data centers could be more scalable and faster to set up if they used OpenFlow.

In Ref. [1], a data center model is created with a large number of nodes to test performance, throughput, and bandwidth. The model included 192 nodes with four regular switches and two core switches with an OpenFlow controller. There was a firewall between the core switches, the OpenFlow controller, and the router. They also used Mininet to prototype their network and test the performance. Mininet is an application that allows researchers to customize an SDN using OpenFlow protocols. Furthermore, they used several tools to analyze their network setup, including Iperf, Ping, PingAll, PingPair, and CBench. These tools allow people to check the possible bandwidth in the network, the connectivity among nodes, the connectivity between the deepest nodes, and the speed in which flows can be changed in a network. Wireshark was also used to view traffic in the network.

After they tested their network using all the tools previously described, they found that OpenFlow would be good to use in data centers because of its ability to reduce switch configuration time as well as to be controlled from a centralized controller. They recommend that data centers wait on OpenFlow to be tested further before changing all their switches to OpenFlow switches. Because their testing was done on a virtual machine using Mininet, it would need to be tested further in a real environment to ensure that all results are realistic.

### SDN Application 6: Mobile Application Offloading

Mobile devices have become increasingly popular both for general consumers and for businesses in recent years. To make mobile devices truly useful in a business environment, they need to securely send data to other servers and other work machines. Because battery life and performance are critical on mobile platforms, any additional software necessary for security would need to be lightweight enough to not impede the functioning of the device.

Two important considerations to make when creating offloading software for mobile devices are privacy and resource sharing [6]. Privacy is important for business applications because people often work on data that need to be kept secure. Some data can be sent among only a few people, whereas other data do not require the same level of security. The ability to determine which data require additional security is important for mobile offloading applications. In addition, resource sharing is important for mobile offloading because it allows mobile devices to exploit the machines that are idle or not fully using their capabilities.

In Ref. [6], they used an enterprise-centric offloading system (ECOS) to address these concerns of security and resource sharing. The controller of the system needed to do most of the computational work to relieve mobile devices of this necessity. ECOS was designed to offload data to idle computers while ensuring that applications with additional security requirements are only offloaded on approved machines. Performance was also considered for different users and applications [6].

ECOS reviews the available idle computing resources to find an appropriate resource to offload data. The machine chosen is required to provide a benefit in either latency or energy savings or both. Furthermore, the machine chosen has to meet the minimum security requirements for the particular application. Three security categories for the applications are user-private data, enterprise-private data, and no-private data. Only certain users can access user-private data, whereas enterprise-private data need only remain within the business environment. No-private data require no security because they can be accessed by anyone [6].

SDN is used to control the network and to select resources. It is a good choice because it is less resource intensive than the other choices.

The selected resources must be able to meet the security requirements listed above. The controller will determine if such a device is available for offloading that meets the security requirements while maintaining energy savings. If there is no such device, data are not allowed to be offloaded from the mobile device. If energy savings are not necessary, then any available resource with enough capacity is used. OpenFlow switches are used so that the controller can regulate the flows.

ECOS was able to provide decision-making controls to offloading while not ignoring security or energy-saving requirements. It is able to use the benefits of SDNs. In Ref. [6], they showed that ECOS was able to offload while considering security requirements without an overly complex scheme. Furthermore, ECOS was shown to scale to accommodate many mobile devices.

### SDN Application 7: Mobile Virtual Machines

Applications running on virtual machines in businesses are becoming increasingly common [7]. These virtual machines allow companies to be more flexible and to have lower operational costs. To extract the full potential from a virtual machine, there needs to be a means of making them more portable. The main issue with making virtual machines portable is the need to maintain the IP address of the virtual machine in the process. To maintain the same IP address, the virtual machine would have to remain in the same subnet or a new scheme would need to be created.

It was found that the current methods of handling virtual machines were not efficient. The choice of solutions that they found in Ref. [7] includes using a mobile IP or a dynamic DNS. The main issue with both of these solutions is that someone has to manually reconfigure the network settings after removing the virtual machine. This limits businesses and data centers from easily porting their virtual machines to new locations [7].

CrossRoads was an application developed by Mann et al. [7] to solve the mobility issue for virtual machines. It was designed to allow the mobility of both live and offline virtual machines. It has three main purposes. The first purpose is to take care of traffic from data centers as well as external users. The second purpose is to make use of OpenFlow with the assumption that each data center uses an

OpenFlow controller. The third purpose is to make use of pseudo addresses for IP and MAC addresses so that they remain constant when porting while allowing the real IP to change accordingly [7].

The basic implementation of their software was to create rules for finding the virtual machines in different networks. The CrossRoads controller would keep track of the real IP and MAC addresses for the controllers in each data center as well as the virtual machines in its own network. When a request is sent for an application running on a particular virtual machine, a request is broadcasted to the controllers. If the controller receives a request for a virtual machine that is not in its table, it then broadcasts the request to the other controllers; the controller who has the real IP address of the virtual machine then sends the pseudo MAC address to the original controller, and the original controller can update its table in case it gets another request in the near future.

To test their implementation, Mann et al. [7] set up a virtual testing environment. This virtual environment consisted of two data centers, which were used to understand how CrossRoads performs in a realistic environment. They ran four 15-min experiments to test the performance of their network. They were able to decrease the address resolution protocol (ARP) latency by 30% while maintaining similar ping times, available bandwidth, and HTTP response time with respect to a default testbed, where the virtual machines are all in the same subnet [7].

## Discussion

SDN has been shown to be a valuable resource in many different types of applications. From Internet research to many data center applications, SDN has found a wide array of useful possibilities. SDN allows users to quickly adapt networks to new situations as well as to test new protocols for production networks. Different network situations can be tested using OpenFlow and SDN.

Table 1.1 shows some differences between the applications previously discussed. As one can see from the table, OpenFlow was used in most of the applications for its versatility. Data centers were also a topic for many of these applications. Data centers continue to become an important part of the Internet and many large companies. The column for mobile applications in Table 1.1 refers to cell phones, tablets,

**Table 1.1**   Differences between Different Applications

| APPLICATIONS | OPENFLOW? | DATA CENTERS? | MOBILE PLATFORM? | CLOUD? | HARDWARE CHANGE? |
|---|---|---|---|---|---|
| Internet research | Yes | Yes/for all that uses the Internet | Yes | No | Maybe |
| Rural connectivity | No | No | No | No | No |
| Changing silicon | No | Yes/could be utilized by data centers and all other users | No | No | Yes |
| Data centers | Yes | Yes | No | No | No |
| Cloud data centers | Yes | Yes | Yes | Yes | No |
| Mobile applications | Yes | No | Yes | No | No |
| Virtual machines | Yes | Yes | Yes | Yes | No |

and other nontraditional media formats rather than laptops and other typical computing platforms. A few of the applications use cloud for their platforms. Hardware changes are difficult to implement in large-scale changes. One of the applications calls for a hardware change to make SDN work better. Hardware changes are often not practical because they require system shutdown during upgrade.

### Conclusion

SDNs have many applications, including researching new protocols prior to implementing them in real networks, increasing connectivity in rural environments, making both cloud-based and regular data centers better, and supporting mobile device offloading. As the Internet continues to grow and becomes available to more and more people, networks will need to adapt to ever-changing circumstances. SDNs allow the data and control planes to be separated to make it easier for improvements.

## References

1. Baker, C., A. Anjum, R. Hill, N. Bessis, and S. L. Kiani. 2012. Improving cloud data center scalability, agility and performance using OpenFlow. *2012 4th International Conference on Intelligent Networking and Collaborative Systems (INCoS)*. IEEE. Bucharest, Romania, pp. 20–27.

2. de Oliverira Silva, F., J. H. de Souza Pereira, P. F. Rosa, and S. T. Kofuji. 2012. Enabling future Internet architecture research and experimentation by using software-defined networking. *2012 European Workshop on Software-Defined Networking (EWSDN)*. IEEE, Darmstadt, Germany, pp. 73–78.

3. Hasan, S., Y. Ben-David, C. Scott, E. Brewer, and S. Shenker. 2013. Enhancing rural connectivity with software-defined networks. *Proceedings of the 3rd ACM Symposium on Computing for Development*. ACM, New York, Article 49, 2 pp.

4. Narayanan, R., S. Kotha, G. Lin, A. Khan, S. Rizvi, W. Javed, H. Khan et al. 2012. Macroflows and microflows: Enabling rapid network innovation through a split SDN data plane. *2012 European Workshop on Software-Defined Networking (EWSDN)*. IEEE, Darmstadt, Germany, pp. 79–84.

5. Boughzala, B., R. Ben Ali, M. Lemay, Y. Lemieux, and O. Cherkaoui. 2011. OpenFlow supporting interdomain virtual machine migration. *2011 8th International Conference on Wireless and Optical Communications Networks (WOCN)*. IEEE. Paris, France, pp. 1–7.

6. Gember, A., C. Dragga, and A. Akella. 2012. ECOS: Leveraging software-defined networks to support mobile application offloading. *Proceedings of the 8th ACM/IEEE Symposium on Architectures for Networking and Communications Systems*. ACM, New York, pp. 199–210.

7. Mann, V., A. Vishnoi, K. Kannan, and S. Kalyanaraman. 2012. CrossRoads: Seamless VM mobility across data centers through software-defined networking. *Network Operations and Management Symposium (NOMS)*. IEEE. Maui, HI, pp. 88–96.

# 2

# AN OPENFLOW NETWORK DESIGN CYCLE

## PEDRO A. ARANDA GUTIÉRREZ AND DIEGO R. LOPEZ

**Contents**

## Introduction

OpenFlow [1] has matured from an academic experiment to a networking tool that is starting to be deployed in real-world scenarios, reflected in an increasing number of use cases for OpenFlow, which have moved from initial exploration and demonstration to actual

**17**

deployments in production environments. Meanwhile, new potential applications of these technologies are being proposed in different areas, from security to transport networks (NWs). In addition to the features offered by OpenFlow itself, another significant factor has been the availability of simple-to-use NW emulation environments, as well as development tools, which have allowed users to gain early hands-on experience.

This reflects the current trend in networking technologies, which have become more softwarized. An NW design cycle that is based on current software (SW) application development practices seems to be appropriate in this context. Research projects, such as 4WARD [2], have explored and confirmed the applicability of SW design methodologies in the design and deployment cycle of telecommunication NWs.

In this chapter, we show how to combine Mininet [3], available OpenFlow development tools, and OpenFlow implementations on commercial off-the-shelf (COTS) equipment to speed up the NW design cycle in OpenFlow environments. We examine the state of the art in SW development tools in general and in OpenFlow specific tools. We then present our current OpenFlow-in-a-box solution as an enabler for proof-of-concept (PoC) environment for OpenFlow use cases. Finally, we examine how the gap between current open-source–based solutions and commercial-grade products can be closed in the OpenFlow ecosystem to produce a full NW design cycle.

We believe that this combination will be useful initially in situations where a fast assessment of OpenFlow applicability and/or a quick evaluation of the features of the resulting solution are required. We also believe that this constitutes the first step toward a paradigm shift on how NWs are conceived, designed, and operated.

**State of the Art**

Software-defined NWs (SDNs) imply a paradigm shift on how NWs are designed and operated. They are the result of a general trend, where hardware (HW) functionality is being replaced by SW implementations. This evolution has been fostered not only by a continuous increase in computing power and competitiveness of SW-based solutions, but also by an evolution in the SW development world toward conceptually sound procedures and easy-to-use tools. Let us analyze the state of the art of SW design frameworks and SW generation tools in general and of these aspects in OpenFlow NWs and components in particular.

*SW Design Cycle*

The design and implementation of a piece of SW, be it a full program or a module in a larger SW system, has not changed much in philosophy from the moment when computer programs started to be written. It follows a cyclic methodology that is similar with a generic system design and implementation cycle. The steps, which are shown in Figure 2.1, are as follows:

1. Design
2. Coding and compiling
3. Unitary tests

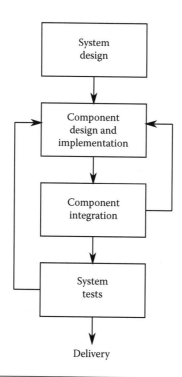

**Figure 2.1**   Systems design and implementation cycle.

To perform unitary tests, different tools are used. The SW component is a system that has to respond in a specific way to different stimuli, and the unitary tests have to make sure that the system responds adequately.

This cycle corresponds to the write-make-test motto, coined by the pioneers of the Unix operating system back in the early 1980s, and all SW development environments apply it, in combination with mechanisms to maintain the successive versions of SW elements as they evolve along the cycle iterations.

Along these cycle iterations, SW debuggers are an essential tool. Debuggers are sophisticated tools that allow the programmer to interact with the program while it is executing on the computer. gdb [4] is one example of an SW debugger. It is a command-line interface (CLI) tool that allows the programmer to load a program, stop the program on demand using breakpoints, and inspect the program state. One of the main advantages of debuggers, such as gdb, is that they map the execution state to the source code of the program. Variables and codes are shown as written by the programmer and not in their binary representation.

A next level in usability is provided by integrated development environments (IDEs). They provide a unique, comprehensive access mechanism to the SW development cycle. They do so by integrating design and coding facilities, version control, and executable code-building options with sophisticated debugging tools. This way, complex building and debugging tasks are hidden from the user, who interacts with the development tools through a graphical user interface (GUI). The integration of an IDE, such as Eclipse [5] and gdb, is shown in ref. [6].

*Component-Based SW Engineering*   Component-based SW engineering (CBSE) [7] is a discipline of SW engineering that was examined in the context of the 4WARD project [2] as a facilitator in the design and development of future-generation NWs.

In general, CBSE provides the basic tools in SW design cycle and enhances its productivity by

1. Abstracting the development of systems as assemblies of components

2. Stressing on reusability, understanding that components are reusable entities and that their development needs to be seen under this perspective
3. Providing a life cycle where systems are maintained and upgraded by customizing or replacing their components

In this context, a component-based architecture (CBA) defines the components, their relationships, and their functionality within the architectural framework defined for the application or applications that are designed using CBSE. Figure 2.1 shows a basic system design cycle using CBSE. During the system design phase, components are identified, and their interaction with the rest of the system is specified. Then, each component is designed, implemented, and tested. Once the component is ready for delivery, it is handed over to a component integration phase, where the full system is assembled. During this phase, design flaws in different components may be identified, and those components need to go back to the component design and implementation phase. Finally, once the full system has been assembled, it also needs to pass an array of system tests. During this phase, further design flaws may be identified, and the components' design and implementation may need improvements. Once the system passes all system tests, it can be delivered. During this process, design patterns and components that are widely used throughout the design cycle are identified and stored for reuse. The 4WARD project identified the need for a design repository (DR) for this purpose.

The Open Services Gateway initiative (OSGi) [8] is a consortium that is specifying a framework to build and maintain Java applications based on modules. Modules are a specific way of understanding the component concept. OSGi proposes a full life cycle for SW modules, as shown in Figure 2.2. This cycle allows programmers and integrators to code, debug, release, and maintain SW modules in Java applications. This SW-building methodology is being used in several companies that include "infrastructure/NW operators, utilities, enterprise SW vendors, SW developers, gateway suppliers, consumer electronics/device suppliers (wired/wireless), and research institutions" as stated in the OSGi Web portal [9].

The Agile SW development methodology [10] has contributed to this overall trend of dynamicity in the design and feedback between

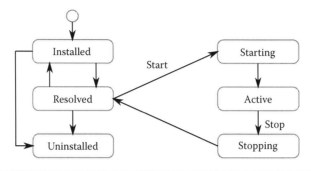

**Figure 2.2** Life cycle of an OSGi module.

the different stages in the SW production cycle. By reducing the scope of each development cycle, it has even allowed more feedback between different development cycles than classic waterfall methodologies [11], which intended to perform just one or, at most, a few iterations of the development cycle.

*OpenFlow Tools*

A basic OpenFlow test environment requires some essential components in both data and control planes. A forwarding node is the basic building element of the OpenFlow NW; it may be a commercial HW-based node or an SW-based implementation. On the other hand, the control plane consists of OpenFlow controllers.

*OpenFlow Switches and Controllers*   Many open-source projects have emerged to help build OpenFlow NWs; both controllers and switches are freely available under different licenses. One of the most popular SW-based OpenFlow switch implementation is the OpenVSwitch (OVS) [12], which is integrated in Linux kernel 3.3; firmwares, such as Pica8 [13] and Indigo [14], are also available.

There are also many open-source OpenFlow controllers for almost every development environment. Nox [15] written in C; Pox [Pox-15], in Python; Floodlight [16], in Java; and Trema [17], for Ruby developers are some of the available controllers. Most of these controllers are part of more-or-less ambitious OpenFlow development frameworks.

*Nox*   Nox [15] was the first reference implementation of an OpenFlow controller. This implementation, referred to as Nox Classic, was programmed in Python and C++. The current implementation is C++ only and, as stated in the Nox Web site, aims at implementing a C++ application programming interface (API) for OpenFlow 1.0, which provides a fast, asynchronous I/O. It is targeted at recent Linux distributions and includes sample components that implement topology discovery, a learning switch, and a NW-wide switch that is distributed over all the OpenFlow switches controlled by an instance of Nox.

*Pox*   Pox [18] is an evolution of the Python OpenFlow interface that has been dropped from Nox and that has been a very fast and convenient entry path to OpenFlow networking and SDN in general. Besides implementing the Python OpenFlow API, Pox provides sample components as Nox does; runs on multiple platforms, including Linux, Mac OS, and Windows; and supports the same GUI and visualization tools as Nox. Benchmarks shown in the Web site confirm that Pox provides an improvement over Nox applications written in Python.

*Java-Based OpenFlow Controllers: Beacon and Floodlight*   Beacon [19] is an OpenFlow controller implementation written in Java, which is integrated in OSGi frameworks (such as Equinox or SpringSource). The Floodlight controller [20] was derived from Beacon, and OSGi integration was stripped to ease the access to developers that are not familiar with it. Both provide different modules, with sample implementations of different NW functions that make it easy for a new developer to start coding. One of the key distinctive features of Floodlight is that it offers a bridge toward virtualization environments based on the OpenStack [21,22] framework.

Both Beacon and Floodlight are reported to be OpenFlow controllers with very high performance [23]. A distinctive feature for Beacon is the integration of the OSGi framework in Beacon, which allows a code bundle to be added, replaced, or removed at run time, without interrupting NW applications that do not rely on it.

*Trema*   Trema is an OpenFlow controller framework. The OpenFlow controller accepts applications written in Ruby and C. In addition to the OpenFlow controller, Trema is shipped with an integrated

OpenFlow NW emulator and OpenFlow traffic analysis tools based on Wireshark [24].

*Simulating OpenFlow on NS-3* NS-3 [25] includes support for OpenFlow switching. It relies on building the OpenFlow SW implementation distribution (OF-SID) [26] and integrates into the simulation environment using NS-3 wrappers. At the point of this writing, the OpenFlow support status in NS-3 is (1) version 1.0 of the OpenFlow protocol and (2) a version of OF-SID that supports MPLS. This version was created by Ericsson. However, the integration with NS-3 is limited because there is no way to pass MPLS packets to the NS-3 OpenFlow switch implementation.

*OpenFlow NW Emulation Environments* Mininet [3] is an NW emulation platform for quick NW testbed setup; it allows the creation of large (up to 1024 nodes) OpenFlow NWs in a single machine. Nodes are created as processes in separate NW namespaces using Linux NW namespaces [27], a lightweight virtualization feature that provides individual processes with separate NW interfaces, routing tables, and ARP tables. Nodes are connected via virtual Ethernet pairs. Mininet supports binding to external controllers and provides a front end for OpenFlow soft switches, such as the Stanford reference implementation and the OVS. Mininet has a full support for OF1.0 and a minimal support for OF1.1 using OVS [28].

*Applications Beyond the Pure OpenFlow Control Plane* In addition, other OpenFlow-related projects mainly focus on additional use cases. FlowVisor [29] provides specific NW virtualization, and RouteFlow [30] investigates how to integrate IP routing features in the OpenFlow control plane.

*Programming Languages* When controlling the NW becomes an SW matter, the natural consequence is to look for high-level languages that provide a level of abstraction powerful enough to simplify programming and achieve more complex behaviors. In this respect, we can consider OpenFlow as similar with a processor instruction set in normal IT, whereas several NW programming languages have already been proposed.

Flow-based modeling language (FML) [31] is a policy language for the Nox controller, which allows operators to declaratively specify NW policies. It provides a high-level declarative policy language based on logic programming. An FML program consists of a collection of inference rules that collectively define a function that assigns a set of flow constraints to each packet.

Procera [32] provides a controller architecture and a high-level NW control language that allows operators to express NW behavior policies, without resorting to the general-purpose programming of an NW controller. Procera is designed to be reactive, supporting how most common NW policies react to dynamic changes in NW conditions. Procera incorporates events that originate from sources other than OpenFlow switches, allowing it to express a policy that reacts to conditions, such as user authentications, time of the day, bandwidth use, or server load.

The Frenetic language [33] is intended to support the following essential features in the context of OpenFlow NWs:

1. High-level abstractions that give programmers direct control over the NW, allowing them to specify what they want the NW to do
2. Modular constructs that facilitate compositional reasoning about programs
3. Portability, allowing programs written for one platform to be reused with different devices
4. Rigorous semantic foundations that precisely document the meaning of the language and provide a solid platform for mechanical program analysis tools

The current Frenetic framework prototype combines a streaming declarative query sublanguage and a functional reactive sublanguage that, together, provide many of the features listed above.

*OpenFlow Debuggers*　Another component for an OpenFlow NW-application development cycle is the debugger. The task of controlling how packets flow in an OpenFlow-based NW is performed by an application running in the OpenFlow controller. Tools to control that this application works as expected and to isolate malfunctioning parts of it are needed. These tools fall into the same category of tools as debuggers in the SW development cycle.

Handigol et al. [34] have proposed an OpenFlow debugger, ndb. This tool mimics the GNU debugger, gdb [4], in the SW development cycle in some aspects. It is a command-line tool that loads an OpenFlow application and allows it to run natively in the OpenFlow controller. It uses breakpoints and backtraces to inspect how an application is behaving. The intended use of ndb are as follows:

1. The debugger loads the NW application.
2. The programmer sets breakpoints at specific points in the NW application, where the state wants to be observed. This normally happens when the NW application did not handle a specific NW packet.
3. The NW application is launched, and a specific NW debugging traffic is launched.
4. The debugger halts the application at a breakpoint.
5. Controller state and test packets that led to the NW application hitting the breakpoint are stored.
6. The debugger hands control to the programmer who has access to the state and the packets that were being processed when the application hits the breakpoint.

As mentioned above, Trema [17] and other OpenFlow controllers go beyond the mere implementation of an OpenFlow controller and provide an OpenFlow development framework. In the specific case of Trema, it includes a Wireshark [24] plugin to dissect OpenFlow packets [35] and a bridge to display OpenFlow events on Wireshark, known as Tremashark [36], as debugging tools. The next step in this development is provided by OFRewind [37]. This tool collects NW traces that can be replayed afterward to detect configuration or SW errors in an NW. In addition, tools to check the different components in the OpenFlow ecosystem. FlowChecker [38] provides a method to check the flow tables within an OpenFlow switch.

A very comprehensive approach to OpenFlow testing should be provided by the final version of NICE [39]. This tool, which is available in a preliminary version [40], is designed to test NW applications on an OpenFlow controller that controls several switches. Techniques, such as code analysis, will let the program determine the classes of packets that receive the same treatment in the controller and generate stimuli that cover the whole OpenFlow application under test. This tool will also create an NW model using an iterative approach to refine the model and cover the application fully.

*OpenFlow Testbeds*    The creation of easy-to-manage, scalable, and programmable OpenFlow-enabled testbeds has been a critical requirement for the growth of the OpenFlow ecosystem. Besides the possibility of setting up emulated testbeds, such as Mininet, Trema, and NS-3, there was a need for more realistic programmable testbed projects, based on NW equipment, as it became available. These test NWs allow end users to test their applications running on the testbed and allow NW equipment developers to perform an early but realistic deployment of their new developments. OpenFlow testbeds have been deployed worldwide and include GENI [41] in the United States, OFELIA [42] in the European Union, and JGN [43] in Japan.

GENI supports at-scale experimentation on a shared, heterogeneous, highly instrumented infrastructure [44]; accessing the facility requires registration and previous approval from the testbed authorities, and then the experimenter will be assigned login information to a Web-based user interface to create and run experiments. OFELIA is intended to create a unique experimental facility that allows researchers to not only experiment on a test NW but also to control and extend the NW itself dynamically. The OFELIA infrastructure is based on a set of interconnected *islands*, each of them managed independently, available to users under a common control framework. To achieve secure remote access, these facilities require the use of encrypted connections (such as OpenVPN in the case of OFELIA). Some apply special policies to the approval procedure. For example, the JGN facility [45] requires that the group proposing an experiment to be conducted on it has, at least, one Japan-based partner.

## OpenFlow in a Box

The introduction of OpenFlow in telco environments was not an easy task. As a new technology, different use cases need to be tested, involving different NW setups and application scenarios for PoC. However, the access and integration of equipment implementing OpenFlow into production and preproduction NW environments was restricted.

Therefore, we needed an alternative for PoC implementations. On the other hand, OpenFlow controllers, such as Trema or Nox, include frameworks with some simulation or emulation capabilities. This feature sets allowed us to kick-start research activities on OpenFlow. However, they are far away of an IDE that would allow an agile NW development cycle. This motivated the development of an environment suitable for early test and able to evolve by successive steps into a full-fledged NW IDE: OpenFlow in a box.

The basic requirement for our OpenFlow-in-a-box system was flexibility because we planned to use it on both functional and initial performance tests. Functional tests can be implemented in an NW emulation environment, whereas initial performance tests imply the use of actual NW equipment. Figure 2.3 shows the OpenFlow-in-a-box concept. The OpenFlow-in-a-box IDE controls different back ends implementing the NW functionality: (1) a local NW emulation environment, for example, our Mininet OpenFlow environment; (2) real NW test environments implemented with COTS HW, which implement SDN nodes; and (3) real NW environments with carrier-grade OpenFlow NW equipment as it becomes available. The current version controls nodes based on the Xen Cloud platform [46], using the Xen [47] hypervisor and the OVS OpenFlow switch. Tests with commercial OpenFlow equipment are planned. Wrapping the emulation and testing environments, a GUI provides users the necessary interactive functionalities to simplify the NW definition and operation tasks.

The emulated environment that we use is Mininet, the NW emulation environment introduced in the previous section, written in Python with a small core of C code. Mininet offers an extensible Python API for NW creation and experimentation. Our GUI controls both the local, emulated OpenFlow NW using Mininet and the remote virtualized environment. Its look and feel is shown in Figure 2.4. We use a mid-level API that allows us to flexibly build custom topologies with no restrictions in NW size, number of links, or ports. We connect to Mininet-emulated hosts through virtual terminals and access switch instances using the dpctl tool to query installed rules or to manually install new flows. The GUI allows us to start per-host services, such as Web or directory naming service (DNS), by means of simple drag-and-drop actions. It also

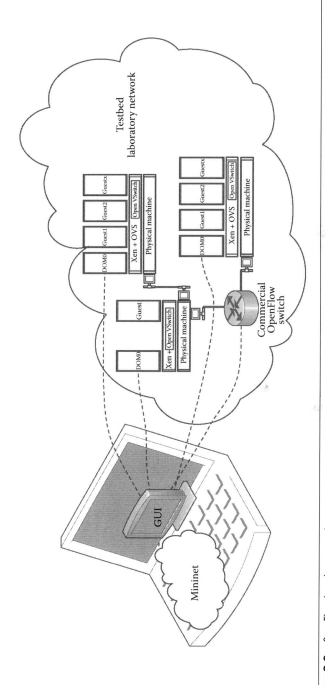

**Figure 2.3**   OpenFlow-in-a-box concept.

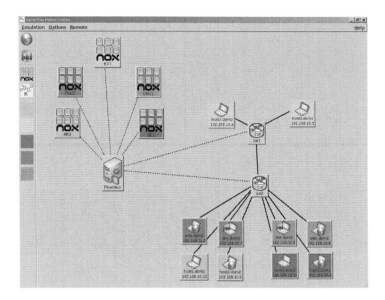

**Figure 2.4**  OpenFlow-in-a-box GUI look and feel.

makes possible the provision of different controllers (currently Nox, Pox, and Floodlight), with possibly different functionalities, as well as the integration of FlowVisor for NW slicing. Initially, NW scenarios are described by means of XML files.

The main objective of the GUI is to model the PoC scenarios for the different use cases. The use cases are expressed as a topology that interconnects the basic functional blocks: (1) OpenFlow controllers, (2) OpenFlow NW sliders, (3) OpenFlow switches, and (4) end hosts that provide the applications that we want to test in the OpenFlow-based scenario. Figure 2.5 shows the concept behind the GUI. It allows us to run different scenarios, which are described in an XML file. The OpenFlow-in-a-box environment loads, parses, and renders the scenarios according to the description file. Different OpenFlow controller types can be used independently from the scenario. Developers can drag and drop the specific OpenFlow controller into the button that implements the OpenFlow controller functionality in the scenario, and an instance of the selected OpenFlow controller will be launched with a scenario-specific configuration (e.g., the OpenFlow applications that are executed in the controller). In addition, different kinds of NW services (NS, in the figure) can be activated on demand. In our current

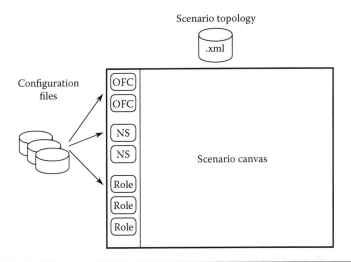

**Figure 2.5**  Mapping scenarios to the GUI. NS, network service; OFC, OpenFlow controller.

implementation, DNS and World Wide Web (WWW) servers can be deployed in the end hosts using a drag-and-drop gesture. These services are also configured using a scenario-specific configuration file. Finally, the GUI allows us to group the end hosts depending on the functions or roles that they implement in a scenario. So, for example, in a PoC of an OpenFlow-driven data center, we can group virtual machines by tenant and by functionality (e.g., machine in the back end, Web, or DNS server in the client's demilitarized zone [DMZ], etc.).

The GUI, which is shown in Figure 2.4, allows a smooth migration to a real HW scenario once it has been tested on Mininet. From the same GUI, we can remotely control our virtualized testbed based on a real HW that consists of multiple hosts with a Xen Cloud platform [46] hypervisor and OVS as OpenFlow switch. Guest virtual machines are created in hosts and are connected via virtual interfaces to the OVS instances. The soft switches are controlled by one or more OpenFlow controllers that reside in a virtual machine; each guest has two interfaces—one connected to the OpenFlow switch and another connected to the management NW. The management NW carries the signaling for the whole scenario, consisting of the OpenFlow control channels from the switches to the FlowVisor and from the FlowVisor to the OpenFlow controllers, as well as the access traffic to the end hosts, which includes control traffic and GUI traffic (including CLI,

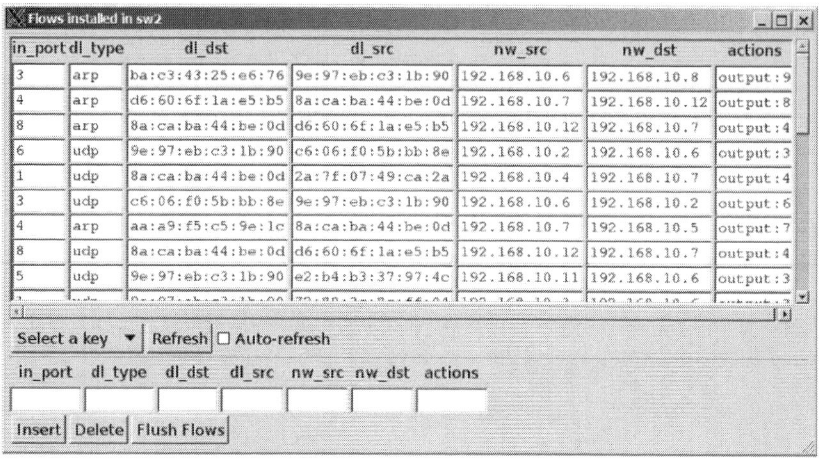

| in_port | dl_type | dl_dst | dl_src | nw_src | nw_dst | actions |
|---|---|---|---|---|---|---|
| 3 | arp | ba:c3:43:25:e6:76 | 9e:97:eb:c3:1b:90 | 192.168.10.6 | 192.168.10.8 | output:9 |
| 4 | arp | d6:60:6f:1a:e5:b5 | 8a:ca:ba:44:be:0d | 192.168.10.7 | 192.168.10.12 | output:8 |
| 8 | arp | 8a:ca:ba:44:be:0d | d6:60:6f:1a:e5:b5 | 192.168.10.12 | 192.168.10.7 | output:4 |
| 6 | udp | 9e:97:eb:c3:1b:90 | c6:06:f0:5b:bb:8e | 192.168.10.2 | 192.168.10.6 | output:3 |
| 1 | udp | 8a:ca:ba:44:be:0d | 2a:7f:07:49:ca:2a | 192.168.10.4 | 192.168.10.7 | output:4 |
| 3 | udp | c6:06:f0:5b:bb:8e | 9e:97:eb:c3:1b:90 | 192.168.10.6 | 192.168.10.2 | output:6 |
| 4 | arp | aa:a9:f5:c5:9e:1c | 8a:ca:ba:44:be:0d | 192.168.10.7 | 192.168.10.5 | output:7 |
| 8 | udp | 8a:ca:ba:44:be:0d | d6:60:6f:1a:e5:b5 | 192.168.10.12 | 192.168.10.7 | output:4 |
| 5 | udp | 9e:97:eb:c3:1b:90 | e2:b4:b3:37:97:4c | 192.168.10.11 | 192.168.10.6 | output:3 |

Select a key ▾ | Refresh | ☐ Auto-refresh

in_port  dl_type  dl_dst  dl_src  nw_src  nw_dst  actions

Insert | Delete | Flush Flows

**Figure 2.6**   Flow table sniffer.

which is accessed through an X terminal). We use Python, fabric [48], and the Xen API to remotely control and collect testbed status information. Once the experiments are verified and optimized in Mininet, the remote testbed allows us to smoothly deploy them in the real HW while maintaining the same control interface.

In addition, the GUI provides us access to the flow tables in the OpenFlow switches. Users can watch in near real time how flow entries are installed and deleted. This feature is key when debugging applications that interact with or extend the OpenFlow controller functionality. Figure 2.6 shows the flow table sniffer GUI.

**The Coming Steps toward an IDE**

The discussion in the previous sections shows the current availability of tools that support an NW development cycle close to those used for SW development. Moreover, it has presented some first steps in the direction of an IDE that supports the application of mainstream SW development techniques to NW design and operation, really encouraging in their results. An integral co-design development environment, addressing NW and application, is a foreseeable result that shall translate into better NW services that are faster to deploy and are more robust and easier to maintain. For achieving this goal, it is essential to pursue two complementary objectives:

1. To develop an IDE that fully supports the full life cycle of an NW application in an SDN
2. To extend the NW realm controlled by the SDN controller in general to other components, such as storage and computation, which are evolving from mere entities deployed at the endpoints of the NW to components integrated in the NW itself

We envisage a full framework to support an SW development and deployment cycle based on a framework equivalent to the OSGi cycle, but without depending on the Java programming language. The paradigm used in the Beacon controller is extremely powerful, and we recognize that the support for dynamic NW applications is critical for a real-world NW deployment. In this regard, building on past experience by the OSGi community, the goal will focus on providing mechanisms that support the whole NW design cycle on an Eclipse framework. Considering its wide adoption, dynamic component approach, and great extensibility, Eclipse seems the most natural support for this endeavor.

Eclipse provides a feature-rich extension development that is already being used to program with different programming languages, such as (1) Java, which comes in the stock Eclipse distribution; (2) Python [49], which can be integrated with a mature plugin; and (3) Ruby [50], for which a plugin is being developed, etc. Moreover, Eclipse has been extended to implement other tools in the SW development cycle, such as unified modeling language (UML-OMG) support, and version control and repository manipulation. Its component-oriented architecture eases the development of projects applying a mix of languages and base technologies, and provides support for tasks related to deployment. Eclipse functionality is essentially based on plugins, which extend the capabilities of the base platform where they reside, as shown in Figure 2.7. This way, it will be possible to integrate plugins dealing with the different stages in the NW development cycle, making them interact with the generic GUI and among them. Moreover, the modular nature of plugins will possibly accommodate the NW development cycle with different application development toolsets as well as adapt it to the different application-NW APIs that will probably exist in the foreseeable future.

This development environment shall also include a DR, as postulated in the 4WARD SW–inspired NW design cycle. This DR will store design invariants and generic components that originate in a

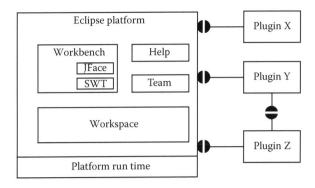

**Figure 2.7**  Eclipse plugin architecture. SWT, standard widget toolkit.

specific NW design and can be reused in the future. We expect this concept to speed up the creation of progressively complex OpenFlow applications in the future.

Regarding the interface for the real networking environment, we intend to develop an interface that will allow us to interact with different cloud-computing infrastructures to explore the applicability of OpenFlow in end-to-end scenarios. In this respect, our next steps are to implement an interface with KVM-based [51] nodes using libvirt [52] to integrate, compute, and store new forms of NW functionality. These plans are inspired by the integration of an interface to OpenStack in Floodlight [22].

As of this writing, we are extending the interface to the OVS to also interact with other HW-based OpenFlow switches. This will allow us to interact with commercial OpenFlow HW as it appears in the market.

In summary, we envisage an OpenFlow IDE that is based on a series of plugins for Eclipse. Because this IDE should cover the whole life cycle of an OpenFlow application, we envisage to integrate available OpenFlow application debugging tools and enhance them with new features to cover the full NW-application development cycle.

## Acronyms

| | |
|---|---|
| **API** | application programming interface |
| **CBA** | component-based architecture |
| **CBSE** | component-based software engineering |

| | |
|---|---|
| **CLI** | command-line interface |
| **COTS** | commercial off the shelf |
| **DMZ** | demilitarized zone |
| **DNS** | directory naming service |
| **DR** | design repository |
| **GUI** | graphical user interface |
| **HW** | hardware |
| **IDE** | integrated development environment |
| **NW** | network |
| **OpenFlow** | OpenFlow |
| **OSGi** | Open Services Gateway initiative |
| **OVS** | OpenVSwitch |
| **PoC** | proof of concept |
| **SAIL** | Scalable and Adaptive Internet Solutions |
| **SDN** | software-defined network |
| **SW** | software |
| **UML-OMG** | unified modeling language |
| **WWW** | World Wide Web |
| **XCP** | Xen Cloud platform |

## Acknowledgment

This work is partly being carried out under the scope of the Scalable and Adaptive Internet Solutions (SAIL) Project under grant agreement number 257448 of the Seventh Framework Programme of the European Union.

## References

1. McKeown, N., T. Anderson, H. Balakrishnan, G. Parulkar, L. Peterson, J. Rexford, S. Shenker et al. 2008. OpenFlow: Enabling innovation in campus networks. *SIGCOMM Comput. Commun. Rev.* 38(2):69–74.
2. Correia, L. M., ed. 2011. *Architecture and Design for the Future Internet.* Signals and Communication Technology Series. pp. 59–87, Netherlands: Springer.
3. Lantz, B., B. Heller, and N. McKeown. 2010. A network in a laptop: Rapid prototyping for software-defined networks. In *Proceedings of the 9th ACM SIGCOMM Workshop on Hot Topics in Networks (HotNets-IX).* pp. 19:1–19:6, New York: ACM.

4. Stallman, R., R. Pesch, and S. Shebs. 2002. *Debugging with GDB: The source level debugger.* Boston: GNU Press.
5. Eclipse. http://www.eclipse.org.
6. Matloff, N., and P. J. Salzman. 2008. *The Art of Debugging with GDB, DDD, and Eclipse.* San Francisco: No Starch Press.
7. Heineman, G. T., and W. T. Councill, eds. 2001. *Component-Based Software Engineering: Putting the Pieces Together.* Boston: Addison-Wesley Longman Publishing Co., Inc.
8. de Castro Alves, A. 2009. *OSGi Application Frameworks.* Manning Publications.
9. About the OSGi Alliance. http://www.osgi.org/About/HomePage. Accessed November 11, 2012.
10. Highsmith, J., and A. Cockburn. 2001. Agile software development: The business of innovation. *Computer* 34(9):120–122.
11. Petersen, K., C. Wohlin, and D. Baca. 2009. The waterfall model in large-scale development. *Product-Focused Software Process Improvement*, pp. 386–400.
12. OpenVSwitch. http://www.openvswitch.org/.
13. Pica8 Open Network Fabric. http://www.pica8.org/solutions/openflow.php.
14. Indigo—Open-Source OpenFlow Switches. http://www.openflowhub.org/display/Indigo/Indigo+-+Open+Source+OpenFlow +Switches.
15. About Nox. http://www.noxrepo.org/nox/about-nox/.
16. Floodlight: A Java-Based OpenFlow Controller. http://www.floodlight.openflowhub.org/.
17. Trema: Full-Stack OpenFlow Framework in Ruby and C. http://www.github.com/trema/.
18. About Pox. http://www.noxrepo.org/pox/about-pox/. Accessed November 10, 2012.
19. Erickson, D. Beacon Home. http://www.openflow.stanford.edu/display/Beacon/Home. Accessed November 18, 2012.
20. Floodlight is an Open SDN Controller. http://www.floodlight.openflowhub.org/. Accessed November 18, 2012.
21. OpenStack: Open-Source Software for Building Private and Public Clouds. http://www.openstack.org/. Accessed November 18, 2012.
22. OpenStack+Quantum. http://www.floodlight.openflowhub.org/quantum-and-openstack/. Accessed November 18, 2012.
23. Controller Performance Comparisons. http://www.openflow.org/wk/index.php/Controller_Performance_Comparisons. Accessed December 18, 2012.
24. Orebaugh, A., G. Ramirez, J. Burke, and L. Pesce. 2006. *Wireshark & Ethereal Network Protocol Analyzer Toolkit (Jay Beale's Open Source Security).* Syngress Publishing.
25. ns3. http://www.nsnam.org. Accessed November 9, 2012.
26. Pelkey, J. OpenFlow software implementation distribution. http://www.code.nsnam.org/jpelkey3/openflow. Accessed December 18, 2012.
27. Mininet at github. http://www.github.com/mininet/mininet.

28. OpenFlow 1.x Plans. http://www.openvswitch.org/development/open flow-1-x-plan.
29. Sherwood, R., G. Gibb, K.-K. Yap, G. Appenzeller, M. Casado, N. McKeown, and G. M. Parulkar. 2010. Can the production network be the testbed? In *OSDI*, pp. 365–378.
30. RouteFlow Project: IP Routing on SDN. http://www.sites.google.com/ site/routeflow/.
31. Hinrichs, T. L., N. S. Gude, M. Casado, J. C. Mitchell, and S. Shenker. 2009. Practical declarative network management. In *Proceedings of the 1st ACM Workshop on Research on Enterprise Networking (WREN '09)*, pp. 1–10, New York: ACM.
32. Voellmy, A., H. Kim, and N. Feamster. 2012. Procera: A language for high-level reactive network control. In *Proceedings of the 1st Workshop on Hot Topics in Software-Defined Networks (HotSDN '12)*, pp. 43–48, New York: ACM.
33. Foster, N., R. Harrison, M. J. Freedman, C. Monsanto, J. Rexford, A. Story, and D. Walker. 2011. Frenetic: A network programming language. *SIGPLAN Not.* 46(9):279–291.
34. Handigol, N., B. Heller, V. Jeyakumar, D. Maziéres, and N. McKeown. 2012. Where is the debugger for my software-defined network? In *Proceedings of the 1st Workshop on Hot Topics in Software-Defined Networks (HotSDN '12)*, pp. 55–60, New York: ACM.
35. OpenFlow Wireshark dissector. http://www.openflow.org/wk/index. php/OpenFlow_Wireshark_Dissector. Accessed November 9, 2012.
36. Chiba, Y., and Y. Nakazawa. Tremashark: A bridge for printing various events on Wireshark. git://www.github.com/trema/trema.gitmaster/ tremashark. Accessed November 9, 2012.
37. Wundsam, A., D. Levin, S. Seetharaman, and A. Feldmann. 2011. Ofrewind: Enabling record and replay troubleshooting for networks. In *Proceedings of the 2011 USENIX Annual Technical Conference (USENIXATC '11)*, p. 29, Berkeley, CA: USENIX Association.
38. Al-Shaer, E., and S. Al-Haj. 2010. FlowChecker: Configuration analysis and verification of federated OpenFlow infrastructures. In *Proceedings of the 3rd ACM Workshop on Assurable and Usable Security Configuration (SafeConfig '10)*, pp. 37–44, New York: ACM.
39. Canini, M., D. Venzano, P. Perešíni, D. Kostić, and J. Rexford. A NICE way to test OpenFlow applications. In *Proceedings of the 9th USENIX Conference on Networked Systems Design and Implementation (NSDI '12)*, p. 10, Berkeley, CA: USENIX Association.
40. Canini, M., D. Venzano, P. Perešíni, D. Kostić, and J. Rexford. nice-of— A NICE way to test OpenFlow controller applications—Google Project Hosting. http://www.code.google.com/p/nice-of/.
41. GENI: Exploring networks of the future. http://www.geni.net.
42. Ofelia: OpenFlow test facility in Europe. http://www.fp7-ofelia.eu.
43. JGN-x Utilization Procedure. http://www.jgn.nict.go.jp/english/info/ technologies/openflow.html.
44. GENI-at-a-Glance. http://www.geni.net/wp-content/uploads/2011/06/ GENI-at-a-Glance-1Jun2011.pdf.

45. JGN: New Generation Network Testbed. http://www.jgn.nict.go.jp/english/procedure/index.html.
46. Xen Cloud Platform. http://www.xen.org/products/cloudxen.html.
47. Barham, P., B. Dragovic, K. Fraser, S. Hand, T. Harris, A. Ho, R. Neugebauer et al. 2003. Xen and the art of virtualization. *SIGOPS Oper. Syst. Rev.* 37(5):164–177.
48. Fabric. http://www.docs.fabfile.org/en/1.4.2/index.html.
49. PyDev. http://www.pydev.org.
50. Williams, A., C. Williams, and M. Barchfeld. 2009. Ruby development tool. http://www.sourceforge.net/projects/rubyeclipse/.
51. Kivity, A. kvm: The Linux virtual machine monitor. 2007. *OLS '07: The 2007 Ottawa Linux Symposium*, pp. 225–230.
52. libvirt—The virtualization API. http://libvirt.org/index.html.

# PART II

# DESIGN

# 3

# IP Source Address Validation Solution with OpenFlow Extension and OpenRouter

## JUN BI

## Contents

## Introduction

Software-defined networking (SDN) has become a promising trend for future Internet development. Current researches mainly focus on flow management, security, and quality of service (QoS) using OpenFlow [1] switch in data center network or campus network. However, the number of OpenFlow application cases to resolve a real problem in a production network is still limited. In the production network, where routers are dominant, some of the challenges in the implementation and deployment of SDN are the integration of existing protocols inside a network device with new protocols, the tradeoff between hardware cost, and deployment profit of network evolution to SDN.

In this chapter, by analyzing the challenges of the current OpenFlow in the production network, we propose three extensions of OpenFlow on FlowTable, control mode, and OpenFlow protocol. Based on these extensions, a commercial OpenFlow-enabled router, named OpenRouter, is designed and implemented using only available and existing hardware in a commercial router.

OpenRouter brings the abilities of control openness; integration of inside/outside protocols; and flexibility of OpenFlow message structure, low-cost implementation, and deployment. We expect that OpenRouter may accelerate the large-scale application and deployment of OpenFlow in the production network.

Currently, the Internet suffers source address spoofing–based attacks based on the observation from the The cooperative association for internet data analysis (CAIDA) data set [2]. Filtering traffic with forged source IP address helps enhance the network security, accountability, and traceability of IP address usage. The source address validation architecture (SAVA) [3] proposed by Tsinghua University discusses the architecture of IP source address validation, dividing it into three levels: local subnet, intra-autonomous system (AS), and inter-AS. For intra-AS, we designed and implemented a method of source address validation named calculation-based path forwarding (CPF), which is based on the idea of centralized computing forwarding path. First, the central controller collects the routing forwarding table and interface table from all the routers in the area. Second, based on the global routing information of the routers, we can calculate the forwarding path from any intra-AS prefix to another and organize it as several triples containing the source IP prefix, the destination IP prefix, and the ingress interface of the current router.

To implement source address validation with OpenFlow architecture is a feasible approach and brings more benefit than traditional network architecture [4]. An OpenFlow application named virtual anti-spoofing edge (VASE) was proposed to improve the source address validation in the access network [5]. VASE uses the OpenFlow protocol, which provides the *de facto* standard network innovation interface, to solve source address validation problem with a global view. Compared with the relatively simple access network, the OpenFlow application in the intra-AS scenario is more challenging. In this chapter, we illustrate a solution of intra-AS IP source address validation named intra-AS IP source address validation application with OpenRouter (InSAVO) [6] with OpenRouter [7] to show that OpenRouter is a feasible and evolvable paradigm for SDN. This application illustrates that OpenRouter not only solves some limits in the current network architecture with SDN features but also gives a tradeoff between software-defined features and evolution cost by means of software image update using the existing hardware of a router.

Another application is a combination of routing and source address validation by having one single algorithm at the OpenFlow controller to generate four-tuple FlowTable entries for both forwarding and filtering. Significant improvements can be found from our evaluation results.

## OpenFlow+ and OpenRouter

### OpenFlow+: An Extension to OpenFlow

OpenFlow is a key technology that enables network innovation. Current researches on OpenFlow are mainly about the applications of OpenFlow switch in campus or in data center. If we could upgrade OpenFlow functionality at legacy commercial layer 3 switches or routers in the production network, then it would make OpenFlow easier to be deployed.

In the production network, layer 3 switches or routers are dominant devices. Many mature network protocols, such as routing protocols, run inside the devices in a distributed control mode. So, there are some challenges of the current OpenFlow to deploy in the production network. First, to adapt all future protocols and different vendors, we need to make FlowTable more open. Second, if a new innovation is mature enough, we need to implement the controller inside the device to improve the efficiency. Third, because it is hard to predefine all the communication requirements between a controller and a device, we need to make the OpenFlow protocol more flexible. Fourth, we need to run OpenFlow in today's router; thus, it will make OpenFlow deployment low cost and deployable.

*FlowTable Extension* Standard and open FlowTable structure, combined with all forwarding elements, currently, is effective to control flows for experiment. However, there are many available forwarding hardware in the production network, such as forwarding information base (FIB), access control list (ACL), and xFlow (NetFlow or sFlow, which depends on vendor implementation). From the view of forwarding behavior, these existing forwarding elements can be

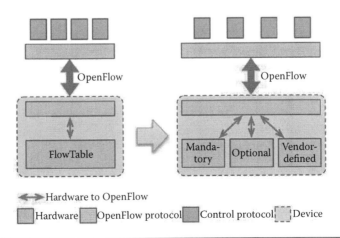

**Figure 3.1** FlowTable extension with mandatory, optional, or vendor-defined forwarding element.

regarded as a type or subset of FlowTable. FlowTable in OpenRouter, as shown in Figure 3.1, can be defined by mandatory, optional, or vendor-defined forwarding element to support not only standard flow forwarding ability by mandatory FlowTable but also existing forwarding ability by optional hardware of FIB or ACL, even enhanced forwarding ability by vendor-defined forwarding ability.

*Control Mode Extension*    The control mode of the current OpenFlow is a centralized control mode that is strong in control efficiency with a global view but is weak in scalability and robust. In OpenRouter, distributed control protocols, such as Open Shortest Path First (OSPF), still keep running inside the device. The information generated by distributed control protocols inside the device, such as routing information, are shared with the outside controller by OpenFlow protocol encapsulation, as shown in Figure 3.2.

*OpenFlow Protocol Extension*    Network data in OpenFlow message are reorganized using type, length, and value (TLV) format. It not only can support transmission of variable-length and optional data but also can achieve the goal of flexible extension of FlowTable structure in the future. According to the content of the current OpenFlow protocol, three TLV types have been defined: network state TLV (e.g., port

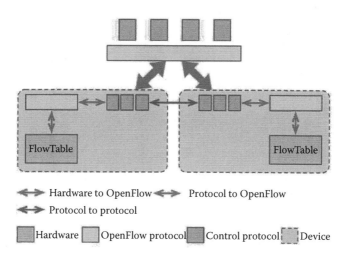

**Figure 3.2**    Control mode extension with sharing data with inside/outside protocols by OpenFlow protocol encapsulation.

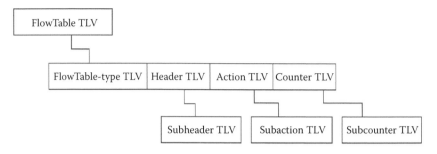

**Figure 3.3** OpenFlow protocol extension with TLV format and newly added FlowTable-type field.

status TLV and routing TLV), network configuration TLV (e.g., interface TLV and queue TLV), and FlowTable TLV. To differentiate the different types of FlowTable in OpenRouter, the "FlowTable Type" field is added to the original FlowTable definition with TLV format, as shown in Figure 3.3. This extension is earlier than the OpenFlow extensible match (OXM) TLV header defined in OpenFlow 1.2, or it can be regarded as an example of OXM.

*OpenRouter: A Software Upgrade on the Legacy Router to Support OpenFlow+*

Based on three OpenFlow extensions, we have designed and implemented an OpenRouter prototype using a commercial layer 3 switch, which architecture is shown in Figure 3.4. An OpenFlow module is embedded into the control plane of OpenRouter. The OpenFlow

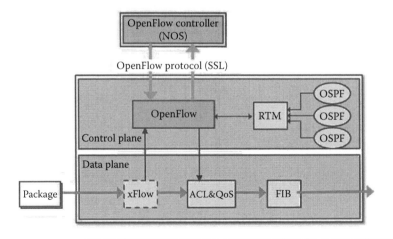

**Figure 3.4** Implementation of an OpenRouter based on DCRS 5980 of Digital China Company; Network Operating System (NOS); Secure Sockets Layer(SSL).

protocol is redesigned and reconstructed with TLV. OpenFlow module communicates with routing table management (RTM) and xFlow modules through two added asynchronous messages and a synchronous message to transmit routing information and sampling packets.

The FlowTable of OpenRouter is implemented using existing forwarding hardware-FIB, ACL, and xFlow, as shown in Figure 3.5:

1. By setting up interfaces with RTM, OpenRouter can send routing table, interface table, and even the routing change information to the controller. We use the TLV format to support these data structures.
2. By setting up interfaces with xFlow (sFlow, NetFlow, or NetStream), OpenRouter can notify sampling packet information to the network operating system (NOS). It also uses TLV to support the data definition.
3. FlowTable is implemented using the existing hardware module of ACL&QoS in the current commercial router.
4. We added several message types in the OpenFlow protocol to support the newly added routing information, sampling information, and so on.

All the extensions can be implemented in the existing commercial router by updating software image instead of replacing the hardware.

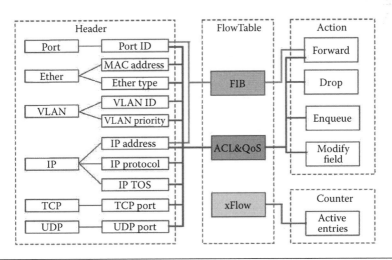

**Figure 3.5** FlowTable structure of an OpenRouter with existing hardware. Virtual local area network (VLAN); transmission control protocol (TCP); user datagram protocol (UDP); media access control (MAC); type of service (TOS).

So, it makes a traditional network device to support OpenFlow in being more cost effective.

In conclusion, the advantages of an OpenRouter are as follows:

1. *Openness:* OpenRouter brings openness to a traditional commercial router with the flexible and customized control abilities offered by OpenFlow.
2. *Integration:* The control protocols inside and outside an OpenRouter can be integrated and exchange network state information.
3. *Flexibility:* OpenRouter enables TLV format to encode and encapsulate network data for variable-length data transmission and flexible extension.
4. *Low cost:* OpenRouter makes use of existing ACL and FIB hardware as a type of FlowTable for low-cost and easy implementation.

*Nox+: An Extension on the Controller Nox to Support OpenFlow+*

Nox is a famous open-source OpenFlow controller. We add some extension based on Nox [8] to support the requirement of CPF, and we call it Nox+. The extension includes the two aspects below.

1. We extended the OpenFlow protocol in Nox to support the routing and sampling information to be received and handled by Nox+. By adding new OpenFlow protocol types such as OFPT_NSM_CHANGES, OFPT_NSM_FULL_TABLE, and OFPT_SAMPLE_PACKET, we can parse the routing information and sampling information packet sent from OpenRouter. We registered new kinds of events and related event handler functions to pass the data up to network application (NAPPs). In addition, we added a module to translate ACL rule data structure into the OpenFlow FlowTable.
2. To physically decouple Nox and NAPP, we designed a socket communication protocol between Nox and NAPP based on the requirement of CPF. The communication data are in TLV format. It can be extended easily.

*OpenFlow Extension in Nox+* As previously mentioned, when introducing OpenRouter, we know that, for the demands of CPF implementation, we extended the OpenFlow protocol to support sending routing information and sampling information. Thus, we added related data handler modules to process such kind of data properly.

First, we added OFPT_NSM_REQUEST, which contains the datapathid of the required router, to represent the routing table request sent from Nox. When OpenRouter with the datapathid receives the OFPT_NSM_REQUEST packet, it will send the OFPT_NSM_FULL_TABLE packet, which contains the whole routing table and interface table of the router, to Nox.

Second, we added the OFPT_NSM_FULL_TABLE type in the OpenFlow protocol and registered the nsm_full_table_event. When Nox+ receives the OFPT_NSM_FULL_TABLE packet, it will parse the type and trigger the nsm_full_table_event event, and then the related event handler function is invoked. In the function, we can parse the data and send them to the NAPP who requires them. In addition, OpenRouters will notify the routing change information and the sampling packet to Nox+. These two kinds of packet are described as OFPT_NSM_CHANGES and OFPT_SAMPLING_PACKET. We added these two kinds of OpenFlow protocol type in OpenFlow. hh and registered related events. When Nox+ receives such two kinds of packet, it will trigger nsm_changes_event and sample_packet_event. Then, the related event handler function will be invoked. In these functions, Nox+ parses the data and sends them to the required NAPP.

Finally, we added the OFPT_ACL_PACKET in the OpenFlow protocol. When a NAPP sends the OFPT_ACL_PACKET packet to Nox+, it will parse the datapathid and tuple-tuple information, including source IP address, destination IP address, and ingress interface. After translating it into FlowTable entry with the OpenFlow protocol type of OFPT_FLOW_MOD, Nox+ will send the OFPT_FLOW_MOD packet to the OpenRouter with the datapathid.

*Communicate between Nox+ and NAPP*   In traditional Nox, it provides a programmatic interface both in C++ and Python upon which developers can develop multiple network applications. The control software of Nox runs on a single-commodity PC. However, integrating multiple applications into a single software will bring too many limitations to the developers, and it may cause computing bottleneck when multiple app's computing burden falls on the shoulder of a single PC.

So, we think that decoupling network control program and network applications physically would be a good choice.

As CPF requires routing table and interface table proactively at any time, we designed the OFPT_NSM_REQUEST packet to support CPF sending such requirement to Nox+. Besides, OpenRouter generates the routing change information and the sampling information in real time and pushed it to CPF proactively. These two kinds of methods of acquiring data from Nox+ are not the same. The former is proactively requiring, and the latter is passively receiving. So, the communication method must support both of the needs.

For the need of requiring a full table, we designed a type of message named Require. When CPF starts up, it sends the Require message to Nox to get the full table. The Require message contains the required content type and datapathid presenting the router of which it wants to get the full table. The Require message is in TLV format.

For the need of passively receiving the routing change information and the sampling information from Nox+, we found that it is very much like the Observer pattern [9], in which the routing change information and the sampling information are the subjects, and the NAPP who is listening to such information is the observer. Every NAPP may be interested in different types of information update. It can subscribe the demanded subject type from Nox+. All kinds of information from all the routers will be sent to Nox+. It must decide to which NAPP it passes the data through. So, Nox+ must maintain an observer list for different kinds of subjects. In the Observer pattern, three kinds of interfaces are provided: notifyObservers, registerObserver, and removeObserver. When different kinds of packets arrive to Nox+ from the routers, they will trigger related events and enter the logic of related event handler functions. We can implement the notifyObservers feature in such event handler functions. To achieve registerObserver and removeObserver, we designed the Register message and the Remove message. NAPP can send the Register message, which contains the observing content type and datapathid describing the router of which it wants to get the updates. When NAPP no longer needs the observed information, it can send the Remove message, which contains the observing content type that it no longer subscribes and the datapathid describing the router of which it no longer needs the updates to Nox+. The Register message and the Remove message are also in TLV format.

For the need of sending data to Nox+ to manage the flows, we designed a type of message named Post. For example, CPF generates ACL rules and sends them to Nox+. We can use the Post message to carry these kinds of data. The Post message contains the datapathid describing the router that it wants to configure and the data structure that can be processed in Nox+. It also uses TLV format for the convenience of extension.

There may be multiple NAPPs in the network. We use the socket select function to have a nonblocking I/O between Nox+, which acts as a server, and several NAPPs, which act as clients.

In Nox+, we maintain an observer list for each event type. Table 3.1 lists the event types and the corresponding observer list variable.

As has been shown above, the observer list is a vector container storing several Observer data structures. The Observer data structure contains four fields:

1. structure App* app, which describes the app's information, such as socket file descriptor.
2. datapathid observe_switch, which describes the observed router's datapathid.
3. int type, an enum type of {PERMENENT, ONCE}, which describes the observer type. If it is a Require message, the type will be set as ONCE; if it is a Register message, the observer type will be set as PERMENENT.
4. int count, which describes the number of packets that should match. When the count decreased to 0, it means that all the responses to the requirement have been sent back completely, and the Observer instance should be removed from the observer list.

**Table 3.1**    Observer List Variables of Each Event Type

| EVENT TYPE | OBSERVER LIST |
|---|---|
| NSM_FULL_TABLE_EVENT | std::vector<Observer> nsm_full_table_observers |
| NSM_CHANGES_EVENT | std::vector<Observer> nsm_changes_observers |
| SAMPLE_PACKET_EVENT | std::vector<Observer> sample_packet_observers |

For example, when Nox+ receives a Request message that contains several datapathids and an event type of NSM_FULL_TABLE_ EVENT, Nox+ will add an observer instance for each datapathid to the related observer list. Then, Nox+ sends the OFPT_NSM_REQUEST message to every router described by each datapathid. When Nox+ receives OFPT_NSM_FULL_TABLE packets from OpenRouters, the handle_nsm_full_table_event() function is invoked. By loop checking every Observer in the NSM_FULL_TABLE_EVENT's observer list std::vector<Observer>nsm_full_table_observers, if the datapathid matches, Nox+ will send the nsm_full_table packet to the socket file descriptor described by the app field in Observer. Because the observer type of the Require message is ONCE, the count will decrease by 1. After count operation, if the count field in Observer equals 0, then Nox+ deletes the Observer entry from the observe list. In our condition, we set the count value to 2 in the nsm_full_table requirement because the nsm_full_table will be divided into routing table packet and interface (address) table packet. It is supposed to finish the sending nsm_full_table data only when Nox+ received two OFPT_ NSM_FULL_TABLE packets. For the Register message, because the observer type is PERMENENT, the count value is set to –1. However times it is matched, as long as the observer type is PERMENENT, the count will not change so as to not remove the Observer instance from the observer list, except for receiving Remove messages.

### InSAVO: Intra-AS IP Source Address Validation Application with OpenRouter

*CPF: An Intra-AS Source Address Validation Method*

Central control mechanism is a good way to validate the IP source address in an AS because the mechanism can get a global forwarding path and then resolve the false positive of filtering information caused by asymmetric routing more than ingress filtering. CPF is a source address validation solution proposed by Tsinghua University.

In CPF, a central controller calculates the forwarding path information based on collected routing and interface information in a campus or enterprise network. Besides, the central controller collects sampling information from several routers, which can also be organized to triple-tuples containing the sampling packet's source IP address, destination

IP address, and ingress interface of the router from which it is collected. By comparing the sampling triple-tuple and filtering rule triple-tuple, we can tell whether the sampling packet's source address is forged or not. As a result, we display the sampling packets with a forged source IP address in the management Web site to show the network status. The network administrator then has a global vision of it and decides whether to configure ACL to the related router. Figure 3.6 shows the architecture of CPF in the current network.

During the implementation of CPF, we used simple network management protocol (SNMP) to collect routing information, xFlow (including sFlow, NetFlow, and NetStream) to collect sampling packet, and Expect tool command language (TCL) script to configure ACL into the router based on the telnet command line. In the current network architecture, we came across some difficulties because of the limitations and differences of the equipment.

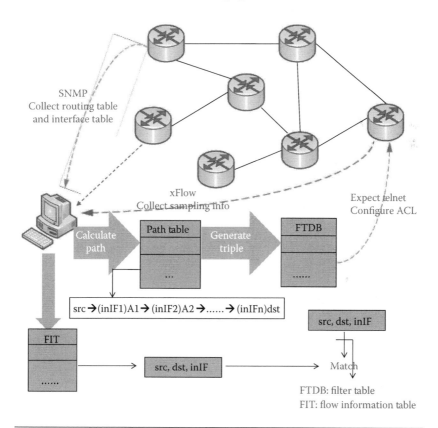

**Figure 3.6**   CPF architecture in the current network architecture.

1. We found that not all the equipment respects the IPv6 SNMP MIB standard strictly, and thus, we spent a lot of time analyzing and resolving the differences.
2. We had to use SNMP polling to collect routing information periodically, which resulted in the false positive of filtering rules because of not getting the routing change information in time.
3. We must implement multiple sampling protocols because of the inconsistencies of different equipment supporting.
4. Using Expect TCL scripts to automatically configure ACL rules with telnet command line is not a good way because it is not easy to handle complex logic.

Moreover, in the current network architecture, using SNMP polling to get the two types of information will result in some false positive of filtering information because of not getting the routing change information in time.

Therefore, we proposed a new intra-AS source address validation named InSAVO, which calculates the forwarding path with OpenFlow architecture. The calculation of the forwarding path makes use of FlowTable in the OpenFlow switch and routing protocols running on an outside controller. It resolves the problems above because OpenFlow (and with its extension OpenFlow+) provides standard interfaces, and the awareness of network status changes will eliminate the false positive in the filtering table.

*InSAVO: CPF as an Intra-AS Source Address Validation Application over OpenFlow*

To deploy InSAVO in the current campus network, we upgrade the router with the OpenRouter software, which not only can provide software-defined abilities to introduce some new network features by extending OpenFlow but also can give a tradeoff between existing hardware and evolution cost by software update inside a network device. In this section, we will illustrate that OpenRouter is a feasible and evolvable paradigm for SDN.

*System Architecture* The system architecture is composed of an OpenRouter, a Nox+ controller, and InSAVO application, as shown in Figure 3.7.

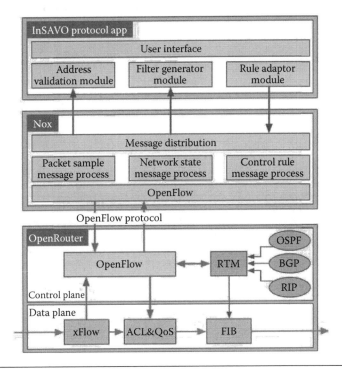

**Figure 3.7** InSAVO architecture; Border Gateway Protocol (BGP); Routing Information Protocol (RIP).

*OpenRouter* OpenRouter can notify the routing change information and the sampling packet message by setting up interfaces with RTMand xFlow (sFlow or NetFlow, which depends on a vendor implementation). The notification messages are encapsulated with OpenFlow protocol to support standard information transmission and network state information sharing with protocols running inside/outside a router. FlowTable is implemented using the existing hardware of ACL&QoS and FIB in OpenRouter to enable software-defined features only updating the software image of a router. The network data structure is redesigned and reconstructed with TLV format to support the user-defined combination and extension of header fields and actions of OpenFlow. The IP address length and match structure of FlowTable is extended to 128 bits to support IPv6 forwarding rules.

*Nox+* Nox+ can receive and process routing information, interface information, and sampling packets from OpenRouter, and then distribute information to the InSAVO protocol app. Nox+ can encapsulate control rules from the InSAVO protocol app to FlowTable format and issue them to OpenRouter. Three new modules are implemented to process sampling packet message, network state message, and control rule message by means of a message-event pattern to improve the efficiency of information process and distribution.

*InSAVO (CPF) App*    The InSAVO protocol app (i.e., CPF) can calculate the forwarding path and the filtering information in the filtering generator module according to the routing and interface information from OpenRouters in an AS. InSAVO can issue filtering information by active control based on the forwarding path and by positive control based on the validation of the sampling packet according to management requirement.

*CPF Application Implementation*    In CPF application, we get all of the data from Nox+ and send ACL rules to Nox+ using unified TLV format.

As shown in Figure 3.8, there are several modules in CPF application:

1. *Communicating with Nox+ module:* This module implements the communication interface with Nox+. After CPF starts to run, it connects to the configured Nox+ IP address and port to set up the socket. Afterward, it will send Request, Register, Remove,

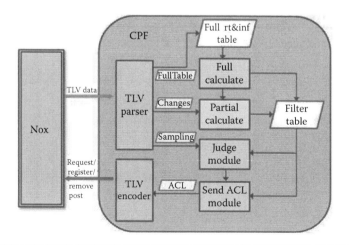

**Figure 3.8**    Implementation of CPF application.

and Post messages to Nox+ using a socket. Besides, it creates a new thread named Recv Thread to receive data from Nox+.

2. *TLV parser and encapsulate module:* The routing, routing change, and sampling information are all encapsulated into the TLV format. The TLV parser module parses these data based on related TLV definitions to organize the data into data structure needed by the calculating module. In addition, the four types of messages, including Require, Register, Remove, and Post, are also in TLV format. TLV-encapsulated module is responsible for encapsulating such TLV packet to send to Nox+.

3. *Calculating filtering rule module:* This is the core of the CPF. It includes the algorithm of calculating the forwarding path. There are two types of calculating: the full calculating method calculates the whole filtering rules based on a full table, whereas the partial calculating method partially calculates the changed part of the filtering rules based on the routing updates.

4. *Sample packet judging module:* This module takes out the received sample packet triple-tuples and compares it with the filtering rule triple-tuple to judge whether the sample packet has forged a source address. Once it decides that the sample packet is a packet with a forged source address, it will write the entry into MySql.

5. *MySql module:* It is the bridge between the CPF front Web site and the backend server. Network administrator can input data in a Web site and acquire a network IP address spoofing status from the Web site. The CPF backend server reads a configuration from MySql and writes calculated data into it.

*Testbed*  We have set up the testbed of an intra-AS source address validation with 10 OpenRouters (DCRS 5980/5950) to evaluate the performance of OpenRouter, as shown in Figure 3.9.

The testbed topology is divided into two parts. The network in Tsinghua University with 10 OpenRouters is configured to run the OSPF protocol as an AS. Nox+ and InSAVO (CPF) are running on two servers. A packet generator is used for generating spoofing packets to server A. A video server is used for generating video traffic to

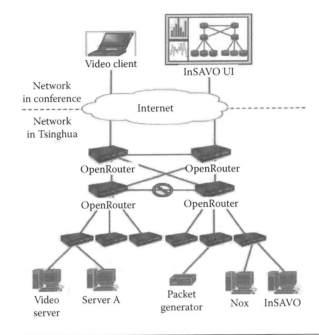

**Figure 3.9**   Intra-AS IP Source Address Validation Application with OpenRouter (InSAVO) testbed; User Interface (UI).

show bandwidth change intuitively caused by spoofing packets. The network in the conference with two laptops is used to show video client and InSAVO user interface.

Some InSAVO core functions can be demonstrated in this testbed:

1. *Source address validation:* The packet generator generates IP spoofing packets to server A. The quality of the video in the video client is declined. After turning on the function of InSAVO, the quality of the video in the video client is recovered. In the user interface of InSAVO, the source of spoofing traffic will be marked in red.

2. *Forwarding path recalculation:* When the link between the two OpenRouters is down, the message of a link down notification can be sent to Nox+ by OpenRouter as soon as possible. The user interface of InSAVO can show the change of network topology and recalculate the forwarding path and the filtering information. This scene shows that InSAVO can respond quickly to network state change with OpenRouter.

*InSAVO Algorithms*

*Full Calculating*   The principle of calculating the forwarding path of a stream (src, dst) is that we can start from the router that owns the source IP prefix src. Then, we find out the routing forwarding entry with the destination IP prefix dst, which is presented as (dst, nexthop, out_inf, route_type). Thus, we get the nexthop IP and the outgoing interface. The field route_type is an enum of {DIRECT, REMOTE}. As CPF collects all the routers' interface table, it can find out the nexthop router and the corresponding ingress interface by searching the router's interface table with nexthop IP. By repeating the above step until we find out a routing forwarding entry with a route_type of DIRECT, we finally calculate the whole forwarding path from src to dst. In each hop, we can generate a related triple-tuple (src, dst, in_inf), which means that, when the traffic with a source IP prefix src and a destination IP prefix dst passes through the current router, the incoming interface must be the in_inf. Otherwise, it is supposed that the packet forges the source IP address.

Based on the above principle, we designed and implemented the recursion calculating the forwarding path algorithm.

1. Based on the routing table and the interface table, create the neighbor table and subnet table. The neighbor table stores each router's neighbor router and the connected interfaces. The subnet table contains the router's subnet prefix.
2. For each router, put all the prefixes in the subnet table into the src_set. For each destination prefix dst, find out the FIB entry (dst, nexthop, out_inf, route_type). If the route_type is DIRECT, exit and go back to step 2 to continue considering the next destination prefix. If the route_type is REMOTE, then search the nexthop address in the neighbor table to find out the nbr entry. Then, get the nexthop router and ingress the interface of the nexthop router based on the nbr entry. Then, go to the nexthop router and to step 3.
3. Generate filter triple-tuple (src, dst, in_inf)(src ∈ src_set) for the current router. Find out the FIB entry with a destination prefix dst (dst, nexthop, out_inf, route_type). If route_type equals DIRECT, exit and jump to step 2; if route_type equals

REMOTE, search the nexthop address in the neighbor table to find out the nbr entry. Get the nexthop router and the ingress interface of the nexthop router from the nbr entry. Then, go to the nexthop router and repeat step 3.

The algorithm of full calculating is described below.

## GEN-FILTER-TABLE (rt)

1. **foreach** sub in rt.subTable
2.     src_set.ADD(sub)
3. **foreach** dst_rt in routers
4.     **if** dst_rt == rt
5.         **continue**
6.     **foreach** dst in dst_rt.subTable
7.         rt_entry←rt.rtTable.FIND-DST(dst)
8.         **if** rt_entry.rtType == REMOTE
9.             n←rt.nbrTable.FIND(rt_entry.nexthop)
10.            **COM-NEXT-ROUTER** (n.rt, src_set, dst, n.nbr_if)

## COM-NEXT-ROUTER (rt, src_set, dst, ifindex)

1. rt.ADD-FILTER-ENTRIES(src_set, dst, ifindex)
2. rt_entry←rt.rtTable.FIND-DST(dst)
3. if rt_entry.rtType == REMOTE
4.     n←rt.nbrTable.FIND(rt_entry.nexthop)
5.     COM-NEXT-ROUTER(n.rt, src_set, dst, n.nbr_if)

*Partial Calculating* In the traditional CPF implementation, we use SNMP polling to get full tables and calculate the filtering tables periodically. This may lead to a false positive of filtering rules because the update is not acquired in time. However, thanks to the function of notifying the routing change information provided by OpenRouter, we can get the routing update in real time and thus can calculate the correct filtering table in time.

As previously mentioned, the Recv Thread in CPF receives data from Nox+, including the routing change information. As the nsm_changes packets may arrive at CPF at any time, what is the most suitable timing to start up a round of partial calculating? In the experiment, we found that a network topology change will produce several nsm_changes packets from several routers, and they will arrive at CPF within a relatively concentrated period. So, we can start a partial calculating thread after figuring out that all the nsm_change packets generated by a network topology change have been received by defining the start and end time points of the period. We use the convergence time of OSPF as the parameter INTERVAL_TIME. When CPF receives a new nsm_changes packet, it means that a new update collecting process starts. Within the update collecting process, if CPF does not receive any nsm_changes packet for INTERVAL_TIME continuously, CPF identifies it as the end time point of this update collecting process and starts to lock the data and run the partial calculating algorithm.

Based on the routing update, we designed and implemented a kind of partial calculating algorithm to generate the filtering change entries. The central idea of partial calculating algorithm is that the change of routing information presented as adding or deleting routing entries (or interface entries) only affects the path of related source–destination prefix pairs. So, we can just focus on the affected source–destination prefix pairs and calculate the added or deleted filtering table entries.

## UPDATE-FILTER-TABLE (rt)

1. **foreach** rt_add in rt.rt_updates_add
2.     dst←rt_add.dst
3.     **foreach** src in rt.subTable
4.        src_set.ADD(src)
5.     **foreach** f in rt.filterTable
6.        **if** f.dst == dst
7.           src_set.ADD(f.src)
8.     n←rt.nbrTable.FIND(rt_add.nexthop)
9.     **COM-NEXT-ROUTER** (n.rt, src_set, dst, n.nbr_if )
10. **foreach** rt_del in rt.rt_updates_del
11.     dst←rt_del.dst

12.     n←rt.nbrTable.FIND(rt_add.nexthop)
13.     **foreach** f in n.nbr_rt.filterTable
14.      **if** f.dst == dst && f.in_if == n.nbr_if
15.        src_set.ADD(f.src)
16.        n. nbr_rt.DELETE-FILTER-ENTRY(f)
17.     nnbr←n.nbr_rt.FIND(dst)
18.     **NEXT-ROUTER-DELETE** (nnbr.rt, src_set, dst)


## NEXT-ROUTER-DELETE (rt, src_set, dst)

1. **foreach** src in src_set
2.     **foreach** f in rt.filterTable
3.      **if** f.src == src && f.dst == dst
4.        rt.DELETE-FILTER-ENTRY(f)
5. rt_entry←rt.rtTable.FIND-DST(dst)
6. if rt_entry.rtType == REMOTE
7.     n←rt.nbrTable.FIND(rt_entry.nexthop)
8.     **NEXT-ROUTER-DELETE** (n.rt, src_set, dst)

For interface table change entries, for example, adding an interface table entry, we can start from the current router and calculate the path from the interface IP prefix to any destination prefix in the area using the recursion calculating the forwarding path algorithm. In deleting an interface entry, we must delete all the filtering table entries containing the interface IP prefix as its source IP prefix or destination IP prefix.

For routing table change entries, for example, adding a routing table entry, we can start from the current router and calculate the path from the possible source to the added routing destination prefix using the recursion calculating the forwarding path algorithm. In deleting a routing table entry, we must go through the old path from the current router toward the destination router and delete the filtering rules with related source-destination prefix pairs of the routers along the old path.

Because the updating process for interface table changes is very similar with the full calculating process, we only describe the updating algorithm for routing table changes. The algorithm is shown below.

*Comparison of Two Algorithms*  In the traditional CPF in the current network architecture, the full calculating process runs periodically. Now, analyze the complexity of the full calculating algorithm. Suppose that there are $n$ routers. The prefix number of router $r_i$ is $m_i$, and the total number of prefix in the intra-AS is $m$. It means that $\sum_{i=0}^{n} m_i = m$. For each router, we call the GEN-FILTER-TABLE function.

In this function, the size of the src_set is $m_i$, and for each destination prefix, we call the COM-NEXT-ROUTER function, in which the average recursive depth is the average path length $\dfrac{n}{2}$. As a result, for each des prefix, the complexity is $\dfrac{n}{2}m_i$. For each router, it is $\dfrac{n}{2}m_i m$. So, for all the routers, the total complexity is

$$\sum_{i=1}^{n} \frac{n}{2}m_i m = \frac{n}{2}m \sum_{i=1}^{n} m_i = \frac{n}{2}m^2.$$

So, every time the full calculating algorithm runs, the complexity is $o(nm^2)$.

In InSAVO, because it can acquire notifications for routing updates (thanks to OpenFlow+), it can just run a partial calculating algorithm when needed. Suppose that there is a network topology change and that CPF receives an interface update entry and an $n$ routing update entry (one for each router). For one interface prefix added, the complexity of calculating the filter rules considering it as source will cost $\dfrac{n}{2}m$ ($\dfrac{n}{2}$ is recursive depth; $m$ is the number of dst prefixes). Considering it as a destination will cost $\sum_{i=1}^{n}\dfrac{n}{2}m_i = \dfrac{n}{2}m$ ($\dfrac{n}{2}$ is recursive depth, starting from every other router; $m_i$ is the size of the src_set). As for the routing update, for each router, there is one routing update change, and we will call the COM-NEXT-ROUTER function. $m_i$ is the size of the src_set, whereas $\dfrac{n}{2}m_i$ is the recursive depth. So, for all the routers, the total complexity is $\sum_{i=1}^{n}\dfrac{n}{2}m_i = \dfrac{n}{2}m$. Sum up the two aspects, and the total complexity of a partial calculating algorithm for one network topology change is $O(nm)$.

Comparing the two algorithms, we can see that the traditional CPF in the polling mode will cost $o(nm^2)$ every polling interval time, whether or not the network topology has changed. However, in InSAVO, it first runs full calculating and runs partial calculating only when it receives updates. When there is no network topology change, it uses much less CPU resource than the traditional CPF. Even when there is an update, as every network topology change will cause $O(nm)$, only when the network topology change rate is bigger than $\dfrac{m}{\text{polling interval time}}$ $\left( m = \dfrac{O(nm^2)}{O(nm)} \right)$, InSAVO will cost much CPU resource than traditional CPF. Even so, however, in such an unstable network, the false positive of filter rules generated by traditional CPF would be very large, thus resulting in very poor filtering performance. Considering the CPU use and the filtering performance of CPF, we can see that InSAVO is much better than traditional CPF.

*Combination of Routing and InSAVO in OpenFlow Architecture*

*A New Algorithm to Generate FlowTable for Both Routing and Filtering* InSAVO is a way to add intra-AS source address validation functions (CPF) in addition to today's intra-AS routing protocols (OSPF, etc.) by OpenFlow technology. In this section, we propose to combine routing and InSAVO together in OpenFlow architecture by generating a forwarding and filtering table (two-dimensional forwarding table) by one single algorithm as an OpenFlow application.

In this new algorithm, we still use Dijkstra, which is used in today's OSPF routing protocol.

## Definition 3.1

Let $Q$ denote the set of all routers $v$ in the network:

$v$ denotes a router in $Q$;

$P(v)$ denotes all IP prefix at router $v$;

$I(p)$ denotes the router interface of prefix $p$;

$I(v1,v2)$ denotes the ingress interface of router $v2$, where $v1$ and $v2$ are connected together with a link;

$O(v1,v2)$ denotes the egress interface of router $v1$, where $v1$ and $v2$ are connected together with a link;

dist[$v$] denotes the distance from the source router $s$ to the destination router $v$;

dist_between($u,v$) denotes the distance between routers $u$ and $v$;

previous[$v$] denotes the previous neighbor of router $v$ on the path from the source to the destination;

following[$v$] denotes the following neighbor of router $v$ on the path from the source to the destination;

FlowEntry($P_1$, $P_2$, $I$, $O$) denotes a FlowTable entry, where $P_1$ denotes the prefix of the source address; $P_2$ the prefix of the destination address; $I$ the ingress interface; $O$ the egress interface; and $F(v)$.add(FlowEntry($P_1$, $P_2$, $I$, $O$)) denotes that the FlowTable entry is added at router $v$.

## Algorithm 3.1

For a source router $s$, we calculate all the paths that start from $s$ and end at all routers in $Q$:

1. For every router $v$ in $Q$, set dist[$v$] = $\infty$ and dist[$s$] = 0; for all prefix in router $s$, generate a four-tuple FlowTable: for all $P_1$, $P_2 \in P(s)(P_1 \neq P_2)$, add FlowEntry ($P_1$, $P_2$, $I(P_1)$,$I(P_2)$) in router $s$.

2. Select a router $u$ from $Q$, where dist[$u$] is the minimal. If dist[$u$] = $\infty$, then go to the end. Otherwise, for all prefixes in $s$ and $u$, add four-tuple FlowTable in all routers on the path from $s$ to $u$: for $P_1 \in P(s)$, $P_2 \in P(v)$, recursively generating FlowEntry($P_1$, $P_2$, $I$(previous[$v$], $v$), $I(v$,following[$v$])) in router $v$, previous[$v$],..., $s$. The recursive algorithm is shown below:

```
node := v
following[node] := undefined
while node ! = s
    for each P₁ in P(s)
        for each P₂ in P(dest)
            if following[node] = undefined
                F(node).add(FlowEntry(P₁, P₂, I(previous[node],
                node), I(P₂)))
```

**else**
 $F$(node).add(FlowEntry($P_1$, $P_2$, $I$(previous[node],
 node), $O$(node,following(node)))))
**end if**
oldnode:= node
node:= previous[node]
following[node]:= node
 **end for**
 **end for**
**end while**
**for each** $P_1$ **in** $P(s)$
 **for each** $P_2$ **in** $P$(dest)
  $F(s)$.add(FlowEntry($P_1$, $P_2$, $I(P_1)$, $O(s$,following[$s$]))))
 **end for**
**end for**

3. For all $v \in$ neighbor($u$), if dist[$u$] + dist_between($u,v$) < dist[$v$], then previous[$v$] = $u$.
4. Repeat steps 2 and 3, until $Q$ is an empty set.

Algorithm 3.1 is used to calculate FlowTable entries when router $s$ is selected as the source node. Then, we can repeat $n$ times (by selecting each router in the network as the source node) to calculate all the FlowTable entries in the whole network ($n$ denotes the total number of routers).

*Evaluation*

Now, we analyze the complexity of this algorithm:

1. The first step in Algorithm 3.1: First, we initialize the dist[]; the complexity is $O(n)$. Then, we construct the four-tuple table. We let $m(s)$ denote the number of prefixes in router $s$; then it costs $O(m(s)^2)$. Because we need to calculate it for all routers in the graph $\sum_{s \in Graph} m(s)^2$, then the total complexity is $O(m(s)^2)$. Let $m$ denote the total number of prefixes in all routers; then the complexity is $O(m^2)$.

2. The second step in Algorithm 3.1: First, we select a router from $Q$ by the following step: selecting the minimal cost, which is $\dfrac{n(n-1)}{2}$ (i.e., $O(n^2)$); then the cost for generating four-tuple table entries [$l \cdot m(s) \cdot m(u)$ entries, where $l$ denotes the number of routers from $s$ to $u$], which is $l \cdot m(s) \cdot m(v_1) + l \cdot m(s) \cdot m(v_2) + \ldots\ldots + l \cdot m(s) \cdot m(v_{n-1})$, i.e., $\displaystyle\sum_{i=1}^{n} l_{s,v_i} m(s) m(v_i) = m(s) \sum_{i=1}^{n-1} l_i m(v_i)$ (where $l_i \leq n$, $v_i$ denotes router $i$).

3. The third step in Algorithm 3.1: The complexity is $O(E)$, where $E$ is the total number of links in $Q$.

4. Then, the total complexity of the second and third steps in Algorithm 3.1 is $O(n^2 + E) + m(s) \displaystyle\sum_{i=1}^{n-1} l_i m(v_i)$. Because $E < n^2$, $l_i \leq n$, and the second and third steps will be repeated $n$ times, then it can be simplified as $O(n^3 + mn^2)$.

5. Then, the total complexity of the algorithm is $O(m^2 + n^3 + mn^2)$. Because $m$ is the total number of prefixes, so $m > n$. Then, the complicity can be denoted as $O(mn^2)$.

Although Algorithm 3.1 brings more complexity, in OpenFlow, we make use of CPU in a server to calculate the FlowTable. We know that the CPUs in a server are much more powerful than a CPU in a router. In our experiment with the Tsinghua campus network topology and prefixes at each routers (a total of 21 routers and 473 prefixes, i.e., $n = 21$, $m = 473$), we used a middle end server with 2.40-GHz CPU and 2-GB memory. The experimental results show that InSAVO costs 4.137 s, whereas Algorithm 3.1 costs 2.929 s. The reason that InSAVO costs more time is because it contains more match operations that costs more CPU time. Therefore, the experimental results show that the OpenFlow-based central calculation is feasible in the real network.

In InSAVO, the total number of generated table entries is the forwarding table entries $n \cdot m$ (generated by OSFP as a distributed routing protocol) and the filtering table entries $n \cdot m^2$ (generated by InSAVO as an OpenFlow application). In Algorithm 3.1, the total number of four-tuple FlowTable entries is just $n \cdot m^2$. We know that the hardware table resource in a router is valuable, so the new approach saves the valuable table resource.

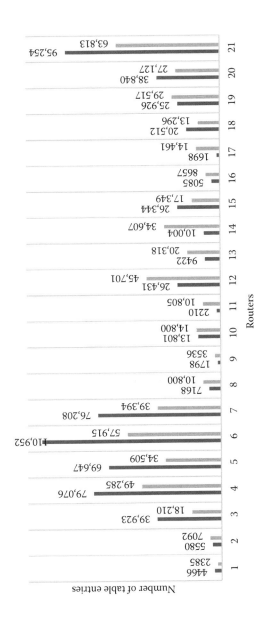

**Figure 3.10** Comparison of table entry consumption of InSAVO and Algorithm 3.1.

We did experiment with the Tsinghua campus network topology and prefixes at each router (a total of 21 routers and 473 prefixes, i.e., $n = 21$, $m = 473$). As shown in Figure 3.10, the dark bar denotes the filtering table generated by InSAVO, and the gray bar denotes the filtering table generated by Algorithm 3.1.

From the experimental result, we can tell the difference between the two approaches. In InSAVO, it costs more filtering tables at some core routers, whereas Algorithm 3.1 generates a more balanced filtering table. The reason is that, in InSAVO, the routing table is affected by some manual configuration and policies. So, if we can combine the forwarding and filtering calculations with a centralized OpenFlow approach, it will generate more optimized FlowTable entries.

Another observation is that, although the theoretical estimation of the filtering table size is $O(n \cdot m^2)$, in reality, in most routers, the table size is quite small, which makes the solution more realistic. However, in some core routers (such as routers 4, 6, and 21), the total number of FlowTable entries is still huge. Therefore, the study of FlowTable compression is still a future research direction.

### Conclusion and Future Work

In this chapter, we illustrate an intra-AS source address validation application to the OpenFlow network architecture.

1. We extended the OpenFlow protocol based on the challenges of the production network. Then, we add OpenFlow extension to a commercial router by software update, and we call it OpenRouter. OpenRouter not only can provide software-defined abilities to introduce some new network features but also can give a tradeoff between existing hardware and evolution cost by software update inside a network device.

2. We applied the OpenFlow extension into the Nox controller (we name it Nox+). We designed and implemented a new communication interface between Nox+ and Application, which can be a generic communication interface between any network operating system and NAPP. Then, we implement the intra-AS source address validation method CPF as an application, and we call it InSAVO. InSAVO listens to and gets the routing

change information in time, and then generates filtering table in OpenFlow table format. The unified OpenFlow interfaces and timely routing update make InSAVO more deployable and efficient than CPF in the traditional network architecture.

3. The core algorithms of InSAVO are introduced. Thanks to the routing updates notification feature provided by it, we design partial calculating algorithm and invoke it only when network update happens. Compared with the traditional polling computing method, it will save much CPU resources and will get better filtering performance. Moreover, a new algorithm that combines both forwarding and filtering calculation by generating four-tuple OpenFlow table entries is introduced. The evaluation in the Tsinghua campus network shows that this algorithm saves more valuable FlowTable entries in the router and that the computation cost is acceptable for a CPU in the server.

Although the evaluation proves that the proposed solution is feasible at the production network, we observed from the experimental results that the size of the FlowTable in some core routers is still explosive. In future works, we will study how to handle the explosive FlowTable. We are also working on the enhancement of the controller software to upgrade it as a complete network operating system, which is called TUNOS [10].

## Acknowledgments

This work is supported by the National High-Tech R&D Program ("863" Program) of China (no. 2013AA010605) and the National Science Foundation of China (no. 61073172).

## References

1. McKeown, N., T. Anderson, H. Balakrishnan, G. Parulkar, L. Peterson, J. Rexford, S. Shenker et al. 2008. OpenFlow: Enabling innovation in campus networks. *ACM Computer Communication Review* 38(2):69–74.
2. Yao, G., J. Bi, and Z. Zhou. 2010. Passive IP traceback: Capturing the origin of anonymous traffic through network telescopes. *ACM Computer Communication Review* 40(4):413–414.

3. Wu, J., J. Bi, X. Li, G. Ren, and K. Xu. 2008. A source address validation architecture (SAVA) testbed and deployment experience. RFC5210.
4. Yao, G., J. Bi, and P. Xiao. 2011. Source address validation solution with OpenFlow/Nox architecture. In *Proceedings of the 19th IEEE International Conference on Network Protocols (ICNP11)*, Vancouver, pp. 19–24.
5. Yao, G., J. Bi, and P. Xiao. 2013. VASE: Filtering IP spoofing traffic with agility. *Computer Networks (Elsevier)* 57(1):243–257.
6. Feng, T., J. Bi, H. Hu, G. Yao, and P. Xiao. 2012. InSAVO: Intra-AS IP source address validation solution with OpenRouter. *31st IEEE International Conference on Computer Communications (INFOCOM12)*, Orlando, FL.
7. Feng, T., J. Bi, and H. Hu. 2011. OpenRouter: OpenFlow extension and implementation based on a commercial router. In *Proceedings of the 19th IEEE International Conference on Network Protocols (ICNP11)*, pp. 141–142, Vancouver, Canada.
8. Gude, N., T. Koponen, J. Pettit, B. Pfa, M. Casado, N. McKeown, and S. Shenker. 2008. Nox: Towards an operating system for networks. *ACM Computer Communication Review* 38(3):105–110.
9. Gamma, E., R. Helm, R. Johnson, and J. Vlissides. 1995. *Design patterns: Elements of reusable object-oriented software*, Addison-Wesley.
10. Feng, T., J. Bi, and H. Hu. 2012. TUNOS: A novel SDN-oriented networking operating system. In *Proceedings of the 20th IEEE International Conference on Network Protocols (ICNP12)*, pp. 1–2, Austin, TX.

# 4

# LANGUAGE AND PROGRAMMING IN SDN/OPENFLOW

## MUHAMMAD FAROOQ AND FEI HU

**Contents**

## Introduction

Conventional networks use special algorithms implemented on dedicated devices (hardware components) for controlling and monitoring the data flow in the network, managing routing paths and algorithms, and determining how different devices are arranged in the network, in other words, finding the network topology. In general, these algorithms and set of rules are implemented in dedicated hardware components, such as Application-Specific Integrated Circuits (ASIC). ASIC are designed for performing specific operations. Packet forwarding is a simple example of this operation. In a conventional network, upon the reception of a packet by a routing device, it uses a set of rules embedded in its firmware to find the destination device as well as the routing path for that packet. Generally, data packets that are supposed to be delivered to the same destination are handled in a similar manner and are routed through the same path irrespective of the data types of different packets. This operation takes place in inexpensive routing devices. More expensive routing devices can treat different packet types in different manners based on their nature and contents. A problem posed by this is the limitation of the current network devices under high network traffic, which poses severe limitations on network performance. Issues, such as the increasing demand for scalability, security, reliability, and network speed, can severely hinder the performance of current network devices because of the ever-increasing network traffic. Current network devices lack the flexibility to deal with different packet types with various contents because of the underlying hardwired implementation of routing rules [1].

A possible solution to this problem is implementing data handling rules as software modules rather than embedding them in hardware. This enables network administrators to have more control over the network traffic and, therefore, has a great potential to greatly improve the performance of the network in terms of efficient use of resources and speed. Such an approach is defined in software-defined networking (SDN). In SDN, data handling is isolated from the hardware, and its control is implemented in a software module called the controller. The basic idea behind SDN is to separate the control of data handling in the networking stack from the hardware and implement it in the software. This results in improved network performance in terms of network management, control, and data handling. SDN is a potential solution to the problems faced by a conventional network and is gaining more acceptance in applications, such as cloud computing. It can be used in data centers and for workload-optimized systems [2].

SDN enables network administrators to deal with data in the network in a more efficient and innovative manner. By using SDN, administrators have the ability to control data flow as well as to alter the characteristics of the switching devices (routing devices) in the network from a central location, with control application implemented as software module, without the need of dealing with each device individually. This enables network administrators to change routing tables (routing paths) in network routing devices. This also gives an extra layer of control over the network data because the administrator can assign high/low priority to certain data packets or allow/block certain packets flowing through the network with different levels of control. As a result, the network traffic can be controlled in an efficient manner and, hence, can be used as a mechanism for traffic load management in networks. A potential application of SDN is in solutions, such as cloud computing, multitenant architecture, etc., because network administrators have more control over the network traffic and can use network resources in a more efficient manner. SDN is less expensive (because of switching of devices used in this approach) and provides more control over network traffic flow as compared with conventional network devices. Several standards are used for the implementation of SDN. One of the most popular and widely accepted standards for SDN implementation is OpenFlow. OpenFlow provides remote control over network routing devices because of its ability to control the routing tables of the traffic in the network. Figure 4.1 illustrates the major difference between SDNs and conventional networks.

As mentioned before, SDNs are more suitable for efficient routing of data packets through paths that have a fewer number of hops or have more bandwidth available. This increases the traffic efficiency of the network. The major advantages of SDNs are listed below.

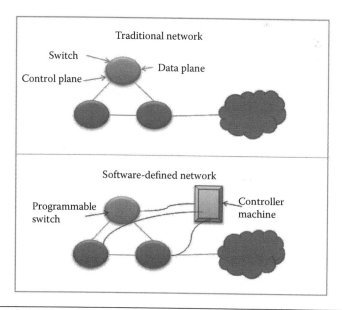

**Figure 4.1** Difference between traditional networks and SDN.

1. *Intelligence and speed:* SDNs have the ability to optimize the distribution of the workload and make end devices more intelligent. This results in high-speed networks and gives more efficient use of the resources.
2. *Patterns in the networks:* In SDNs, the administrators have remote control over the network and can change the network characteristics, such as services and connectivity, based on the workload patterns. This enables administrators to have more efficient and instant access to the configuration modifications based on the network status.
3. *Multitenancy:* The concept of SDN can be expanded across multiple partitions of the networks, such as the data centers and the data clouds; this enables different players of the network to interact with each other in a more secure and efficient manner and provides the administrator with better tools for monitoring.
4. *Virtual application networks:* Virtual application networks can be implemented by network administrators using the Distributed Overlay Virtual Network (DOVE), which helps in the transparency, automation, and better mobility of the network loads that have been virtualized.

**OpenFlow Concept**

There are several protocol standards on the use of SDN in real-life applications. One of the most popular protocol standards is OpenFlow (Figure 4.2) [2]. OpenFlow is a protocol that enables the implementation of the SDN concept in both hardware and software. This approach is used by scientists to make new protocols and test their ability in network performance and analyze it in real network scenarios. An added feature is that a scientist can use the existing hardware to design new protocols and analyze their performance. Now, it is becoming a part of commercially available routing devices (routers) and switches as well.

In conventional routing devices, both the packet forwarding algorithms (datapath) and the routing algorithms (control path) are implemented by hardware modules on the same device; however, OpenFlow protocols decouple these operations from each other.

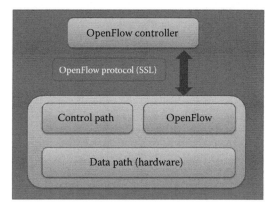

**Figure 4.2** OpenFlow architecture.

An isolated control algorithm, the controller, is used to control the routing algorithms and is implemented as a separate standard server, whereas the datapath remains as a routing hardware component. The OpenFlow protocol is the main control protocol in the new devices, which are termed as the OpenFlow switch and the OpenFlow controller. They use the OpenFlow protocol for communications among different devices and for carrying out operations, such as packet reception, packet transmission, and modifications in the routing tables, etc.

Initially, the datapath of the OpenFlow routing devices has an empty routing table with many fields. This table contains several packet fields that are searched, as well as an action field that contains the code for different actions, such as the destination of different ports (receiving or transmission), packet forwarding or reception, etc. This table is populated based on the new data packets. When a new packet is received, which has no matching entry in the data flow table, it is forwarded to the controller to be taken care of. The controller is responsible for packet handling decisions, for example, a new entry on how to deal with this and similar packets received in the future, is added into the data flow table [3].

SDN has the capability of programming multiple switches simultaneously, but still, it is a distributed system, and therefore, it suffers from conventional complexities, such as dropping packets, delaying control packets, etc. Currently used platforms for SDN, such as Nox and Beacon, enable programming, but it is still hard to program them in a low level.

## Language Abstractions for SDNs

SDN uses an event-driven paradigm, where the software application reacts when certain events take place (Figure 4.3) [4,5]. It uses a set of rules embedded in the routing device or switches when packets are received; however, there are no rules present in the data flow tables on how to forward them.

One of the problems is the decoupling of the control function into two parts: (1) the controller with the program and (2) the set of rules implemented on the routing devices. This has an implication of making the programmer worry about the low-level details that also include the switch hardware. The NetCore programmers write the specification that captures the intended forwarding behavior of the network instead of write programs dealing with the low-level details, such as the events and the forwarding rules of the network. This enables interactions between the controllers and switches. A compiler transforms these specifications into code segments for both controllers and switches. A prominent feature of the NetCore allows descriptions of the network rules and policies in terms of the simple specifications, which cannot be implemented or realized directly in the switches. Another important fact about NetCore is that it has a clear formal set of rules that provide a basis for reasoning about programs [4].

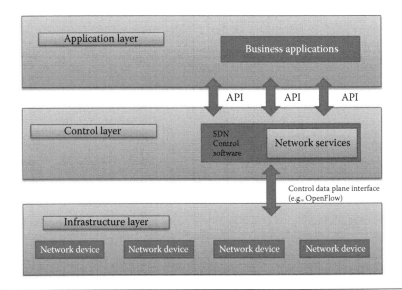

**Figure 4.3**    Programming of the SDN and language abstraction.

*Network Query Abstractions*

In SDNs, each switch stores counters for different forwarding rules. They are for the counts of the total number of packets and data segments processed using those rules. For traffic monitoring, the controller has the ability to check the different counters associated with the different sets of forwarding rules. This allows programmers to monitor the fine details of implementation in the switches. This is a tedious job and makes the program complicated. Therefore, an added level of abstraction will help the programmers. To support applications whose correct operation involves a monitoring component, Frenetic includes an embedded query language that provides effective abstractions for reading network state. This language is similar with structured query language (SQL) and includes segments for selecting, filtering, splitting, merging, and aggregating the streams of packets flowing through the network. Another special feature of this language is that it enables the queries to be composed of forwarding policies. A compiler produces the control messages needed to query and tabulate the counters in the switches.

*Consistent Update Abstractions*

Because SDNs are event-driven networks, some programs in SDNs need to change their policy from one state to another because of the changes in the network topology, failures in the network, etc. An ideal solution is the automatic update of all the network switches, but practically, it is not possible to implement. A possible solution is to allow some level of abstraction and then propagate these changes from one device to another. An example is the per-packet consistency, which ensures that each packet going through the network uses just one version of the policy (instead of a combination of both the old and the new policies). This preserves all properties that can be expressed in terms of individual packets and the paths that they take through the network—a class of properties that subsumes important structural invariants, such as basic connectivity and loop freedom, as well as access control policies. Going a step further, per-flow consistency ensures that sets of related packets are processed with the same policy. Frenetic provides an ideal platform for exploring such abstractions, as the compiler and run-time system can be used to perform the tedious bookkeeping related to implementing network updates [6].

### Languages for SDNs

Conventional computer networks are responsible for carrying out several operations, such as monitoring of network traffic, routing of data in the network, access control in the network, and server workload balancing. Apart from this, computer networks are made up of several different types of devices, such as combination of routers, switches, firewall, repeaters, etc. This heterogeneous nature of different modules present in the networks makes the network management a difficult and complicated task. A simple solution is provided by SDN by achieving the task of network resource management using a simple and neat interface among several different devices and the software module used to control them. As previously mentioned, OpenFlow is an innovative idea where a standard protocol is used for changing the behavior of devices (programming them) using low-level application programming interface (API) that mimic the underlying hardware components. SDN requires a higher level of abstractions to create applications. An example is the Frenetic project, where the goal is to provide a simple and higher level of abstraction for making applications with three main areas in mind: (1) monitoring of traffic in a network, (2) managing (creation and composition) packet forwarding policies in the network, and (3) ensuring the consistency in updating those policies [5]. By providing these abstractions, the job of the programmer becomes easy and efficient in creating and managing new applications for SDN without a need of worrying about the low-level details.

### Querying Network State

The Frenetic project uses a language that supports a high-level query scheme for subscribing to information streams. It gives information about the state of the network, including traffic statistics and topology changes. The run-time system is responsible for managing the polling switch counters, gathering statistics, and reacting to the happening of events. In the Frenetic project, the specification of the packet forwarding rules in the network is defined by the use of a high-level policy language that makes it really simple and easy for programmers. Different modules can be responsible for different operations, such as routing, discovery of the topology of the network, workload balancing, access control, etc. This modular design is used to register each module's task with the run-time system, which is responsible for the composition, automatic compilation, and optimization of the programmer's requested tasks.

*Reconfiguring the Network*

To update the global configuration of the network, the Frenetic project provides a higher level of abstraction. This allows programmers to configure the network without physically going to each routing device to install or change the packet forwarding rules in each table. This process is very tedious and is prone to errors. The run-time system makes sure that, during the updating process, only one set of rules is applied to them: either the old policy or the new policy, but not both of the rules. This makes sure that there are no violations for the important invariants, such as connectivity, control parameters of the loops, and access control, when the OpenFlow switches from one policy to another [3].

The Frenetic language project is an aggregation of simple yet powerful modules that provide an added level of abstraction to the programmer for controlling the routing devices from the central application. This added layer of abstraction runs on the compiler and the run-time system, and is vital for the fast and efficient execution of the code.

## Virtualization Layer for SDNs

As mentioned before, the virtualization layer helps in the development and operation of the SDN slice on the top of shared network infrastructures. A potential solution is the concept of AutoSlice [7]. This provides the manufacturer with the ability to redesign the SDN for different applications while the operator intervention is minimized. Simultaneously, the programmers have the ability to build the programmable network pieces that enable the development of different services based on the SDN working principles.

Network virtualization enables the development of a viable and working solution toward the development and functionality of the small segments of the network slices on the top of a common network infrastructure. SDNs can combine different programmable routing devices such as switching hardware by facilitating the application of the network services, such as centralized control, network-wide availability, and visibility. These features of SDNs enable the network clients to have control over their segments of slices, implementation of custom forwarding decisions, security policies, and configuration of access control, as needed.

FlowVisor is considered to be a fundamental building block for SDN virtualization and is used to partition the data flow tables in switches using the OpenFlow protocol by dividing it into flow spaces. This results in hard switches that can be manipulated concurrently by several software controllers. Nevertheless, the instantiation of an entire virtual SDN (vSDN) topology is nontrivial because it involves numerous operations, such as mapping vSDN topologies, installing auxiliary flow entries for tunneling, and enforcing flow table isolation. Such operations need a lot of management recourses and planning of the network infrastructures. The goal is to develop a virtualization layer that is called the SDN hypervisor. It enables the automation of the deployment process and the operation of the vSDN topologies with the minimum interaction of the administrator. vSDN focuses on the scalability aspects of the hypervisor design of the network.

In Ref. [8], an example is presented in which a network infrastructure is assumed to provide vSDN topologies to several tenants. The vSDN of each tenant takes care of several things, such as the bandwidth of the link, its location, the switching speed (capacity), etc. The assumption in this example is that every tenant uses switches that use OpenFlow protocol standards, with a flow table that is partitioned into several segments. The proposed distributed hypervisor architecture has the capability of handling a large amount of data flow tables for several clients.

There are two very important modules in its hypervisor: management module (MM) and multiple controller proxies (CPX). These modules are designed in such a manner that it distributes the load control over all the tenants. Upon the receipt of a request, MM inquires the vSDN about the resources available in the network with every SDN domain and then accordingly assigns a set of logical resources to each CPX. Next, each CPX initializes the allocated segment of the topology by installing flow entries in its domain, which unambiguously bind traffic to a specific logical context using tagging. Because the clients are required to be isolated from each other, every CPX is responsible to do a policy control on the data flow table access and make sure that all the entries in these tables are mapped into segments that are not overlapping. CPX are responsible for controlling the routing switches. Moreover, CPX take care of all the data communication between the client controller and the forwarding plane. A new entry into the switch has to follow certain steps. First, the proxy creates a control message for the addition of a new entry into the switch flow table in such a manner that all references (addresses) to memories are replaced by the corresponding physical entries, and corresponding traffic controlling actions are added into the packet. The proxy is responsible for maintaining the status of each virtual node in a given SDN. As a result, the CPX has the ability to independently transfer virtual resources within its domain to optimize interdomain resource allocation. The MM portion is to optimize global parameters. The transport control message translation is used to enable the tenants to have access to the packet processing set of rules within a specific SDN layer without having to disturb simultaneous users.

If there are a number of clients in a network, we need a large number of flow tables in the memory of a routing switch. The job of CPX is to make sure that all the flow tables are virtually isolated, all packet processing takes place in the correct order and in a timely manner, and all the actions are carried out in case a connected group of virtual nodes is being mapped to the same routing device. In the OpenFlow routing devices, there is a problem on the scalability of the platform because of the large flow table size. There could be a large number of entries in the flow table. To deal with such situation, an auxiliary software data (ASD) path is used in the substrate network. For every SDN domain, an ASD path is assigned. It has a software routing switch that is running on a separate server. The server has enough memory to store all the logical flow tables that are needed by the corresponding ASD path as compared with the limited space on the OpenFlow routing devices. Although the software-based datapath (as in the SDN) architecture and commodity have achieved recent advantages, there is still a huge gap between the software-defined OpenFlow protocol and the actual hardware components. Hardware-implemented modules have a higher capacity [6].

To overcome these limitations, the Zipf property of the aggregate traffic [9], i.e., the small fraction of flows, is responsible for the traffic flowing through the network. In this technique, ASD paths are used for handling heavy data traffic while only a very small amount of high volume traffic is cached in the dedicated routing devices.

## Discussion

There are still questions on how to make the SDN more efficient, how to optimize it across all the network sets, and how to define the tradeoffs between different implementations. There is a need to have a quantitative approach for evaluating the performance and efficiency of the SDNs. Controllers are responsible for managing the packet forwarding tasks. There are several practical applications and commercial applications/products based on SDN. There is still the question of how to quantify and test their performance and what are the best tradeoffs in the networks.

Some of the important parameters while considering the performance of the SDN are scalability, availability, and latency. These issues are vital and must be addressed before choosing SDN for any application. The situation about the SDN design is made complicated because of several factors. A few of them are listed below [10].

1. *Topologies vary:* In different networks, there are different numbers of parameters, such as the number of edges, number of nodes, distance between the nodes, connectivity of the nodes, etc. Therefore, it is a very interesting research area to find out how to generate reliable network topologies.

2. *Finding relevant metrics:* It is a challenging topic to decide the proper metrics for the SDN performance evaluation. Which metrics or factors are the most important in the evaluation of the performance of the SDN, for example, is having a low average cost among a large number of sets less important than having a low delay among them?

3. *Combining metrics:* A good solution is to combine several factors (parameters) and use them for evaluation. However, the question of how to combine different criteria or come up with a good solution remains the same.

4. *Computational complexity:* Another great problem is the optimization of each factor in consideration. This poses an non-deterministic polynomial-time hard (NP-hard) problem for each metric. Some of these factors/metrics include, but are not limited to, the availability of resources, how fair the distribution of the resources is, and the latency and congestion in different channels.

5. *Design space size:* There are several options to implement the scalability and fault tolerance across multiple nodes in the network. Some of these parameters are the number of controllers, where to place these controllers, the number of state replicas, how to distribute the processing method, and how many switches are to be connected to each controller in the SDN.

Most of these factors can be addressed in different manners, and they can be addressed by repeating analysis on a large number of different topologies to find out the embedded trends. Other factors can be addressed by the use of some approximation algorithms, and the rest could be addressed through simplified models of distributed system communication.

# References

1. Casado, M., M. J. Freedman, J. Pettit, J. Luo, N. Gude, N. McKeown, and S. Shenker. 2009. Rethinking enterprise network control. *IEEE/ACM Transactions on Networking* 17(4).

2. McKeown, N., T. Anderson, H. Balakrishnan, G. Parulkar, L. Peterson, J. Rexford, S. Shenker et al. 2008. OpenFlow: Enabling innovation in campus networks. *SIGCOMM CCR* 38(2):69–74.

3. Kudou, H. and B. C. Victoria. 2011. Effects of routing granularity on communication performance in OpenFlow networks. *2011 IEEE Pacific Rim (PacRim) Conference on Communications, Computers, and Signal Processing.* IEEE. pp. 590–595.

4. Monsanto, C. and A. Story. 2013. *Language Abstractions for Software-Defined Networks.* Communications Magazine, IEEE, vol. 51, no. 2, pp. 128–134.

5. Foster, N., A. Guha, M. Reitblatt, A. Story, M. J. Freedman, N. P. Katta, C. Monsanto et al. 2013. Languages for software-defined networks. *IEEE Communications Magazine* 51(2):128–134. doi:10.1109/MCOM.2013.6461197.

6. Kotani, D., K. Suzuki, and H. Shimonishi. 2012. A design and implementation of OpenFlow Controller handling IP multicast with Fast Tree Switching. *2012 IEEE/IPSJ 12th International Symposium on Applications and the Internet (SAINT)*. IEEE. Izmir, Turkey, pp. 60–67.

7. Bozakov, Z. and P. Papadimitriou. 2012. AutoSlice: Automated and scalable slicing for software-defined networks. *Proceedings of the 2012 ACM CoNEXT Student Workshop*. ACM, New York, NY, USA, 3–4.

8. Schaffrath G., C. Werle, P. Papadimitriou, A. Feldmann, R. Bless, A. Greenhalgh, A. Wundsam, M. Kind, O. Maennel, and L. Mathy. 2009. Network virtualization architecture: Proposal and initial prototype. In Proceedings of the 1st ACM workshop on Virtualized infrastructure systems and architectures (VISA '09). ACM, New York, NY, USA, 63–72.

9. Sarrar, N., S. Uhlig, A. Feldmann, R. Sherwood, and X. Huang. 2012. Leveraging Zipf's law for traffic offloading. *SIGCOMM Comput. Commun. Rev.* 42, 1, 16–22.

10. Heller, B. *Quantitatively Evaluating (and Optimizing) Software-Defined Networks.* http://changeofelia.info.ucl.ac.be/pmwiki/uploads/SummerSchool/Program/poster_023.pdf.

# 5

# Control and Management Software for SDNs

## Conceptual Models and Practical View

NATALIA CASTRO FERNANDES AND
LUIZ CLAUDIO SCHARA MAGALHÃES

Contents

**Architectural View**

Software defined networks (SDNs) are based on the separation of data and control planes. The control plane abstraction allows an easy deployment of network applications based on a central view of the network and also allows an unprecedented high-level view of network configuration. The network operating system, which is the software inside the network controller, provides a simple programming interface and a topological view of the network, which simplifies the tasks of network engineers.

Because OpenFlow is the main SDN implementation, we detail in this chapter the OpenFlow architecture. The key idea of an SDN is to split the network forwarding function, performed by the data plane, from the network control function, performed by the control plane. This allows a simpler and more flexible network control and management and also network virtualization.

In OpenFlow, the data and control planes are executed in different network elements. Figure 5.1 shows the OpenFlow general architecture view. The network controller communicates with the OpenFlow switches using the OpenFlow protocol through a secure channel.

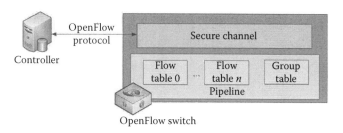

**Figure 5.1** OpenFlow general architecture view.

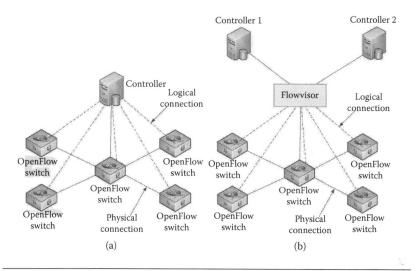

**Figure 5.2**   Typical OpenFlow networks. (a) Without virtualization. (b) Using virtualization.

Using this connection, the controller configures the forwarding tables of the switch.

Figure 5.2a shows an example of a typical OpenFlow network using a single controller. Figure 5.2b shows the use of virtualization in the same network. OpenFlow provides a proper architecture for network virtualization because of the use of a shared flow table. To allow different controllers to access the shared flow tables, FlowVisor is used. Further details about network virtualization using OpenFlow are given in Control, Monitoring, and Virtualization Primitives.

The OpenFlow protocol defines the communication among the controller and the OpenFlow switches. This communication channel is used to send configuration and collect monitoring data from the network. Indeed, every time a packet does not match any of the flow rules installed in the switch, the packet header is forwarded to the controller. Based on the received data, the controller configures the switch to correctly forward the packet. In the rest of this section, we detail the OpenFlow control and data planes, the communication channels, and the network virtualization.

*Control Plane*

In OpenFlow networks, the control plane is executed inside the controller, whereas the data plane is executed inside the switches. The control plane can be executed in a centralized or distributed way, as shown in Figure 5.3, but the usual model is to use a centralized control plane.

The communication channel is used to link these planes, forwarding data among them. Using the OpenFlow protocol, the controller can add, update, and delete flow entries.

The control plane builds a network view based on the data received from the network. First, the controller identifies the nodes and links in the network using the Link Layer Discovery Protocol (LLDP). To do so, the controller iterates over all switch ports of the network and sends out LLDP packets periodically. When a switch receives an LLDP packet, it forwards the header to the controller because the match rule of this flow is not set. Therefore, the controller can infer the link-level connectivity by combining the data of each received LLDP packet. The controller exposes the topology to the applications through an interface that is controller specific.

The controller also presents an interface for programming the network. The network operating system consists of a programming interface coupled to the network view. The key idea is that the network programming interface works in a way similar to the system calls in machine operating systems. In a nutshell, OpenFlow enables transparency and flexibility to the network by providing an open interface to a black box networking node.

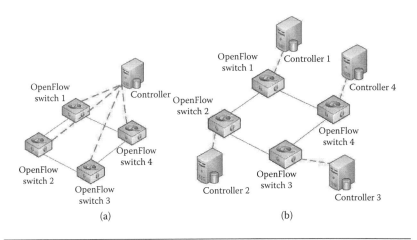

**Figure 5.3**  OpenFlow network control models. (a) Centralized model. (b) Distributed model.

The controller programming interface depends on the primitives available on the data plane. These primitives are also closely related to the switch components. Therefore, we will present briefly the major switch components and, after that, tie those in with the main control and monitoring primitives (based on the OpenFlow 1.1 specification).

*OpenFlow Data Plane*

An OpenFlow switch is composed of one or more flow tables and one group table OpenFlow [1]. Each table is composed of a set of flow entries, whereas a flow entry consists of a set of match fields, counters, and instructions, as shown in Figure 5.4.

Match fields consist of the ingress port and packet headers and, optionally, metadata specified by a previous table. The match fields can be completely specified or wildcarded, which allows the administrator to group sets of flows.

The counters are values that store statistics. A list of the specified counters in OpenFlow 1.1 is in Table 5.1.

Instructions are activities over an action set, which must be performed whenever a packet of that flow arrives. Table 5.2 contains a list of the instructions in OpenFlow 1.1. An action set presents, at most, one action of each type. The action types are in Table 5.3. We can observe that some actions are required, which means that they are implemented in all OpenFlow 1.1 switches, whereas others are optional.

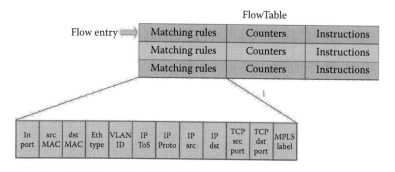

**Figure 5.4** Flow entry structure, which contain instructions, counters, and a matching rule that consists of a set of matching fields. ToS, type of service.

**Table 5.1**    List of Counters of OpenFlow 1.1

| COUNTER | TYPE | BITS |
|---|---|---|
| Reference counter | Per table | 32 |
| Packet lookups | Per table | 64 |
| Packet matches | Per table | 64 |
| Received packets | Per flow | 64 |
| Received bytes | Per flow | 64 |
| Duration (s) | Per flow | 32 |
| Duration (ns) | Per flow | 32 |
| Received packets | Per flow | 64 |
| Transmitted packets | Per flow | 64 |
| Received bytes | Per flow | 64 |
| Transmitted bytes | Per flow | 64 |
| Received drops | Per flow | 64 |
| Transmitted drops | Per flow | 64 |
| Received errors | Per flow | 64 |
| Transmitted errors | Per flow | 64 |
| Received frame alignment errors | Per flow | 64 |
| Received overrun errors | Per flow | 64 |
| Received CRC errors | Per flow | 64 |
| Collisions | Per flow | 64 |
| Transmitted packets | Per queue | 64 |
| Transmitted bytes | Per queue | 64 |
| Transmitted overrun errors | Per queue | 64 |
| Reference count | Per group | 32 |
| Packet count | Per group | 64 |
| Byte count | Per group | 64 |
| Packet count | Per bucket | 64 |
| Byte count | Per bucket | 64 |

*Source:* OpenFlow. 2011. OpenFlow Switch Specification—
Version 1.1.0 Implemented (Wire Protocol 0×02).

**Table 5.2**    List of Instructions in OpenFlow 1.1

| INSTRUCTION | MEANING |
|---|---|
| Apply-actions, *action(s)* | Apply the action immediately, without changing the action set. It can modify the packet or execute multiple actions of the same type. |
| Clear-actions | Clear the action set immediately. |
| Write-actions, *actions(s)* | Merge the specified actions in the current action set. If an action is already in the set, it is overwritten by the new one. |
| Write-metadata, *metadata/mask* | Write masked metadata in the metadata field. |
| GoTo-table, *table* | Specify the next table in the processing pipeline. |

**Table 5.3** List of Actions in OpenFlow 1.1

| ACTION | MEANING | REQUIRED |
|---|---|---|
| Output | Forward packets to the specified physical port(s) or to the virtual ports: ALL, CONTROLLER, TABLE, or IN_PORT (ALL = send the packet to all ports except the input port; CONTROLLER = send the packet to the controller; TABLE = send the packet to be processed from the first flow table; IN_PORT = send the packet to the input door). | Yes |
| Output | Forward packets to the virtual ports: LOCAL, NORMAL, or FLOOD (LOCAL = send the packet to the local networking stack of the switch; NORMAL = process the packet as a non-OpenFlow switch; FLOOD = flood the packet as a non-OpenFlow switch). | No |
| Set-queue | Set an output queue to the flow for providing QoS. | No |
| Drop | Drop the packet. | Yes |
| Group | Process the packet through a specified group. | Yes |
| Push-tag | Push header fields, such as VLAN headers and MPLS headers, in the packet. | No |
| Pop-tag | Pop header fields, such as VLAN headers and MPLS headers, from the packet. | No |
| Set-field | Modify the value of a specific header field, such as the TTL, MAC address, VLAN ID, etc. | No |

The packet is processed in OpenFlow switches in the OpenFlow pipeline, described in Figure 5.5.* The flow tables of an OpenFlow switch are numbered from 0 to *n*. The pipeline processing always begins on table 0. Other tables will be used depending on the result of the match in the first table. If there is a match, which means that the rule for that flow is set, then the instructions of that flow are applied over the action set. The instructions modify the action set by adding or deleting actions, such as packet forwarding, packet modification, group table processing, and pipeline processing.

Pipeline processing is performed by explicitly directing the packet to another flow table using the GoTo instruction. In this case, the instructions to modify the packet header and update the match fields are performed before the new match is executed. In case of match success in the next table, another set of instructions will be applied to

---

* OpenFlow switches are classified into OpenFlow-only and OpenFlow-hybrid. OpenFlow-only switches uses only the OpenFlow pipeline, whereas OpenFlow-hybrid can process packets through the OpenFlow pipeline or as an ordinary switch. The classification of the packet flows to ordinary switch or OpenFlow switch is specified by each manufacturer of OpenFlow-hybrid switches. From now on, we describe the OpenFlow pipeline, which is common to both OpenFlow switch types.

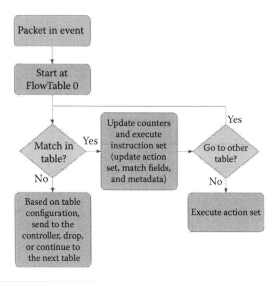

**Figure 5.5**   Detailed pipeline scheme of OpenFlow switches 1.1.

the action set. Before sending the packet to the output port(s), every instruction in the action set is performed. If no match is found in a table, the switch response will depend on the table configuration. The default action is to send the packet to the controller, but it is possible to configure the switch to drop the packet or send it to the next table in the pipeline. A more detailed specification of the flow match in each table is in Figure 5.6. Figure 5.7 shows an example of pipeline processing in OpenFlow switches.

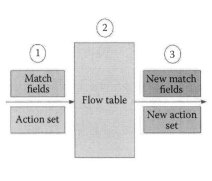

1) – Find the highest priority matching flow entry, based on the ingress port, packet headers, and metadata.
2) – If there is a match, apply instructions, creating the new match fields and the new action set:
a) Apply action instruction, which modify the packet and update the match fields.
b) Update action set, which means to clear actions and/ or write action instructions.
c) Update metadata.
3) – Send the new match fields and the new action set to the next table.

**Figure 5.6**   Per-table processing in the OpenFlow pipeline.

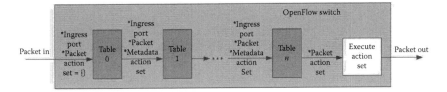

**Figure 5.7**   Example of processing pipeline in OpenFlow.

In addition to the *n* ordinary tables, an OpenFlow switch also presents a group table, which specifies actions to a group of flows. This table has a special description to allow the use of broadcast, multicast, and anycast primitives. A group table entry consists of the fields specified in Figure 5.8.

The group identifier is a 32-bit field that specifies the group number. The counters work in a way similar with the counters of ordinary forwarding tables, storing group statistics. The action buckets are a set of instructions associated to the group. It corresponds to a set of actions that modify the packet header and/or send the packet to the output port. The administrator can assign different action buckets to each group. The group type describes how the group works in terms of the use of the action bucket list. Hence, this field can assume the following values:

1. **all:** This is used for broadcast and multicast. The packet flows through all the specified action buckets.
2. **select:** This is used for anycast. Only one of the action buckets will be used. The action-bucket selection algorithm is out of the scope of OpenFlow, being chosen by the switch manufacturer. Usually, it is implemented using hash functions or in a round-robin manner.
3. **indirect:** This is similar with the group type "all," but uses a single action bucket.
4. **fast failover:** This is used to recover fast from failures. In this case, a set of action buckets are specified, and the flow is forwarded to the next action bucket. In case of a failure, for

| Group identifier | Group type | Counters | Action buckets |
|---|---|---|---|

**Figure 5.8**   Group table entry.

instance, the link of the output port of the first action bucket is down, so the packets are redirected to the next action bucket, without requiring a round trip to the controller.

*OpenFlow Protocol and Communication Channel*

The OpenFlow protocol allows the use of different groups of messages:

1. **Controller-to-switch messages**, which are sent from the controller to the switches to manage or inspect the switch state. These messages allow the controller, among others, to configure the switch, modify states, and modify flow entries.
2. **Asynchronous messages**, which are generated by the switch without the intervention of the controller because of asynchronous events in the network. Examples include the arrival of a new flow in the switch, the expiration of a flow, and errors.
3. **Symmetric messages**, which are generated in either direction (from controller to switch or from switch to controller), but without a solicitation. Examples are hellos, echoes, and experimenter/vendor messages.

The communication channel is used to exchange the OpenFlow protocol messages among the controller and the switches. The message exchange is performed through a transport layer security (TLS) connection. Hence, switches and controllers must mutually authenticate exchanging certificates.

*Control, Monitoring, and Virtualization Primitives*

The control and monitoring primitives are defined by the OpenFlow switch capabilities. The monitoring primitives consist of the set of primitives used to read the value of the counters from each switch (Table 5.2). The control primitives are defined based on the set of actions that the switch can perform, combined with the switch pipeline capabilities. Hence, the control logic is designed based on the instruction set and the action set. Based on the instructions and actions, the network administrator can define how the packet should be processed, specifying the table chain that will be followed by a packet and in which moments the actions should be executed or rewritten.

The virtualization primitives are defined based on the network virtualization model. OpenFlow networks are virtualized by inserting an additional plane between the control and the data planes. Currently, FlowVisor is the tool used to create this extra plane [2].

A simple virtualization model for OpenFlow networks would be to define separate sets of applications in a single controller, each set controlling its own network slice. The main disadvantage of this approach is that there is no isolation in the control plane. One cannot guarantee that the applications of one slice will not reduce the performance of the applications in another slice because all of them share the same controller structures and are run over the same hardware without any protection. Moreover, there is no way to guarantee that a set of applications will work only on its specified slice. Hence, we also cannot guarantee the isolation in the data plane [3].

FlowVisor guarantees the isolation in both control and data planes. First, it allows the creation of many controllers in the network. Second, it ensures that one controller cannot interfere in the other controller's virtual data plane. These features are achieved because FlowVisor works as a proxy between the controllers and the switches, as shown in Figure 5.9. Hence, this network virtualization software multiplexes

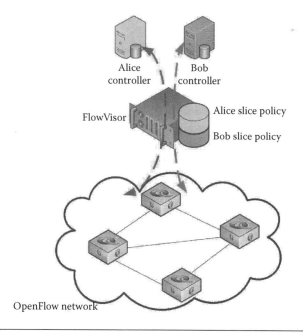

**Figure 5.9**   Logical interconnection among OpenFlow switches and controllers.

the messages to and from the switches according to the specification of the network slice. Therefore, if a controller sends a message to configure the data plane of a switch, it will send this message to the FlowVisor. FlowVisor, then, checks whether the flows affected by the controller message belong to the controller's network slice. If so, it forwards the message to the corresponding switches.

When virtualizing the OpenFlow network, all the switches are configured, pointing to FlowVisor as the controller. When a message is sent from the switches to the controller, FlowVisor intercepts this message to check which controller(s) should receive that message.

The network slice defines the resources of each controller. In FlowVisor, the features that define a slice consist of

1. *Bandwidth:* Each slice can receive a limited bandwidth to guarantee that all the slices will have, at least, a guaranteed minimum bandwidth. This feature can be implemented only if the data plane supports quality of service (QoS) primitives.
2. *Topology:* The virtual topology viewed by a controller may not include all the switches and links in the network. This partial network view can be guaranteed if the virtualization plane filters logical link control (LLC) messages received by each physical switch.
3. *Traffic:* The virtualization plane must determine which set of flows belong to each control plane to guarantee a consistency in the network.
4. *CPU:* The switch CPU is shared among all the control planes. Therefore, it is important to guarantee that it is not overloaded by one slice, reducing the performance of the other slices. Hence, the virtualization plane must infer CPU use and to control CPU sharing among the slices.
5. *Forwarding tables:* There is a physical limit to the number of flow entries in an OpenFlow switch. Therefore, the virtualization plane must also ensure that a virtual network will not use more entries than the threshold specified for that slice.

These are the basic features for creating virtual networks in OpenFlow. By controlling these features, it is possible to create isolated virtual network slices.

## SDN Controllers

There are many flavors of SDN controllers, which were developed with different objectives. Here, we describe some of the most relevant controllers, including the first proposals and the currently most popular software, highlighting their motivations and functionalities.

### Intelligent Route Service Control Point

The Intelligent Route Service Control Point (IRSCP), also called the Routing Control Platform (RCP), is a non-OpenFlow SDN controller [4,5]. This controller selects interdomain routes, configuring the routers that belong to its autonomous system (AS). Because IRSCP controls forwarding rules, it is considered as an SDN with an architecture different from that proposed by OpenFlow. In IRSCP, the Border Gateway Protocol (BGP) is used as the control protocol, and the data plane consists of the IP forwarding tables. Indeed, this architecture is less flexible than OpenFlow because the control is restricted to the IP layer and the use of BGP.

IRSCP was proposed to overcome the complexity of the Internet's routing infrastructure. The main routing limitations that it addresses, such as forwarding loops, signaling partitions, and scalability issues, arise from the distributed path selection performed by routers. The key idea is that the distributed interdomain routing should be executed in centralized servers instead of in BGP routers. Each AS has an IRSCP server, which talks to other IRSCP servers placed in different ASs, exchanging reachability data and selecting paths for the routers. The general IRSCP architecture is shown in Figure 5.10 [6].

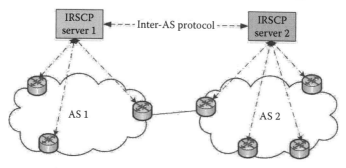

**Figure 5.10**   IRSCP general architecture, in which the edge routers are connected to the IRSCP server, responsible for the interdomain routing.

The main advantages of approaches such as IRSCP are simpler traffic engineering, easier policy expression, and more powerful troubleshooting than BGP. It also preserves the AS autonomy for choosing routes, which is an important feature in the Internet. IRSCP also proposes a way to migrate from BGP to this new proposal. Indeed, the authors point out that the use of IRSCP even in a single AS would bring many benefits for that AS.

*Secure Architecture for the Networked Enterprise and Ethane*

Secure Architecture for the Networked Enterprise (SANE) and Ethane were precursors of the controllers for OpenFlow networks [7]. These controllers aimed to reduce costs and make network management in enterprise networks easier. Indeed, up to these developments, enterprise networks usually required manual configuration for many functionalities and, therefore, presented security issues, which causes frequent network downtimes [8]. The key idea introduced by SANE and Ethane was that, instead of building a complex management plane, it would be interesting to change the network architecture to make it more manageable [9].

Defined as a clean slate approach, SANE [10] was developed with the objective of building a secure enterprise network. Based on the idea of a central control, SANE proposes that all access control decisions should be performed by a logically centralized server, which uses high-level access control policies to create encrypted source routes. SANE was developed only as a prototype, but it provided the first insights for building Ethane.

Because Ethane was based on experiences with other controllers, it was more mature than the other proposals. This allowed its implementation and testing in the production network of the computer science department of Stanford, which is composed of more than 300 nodes. Based on this operational experience, researchers proposed changes and extensions to the original design of Ethane.

The basic principles of Ethane are as follows:

1. Policies with high-level names govern the network. Hence, instead of using low-level names, such as dynamically allocated addresses, the network should use the proper name of

the entities involved in the network, for instance, users and hosts. This kind of policy declaration is easier to describe and debug.

2. Routing must consider policies because they have implications on the type of communication between network entities, which route packets should follow, traffic priorities, and also the need of using middle boxes. Hence, considering routing and policy management in different stages will probably reduce network efficiency.

3. Because security is a main issue in current networks, the network core should enforce a trustful binding between a packet and its source/entity. In current transmission control protocol (TCP)/IP networks, the packet origin is specified as the source IP address, which can be dynamically updated. Moreover, as the network core usually does not verify the packet origin, malicious entities can spoof the source address. The use of high-level names in the packet level allows the identification of the real source entity of any data in transit, which include the host and the user. For keeping this binding consistent, additional mechanisms for tracking service and device mobility are also required in Ethane.

To accomplish these basic principles, Ethane adopted a centralized control architecture, which is not proper for large-scale networks but fits well for enterprise network management. The use of replication techniques are enough to guarantee resilience to the faults in the controller.

*OpenFlow Reference Controller*

The OpenFlow reference controller, also called the OVS controller, is a simple controller that is distributed with OpenFlow. Its main purpose is to provide the simplest reference implementation of an OpenFlow controller. Usually, the OVS controller is used to check OpenFlow installations, verifying basic functions, such as creating L2 media access control (MAC)-learning switches or hubs. Indeed, this controller can manage multiple OpenFlow switches using the OpenFlow protocol.

*Nox*

Nox is currently one of the main OpenFlow controllers. Nox introduced the idea of a network operating system (NOS), which provides a uniform and centralized interface to program the entire network. The idea is that a centralized network control and management is simpler to perform and more efficient than a distributed approach [11].

In Nox, the control and management functions are developed as applications of the network operating system, as shown in Figure 5.11. Hence, control functions, such as routing, spanning tree, data monitoring, load balance, etc., are provided as applications that run over Nox.

The Nox interface automatically provides a network view that includes the OpenFlow network topology. Based on this topology, the control and management algorithms can perform their duties. Nox also provides an interface for collecting monitoring data on switches and flows. However, the application developer still has to decide with which granularity the monitoring data are acquired from the switches.

The full basic set of OpenFlow actions are provided by Nox. This basic set includes operations, such as forwarding to one or more output interfaces, forwarding to the controller, dropping the packet, and modifying packet header fields. Events on the network, such as a

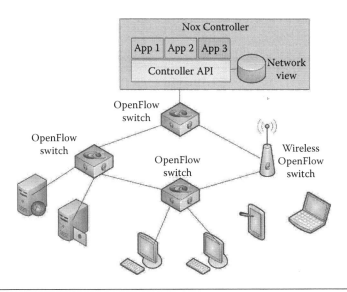

**Figure 5.11**    Nox controller in an OpenFlow network.

connection of a new node, link loss, and arrival of a packet from an undefined flow, are notified to the controller, which calls the applications that are bound to that type of event.

Nox can set flows proactively or on demand. In the on-demand approach, when a packet of a nonspecified flow arrives at the switch, the switch forwards the packet header to the controller. The controller, based on the received data, sets the adequate flow matching rule on the switch. In the proactive mode, the controller can preset matching rules for known flows or create wildcard rules for these known flows. The idea is to reduce the delay to set a flow. Indeed, the controller should optimize the network performance by balancing both methods.

Initially, Nox was developed in C++ and Python, but this implementation had some performance issues. The code was rewritten, and a version of Nox was released, written exclusively in C++. Python developers can use Pox, which is a version of Nox written in Python. Other controllers were also developed based on Nox, such as Jaxon, which is a Java-based OpenFlow controller that provides a Java thin interface that interacts with Nox. Simple network access control (SNAC) is another controller derived from Nox, which provides a Web interface that uses a flexible policy-definition language. SNAC makes it easier to configure devices and monitor the network.

*Onix*

Onix is an SDN controller that runs on a cluster of one or more physical servers [12]. Each server may run multiple instances of Onix. This controller provides a programmatic access to the network, which allows the cluster of servers to set and get the network state. One of the main advantages of Onix is that the network control plane can be implemented as a distributed system.

The main differences of Onix from Nox are that Onix is more reliable, and scales and provides a more flexible programming application programming interface (API). Onix, on the other hand, presents an interface that efficiently handles failure events on the network. Indeed, one of the Onix instances is responsible for disseminating network state to other Onix instances in the cluster. Scalability is achieved by using a cluster of servers instead of a single server, as in Nox. The servers in the cluster communicate and concurrently control the network. In terms of programming flexibility, Onix allows designers to implement distributed control applications. In Nox, applications must be redesigned to work in a central mode.

Onix provides an API for implementing network control logic. Onix does not collect and show network behavior but provides primitives for accessing the network state. The control platform implemented in Onix acquires data from network nodes and distributes it through the appropriate instances of Onix in the cluster. This platform also coordinates the distribution of control data generated by Onix instances to the network nodes.

One important characteristic of Onix is that the proposed platform does not restrict the choice of protocols used to control and manage the network nodes. OpenFlow is one of the possible protocols that can be used with Onix. In that case, it is also possible to use additional services in parallel, such as the configuration database protocol proposed by OpenVSwitch. Additional services are used to collect network data that are not available through OpenFlow.

Onix tracks network state by storing it on a data structure called the network information base (NIB), which is similar with the routing information base (RIB) used by routers. The NIB is the key module of the control model proposed by Onix. Instead of just storing prefixes and destinations as RIB does, NIB stores a graph describing the network topology. Control applications running over Onix read and write on the NIB, whereas Onix control platform replicates and distributes these data to guarantee the resilience and scalability of the controller.

*Beacon*

Beacon is a Java-based OpenFlow controller. This allows it to be run on different platforms, even on Android-based smart phones. This controller was developed as an event-based controller. It also supports threaded operation to increase the control plane performance.

One of the main advantages of Beacon is its dynamic nature. This controller allows applications to be started, stopped, and installed on run time. In contrast to other proposals, these operations do not affect nondependent applications. Hence, the controller can be modified during run time without rebooting the switches on the network.

Because of its advantages, Beacon is the base for an enterprise-class controller, called Floodlight. Floodlight is designed to guarantee high performance while controlling OpenFlow networks.

*Other Controllers*

There are many other OpenFlow controllers such as Maestro, Trema, network development and deployment initiative - open exchange software suite (NDDI-OESS), Ryu, and NodeFlow. They present small differences when compared with the controllers presented on this text. This multiplicity shows the diversity of the OpenFlow community and the fast evolution of technology. Hence, it is expected that, in a few years, the set of the main OpenFlow controllers will contain a variety of strategies to increase performance, reliability, and scalability. The concepts and requirements of how to control an SDN network, however, remain.

## Hybrid Control Approaches

The OpenFlow architecture brought flexibility to network management, allowing flow-level control in every switch in the network. The use of a centralized controller allows the use of simpler algorithms, which are built based on a complete network view. Compared with current packet switching, OpenFlow brings many advantages. For instance, the central controller supervises all flow-level decisions, applying policies in an easier way. The global, complete view of network flows also allows the implementation of simpler load-balance algorithms. Moreover, because OpenFlow moves the network intelligence to a central node, new advances do not imply changes on the firmware or hardware of switches, but only on the software inside the network controller. Nevertheless, these advantages come with costs. OpenFlow controllers must take part in every flow setup, which presents scalability issues and increases the delay for setting a flow and the buffer demand on OpenFlow switches.

Although the OpenFlow architecture allows installing matching rules for all flows in advance, similar with current routing schemes, this strategy is not advantageous. Current routing schemes are based exclusively on destination addresses, whereas OpenFlow switching can be based on an extensive set of header fields. Hence, the number of matching rules for flows can be much greater than the number of routes currently used in active network devices. The use of wildcard rules can reduce this problem by reducing the number of installed matching rules. This approach, however, presents two main issues. First, by using wildcard rules, the network administrator loses the per-flow control and monitoring. Second, wildcard rules are placed in the ternary content-addressable memory (TCAM) of OpenFlow switches, which is restricted because of costs and power consumption. Hence, the hardware of an OpenFlow switch allows the instantiation of much more exact-match rules than wildcard rules. That is why current solutions use the redirection of the first packet of each flow to the controller to create an exact-match rule.

Because of these limitations, the use of new approaches in the flow control is being studied. The main point being criticized is that OpenFlow controllers bind two different and important functions: maintenance of the network view and central flow control. Experiments show that a central network view positively impacts the design of algorithms, but putting the controller in the critical path for setting up all flows presents scalability issues.

We now present two proposals for reducing the role of the controller to increase network performance and scalability: Distributed Flow Architecture for Networked Enterprises (DIFANE) and Devolved OpenFlow (DevoFlow) [13,14]. The key point of this kind of proposal is the way to establish thresholds to controller actions without over-stressing the data plane on switches.

*Distributed Flow Architecture for Networked Enterprises*

DIFANE proposes the idea of keeping all traffic in the data plane. Hence, it avoids redirecting the first packet of a flow to the controller. Instead, in DIFANE, the nonmatched traffic is forwarded to special switches called authority switches. These special switches store flow rules and configure the other switches. In DIFANE, the controller's role is reduced to partitioning the set of matching rules among the authority switches to maintain scalability.

The architecture of DIFANE is described in Figure 5.12. The controller, based on the network view, distributes the forwarding rules to the authority switches. When the first packet of a flow reaches an ingress switch, the switch checks whether there is any rule that matches the packet header. In case there is a rule, the packet is forwarded. Otherwise, the packets are encapsulated and forwarded to its authority switch, which decapsulates, processes, reencapsulates, and forwards the packets to the egress switch. To avoid redirecting all flows to authority switches, DIFANE provides a mechanism called rule caching. Using this mechanism, authority switches cache rules in the ingress switch to guarantee that this switch will forward traffic directly to the egress switch. It is important to notice that there is no delay to set a flow rule on DIFANE. The flows are forwarded by the authority switch instead of being kept on the original switch while the matching rule was being created, as the previous model using only a controller would do. Hence, the ingress switch has no need for extra buffer space for storing data while the controller processes and sets the new flow rule.

The implementation of DIFANE does not require changing the OpenFlow firmware on ordinary switches. It requires, however, that authority switches include a small part of the control plane.

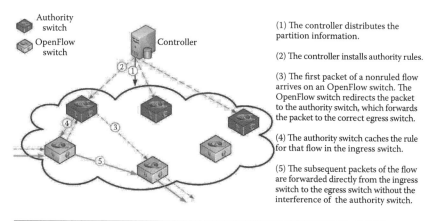

(1) The controller distributes the partition information.

(2) The controller installs authority rules.

(3) The first packet of a nonruled flow arrives on an OpenFlow switch. The OpenFlow switch redirects the packet to the authority switch, which forwards the packet to the correct egress switch.

(4) The authority switch caches the rule for that flow in the ingress switch.

(5) The subsequent packets of the flow are forwarded directly from the ingress switch to the egress switch without the interference of the authority switch.

**Figure 5.12**   DIFANE architecture, in which the control plane is distributed among the central controller and the authority switches.

Therefore, DIFANE introduces two main ideas:

1. The controller monitors the network view and calculates the flow rules. Afterward, it distributes the rules among a set of special switches called authority switches.
2. All data traffic is restricted to the data plane in the switches through the use of wildcard rules. These rules should be simple to match the capabilities of commodity flow switches, guaranteeing that no change in the data plane is required to implement OpenFlow.

*Devolved Flow*

The DevoFlow proposal, similar with the DIFANE proposal, is based on breaking the coupling between centralized flow control and centralized network view. In DevoFlow, part of the control plane is returned (devolved) back to the switches to reduce the interaction between the switches and the controller. This preserves the central network view without affecting the network scalability.

To reduce the interaction between switches and controllers, DevoFlow makes an extensive use of wildcard rules. The idea is that the controller does not need to interfere on the setting of all control rules. Instead, the controller should handle only security-sensitive flows and significant flows, which can impact the global QoS and congestion control. All other flows should be handled by the switch because they do not hinder policy application in the network. Hence, some flows are handled individually, whereas others are automatically forwarded through wildcard rules. However, the central controller must see all the flows set in the switches through the OpenFlow monitoring module. This is important to guarantee that the network view is consistent and also to allow the controller to manage network performance. The use of wildcard rules, however, would group flow statistics, reducing the granularity of the monitoring data.

In sum, the main design principles introduced by DevoFlow are as follows:

1. Keep centralized control over important things, which include security and QoS, but give control back to the switches on ordinary decisions.

2. Reduce the required bandwidth between the controller and the switches by significantly reducing the number of packets that leave the data plane.

In DevoFlow, the role of the OpenFlow switch is slightly modified by including some control functions inside the switch. Indeed, there are two new mechanisms: rule cloning and local routing actions.

Rule cloning is implemented by extending the action field of OpenFlow rules with a CLONE flag. If the flag is set, the switch creates new rules based on the wildcard rule. Hence, if a packet matches the wildcard rule, the switch creates a new rule, with all fields specified for that flow, using the actions defined in the wildcard rule. Otherwise, if the CLONE flag is clear, the switch follows the standard OpenFlow behavior. Rule cloning guarantees that the use of wildcard rules does not masquerade per-flow statistics. Hence, the controller still saw the specific counters of each flow, even if only a wildcard rule was preset in the switches.

The local routing actions include a set of tools:

1. *Multipath support:* A wildcard rule may be configured with several output ports. When the multipath support is active, the switch selects one of these ports for each new flow derived from it. This allows flow-level load balancing in the network.*
2. *Rapid rerouting:* A switch has a backup path in case the designated output port fails. Keeping this decision local significantly reduces delays when recovering faults because the switch does not wait for the controller to configure new routes.†
3. *Triggers and reports:* The switches can set triggers, such as counters, to each flow. Once a trigger condition is met, the switch sends a report to the controller.

*Discussion*

There are advantages in making the initial scheme proposed by the OpenFlow development team flexible. Changing the OpenFlow switch, however, causes compatibility problems with controllers and

---

* This function is now available in OpenFlow 1.1 through the use of the group table.
† This function is now available in OpenFlow 1.1 through the use of the group table.

current equipment. Therefore, the main idea here is to analyze which are the ideal barriers to the control plane to improve future specifications of OpenFlow or other SDNs.

### Wireless and Optical SDN Models

Wireless and optical networks present particularities other than the ones we address in ordinary wired networks. In this section, we discuss some of these particularities and discuss some proposals.

#### Optical SDN Models

One of the main problems when managing optical networks is that IP and transport layers are independently operated. This increases costs on both equipment and network operation and also increases network complexity [15]. SDNs can circumvent this problem by unifying transport and IP control planes.

An important constraint of the current optical network is that these two layers do not dynamically interact. Indeed, IP routers are typically interconnected via static circuits created with the transport layer. Therefore, for the IP routers, these circuits are fixed links. This prevents the carriers from benefiting from dynamic circuit switching in the network core. If IP routing and dynamic circuit switch could be performed together, the network efficiency would increase.

OpenFlow is a technology that meets the requirements for mixing IP and transport control in a single packet and circuit network. One of the first study cases in this area implemented a unified control plane for IP-/Ethernet- and TDM-switched networks [15]. This implementation was a simple proof of concept, but it created and modified L1/L2 flows on demand in addition to performing network congestion control. Among the developed functionalities are the preprovisions; a synchronous optical networking (SONET)/synchronous digital hierarchy (SDH) Virtual Concatenation Group (VCG), which works as virtual ports connecting the Ethernet packet flows with the SONET/SDH circuit flows; the dynamic creation of virtual local area networks (VLANs); the mapping of VLANs into the VCGs; and the emission of SONET signals, which are also mapped in the VCGs. In this proposal, the VCG performed the dynamic linking between the packet flow and the transport circuit.

Currently, there are a lot of efforts for enabling protocols, such as multiprotocol label switching (MPLS) and generalized multiprotocol label switching (GMPLS), over OpenFlow. The current OpenFlow version already supports the use of MPLS, QoS bits, and the modification of the VLAN header. Nevertheless, there are still other characteristics of optical networks that were not addressed, but that are the focus of next versions, to allow the creation of fully functional carrier grades using OpenFlow [16].

*Wireless SDN Models*

Wireless SDN networks also present peculiarities because the control of a wireless network involves the knowledge about the node mobility and the radio interference.

One of the first approaches for controlling wireless SDNs was OpenRoads [17]. This is an open-source platform for providing the basis for innovation on OpenFlow wireless networks. For this reason, OpenRoads architecture includes Wi-Fi and worldwide interoperability for microwave access (WIMAX) support. In addition, this architecture includes extra modules to monitor the wireless medium through simple network management protocol (SNMP).

The OpenRoads architecture was developed to specify the wireless OpenFlow. The proposed architecture consists of three layers, as shown in Figure 5.13. The first layer, called the flow layer, is responsible for managing the flow tables in the access points in the usual OpenFlow way. This table also collects medium-specific parameters, such as the channel, power level, and service set identifier (SSID), through the SNMP. The following layer is the slicing layer, which is responsible for network virtualization. The OpenFlow tables in the switches are shared using FlowVisor in the usual OpenFlow way. Nevertheless, FlowVisor did not slice the access through SNMP. Therefore, OpenRoads includes an SNMP demultiplexer, which performs the network slicing for the SNMP traffic. The last layer is the control layer, which corresponds to a typical OpenFlow control plane. Specifically, OpenRoads was developed over Nox.

OpenRoads also includes some applications for controlling the wireless network called mobility managers. One of the mobility managers

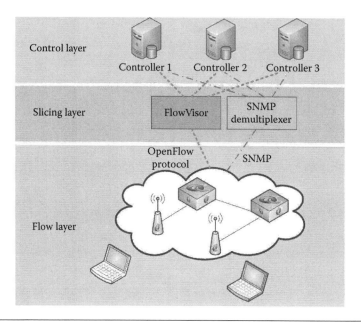

**Figure 5.13**    OpenRoads architecture, which consists of three layers.

is the informed handover, in which the client receives information to improve his/her handoff decisions. Another mobility manager is the n-casting, in which flows are multicasted through different paths to reduce packet losses. Another interesting application is Hoolock, a mobility manager for performing vertical handovers, exploiting multiple client interfaces.

Other applications are being developed to increase wireless network performance, such as the control of the client-AP association according to the IP characteristics, the implementation of protocols to increase the quality of service, home network control, etc. The main point here is that the use of SDN increases the control over the wireless network, allowing the network manager to create new applications that increase network performance.

**Case Studies**

Currently, there are plenty of demonstrations using OpenFlow. These demonstrations include topics, such as load balancing, node mobility, use of optical networks, etc. We now show small examples of how to perform simple network actions using the OpenFlow control plane.

In addition, we also describe some of the main OpenFlow-based test-beds, which can be used by the readers for experimental research.

*Simple Application Examples*

In this section, we provide an overview of how to perform usual L2 switching, routing, load balancing, and traffic filtering using OpenFlow.

*L2 Switching*  An L2 switching application is easily developed under OpenFlow. The key idea here is to define which header fields should be inserted in the wildcard. Because we are using only the layer 2 headers, the IP and TCP headers must be masked. Hence, the controller must set flows specifying only the fields highlighted in Figure 5.14.

The L2 switching application must take into consideration the construction of the spanning tree and the address resolution protocol (ARP) protocol. Based on the packet-in events, the controller observes the ARP requests and ARP replies, which are the bases for setting the L2 flows.

*Routing*  Currently, routing is performed by the routers in a distributed way. Moreover, routers perform best-match searches in the forwarding tables to choose the best route to the packet. When the best route is found, the time to live (TTL) field is decremented, and the packet is forwarded.

In OpenFlow, the packets are forwarded according to the first matching flow in the flow table. Hence, a different scheme must be implemented if hardware support for the best-match search is not available in the switch. In this case, the controller application should calculate the routing tables and store it locally. Whenever a new flow

Layer 2 switching
(MAC/VLAN)

| In port | src MAC | dst MAC | Eth type | VLAN ID | IP ToS | IP Proto | IP src | IP dst | TCP src Port | TCP dst Port | MPLS label |
|---------|---------|---------|----------|---------|--------|----------|--------|--------|--------------|--------------|------------|
|         |         |         |          |         |        |          |        |        |              |              |            |

**Figure 5.14**  Nonmasked fields when developing a simple switch application.

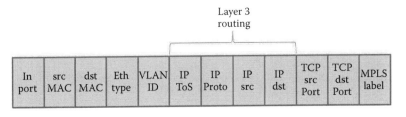

**Figure 5.15**   Nonmasked fields when developing a simple routing application.

arrives in the switch, the switch will send the packet header to the controller. The controller will then perform the best match locally and will set a flow in the switch, including actions to decrement the TTL and to forward the packet to a specific output door. This flow must include all the fields highlighted in Figure 5.15. The use of IP masks also depends on the hardware support. More efficient schemes can be developed if we use the processing pipeline to implement searches similar with the best match.

*Load Balancing*   Load balance applications must collect data about the network state and dynamically reconfigure the network paths. The control plane collects the monitoring data of the switches because the switch counters store information, such as the number of packets successfully forwarded and the number of dropped packets. The network state is the input for the algorithm to reconfigure flows. Based on these data, it is possible to discover the overloaded links.

OpenFlow 1.1 presents some tools that can be used to dynamically reconfigure flows. First, it is simple to recalculate paths of flows already set in the switches because of the control plane central network view. Indeed, it is possible to migrate packet flows without losing packets. The idea is to build a new path before destroying the current path. Another important tool for load balancing algorithms is the group table. The group table allows the administrator to set anycast groups, which is also a powerful tool for performing load balancing. Hence, the control plane monitors the links, chooses new paths, and migrates flows without incurring packet losses.

*Traffic Filtering*   Traffic filtering can be performed using the drop primitive. A simple traffic filtering application would send the first packet of all flows to the controller. Based on the header field data, the controller can configure the forwarding rules in the switches or configure a drop rule to that flow. This could be used to filter IP and MAC addresses, as well as applications, based on the TCP port.

Content-based filtering can also be performed by using middle boxes. In this case, the controller would forward the suspect flows to a middle box. This middle box would run any traffic filtering application, such as IPtables, to verify the packet content. Afterward, the middle box could deliver the packet to an OpenFlow switch or drop the packet.

### OpenFlow-Based Testbeds

OpenFlow-based testbeds are used to test new applications using OpenFlow. Nowadays, there are some open testbeds that can be used for academic research. We now detail some of the most important OpenFlow-based testbeds. These testbeds are based on network virtualization, and then, a network slice is provided to each experimenter.

*Global Environment for Network Innovations*   The Global Environment for Network Innovations (GENI)* is a wide testbed for developing future Internet experiments. This testbed is an initiative of the National Science Foundation–Directorate for Computer and Information Science and Engineering (NSF CISE) of the United States and is being developed since 2005. GENI is a long-term project that aims to create a huge testbed by federating previous initiatives, such as PlanetLab, Internet 2, LambdaRail, Emulab, etc. The development of the testbed is performed in spirals to create and evolve tools to guarantee a proper environment for the researches. In sum, GENI is defined as a virtual laboratory to explore new network architectures and applications at scale.

GENI provides OpenFlow-based slices inside the testbed. Indeed, there are already many OpenFlow switches of different manufacturers in this testbed comprising the core of the LambdaRail and Internet 2 networks. The network slices are created through virtualization, using FlowVisor.

---

* Available at http://www.geni.net/.

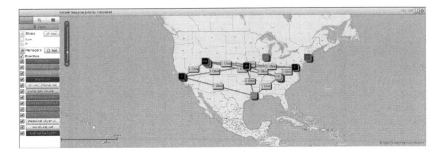

**Figure 5.16**    Resource map of the GENI testbed in the United States obtained with the ProtoGENI Flack tool. (From http://www.protogeni.net/flack, accessed January 2013.)

A general view of the current resources available in GENI in the United States is shown in Figure 5.16.

*OpenFlow in Europe: Linking Infrastructure and Applications*    OpenFlow in Europe: Linking Infrastructure and Applications (OFELIA)* is an OpenFlow testbed developed in Europe under the FP7 ICT Work Program. The idea is to provide experimenters with a testbed in which they can dynamically control and extend the network through the use of OpenFlow.

Because of the use of virtualization, each experimenter receives a network slice. Each slice consists of the following:

1. Virtual machines, which represent the servers and the clients in the network
2. A virtual machine to run the OpenFlow controller
3. A virtual OpenFlow network

The experiments in OFELIA are performed using a control framework called the OFELIA Control Framework (OCF). OCF provides tools to authenticate and access the testbed, allocate the slice, configure the network, instantiate applications in the controller and on other virtual machines, and monitor the experiment results. It is also the basis for the federation of different islands running OFELIA. The federation of the testbeds is important to create the environment to test applications and architectures that demand large-scale networks.

---

* Available at http://www.fp7-ofelia.eu/.

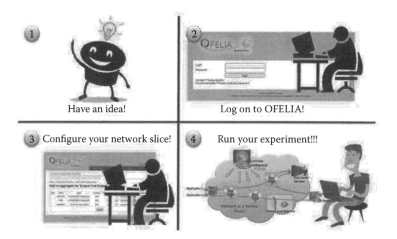

**Figure 5.17**   Using the OFELIA facility. (From http://www.fp7-ofelia.eu/.)

To access the OFELIA testbed, the experimenter must create an OFELIA account, set up a virtual private network (VPN) connection, create a project, define the required resources, and, finally, perform the experiment, as shown in Figure 5.17.

*Future Internet Testbeds Experimentation between Brazil and Europe (FIBRE)*
FIBRE* is a testbed under development as a cooperation between Brazil and Europe, whose main objective is to build a testbed for future Internet architectures. The key idea is to build a federated testbed, which combines wide area network (WAN) and wireless networks. The control of the WAN testbed is performed using OpenFlow. The wireless networks can be optionally controlled using OpenFlow.

As a GENI, FIBRE federates existing testbeds and newly created ones. Hence, it deals with different control frameworks to set up experiments, which include OCF, orbit management framework (OMF), and ProtoGENI.

The access to the testbed is performed in a way similar to OFELIA. The user authenticates himself and obtains credentials to reserve resources of different islands. The machine and network virtualization are, again, key tools to guarantee that many researchers can access the testbed simultaneously without interfering with each other.

---

* Available at http://www.fibre-ict.eu/.

# References

1. OpenFlow. 2011. OpenFlow Switch Specification–Version 1.1.0 Implemented (Wire Protocol 0×02).
2. Sherwood, R., M. Chan, A. Covington, G. Gibb, M. Flajslik, N. Handigol, T.-Y. Huang et al. 2010. Carving research slices out of your production networks with OpenFlow. *SIGCOMM Comput. Commun. Rev.* 40(1):129–130.
3. Fernandes, N. C., M. D. D. Moreira, M. Moraes, L. H. G. Ferraz, R. S. Couto, H. E. T. Carvalho, M. E. M. Campista, L. H. M. K. Costa, O. C. M. B. Duarte. June 2011. Virtual networks: Isolation, performance, and trends. *Annals of Telecommunications - annales des télécommunications,* 66(5–6):339–355.
4. Caesar, M., D. Caldwell, N. Feamster, J. Rexford, A. Shaikh, J. van der Merwe. 2005. Design and implementation of a routing control platform, in *Proceedings of the 2nd Conference on Symposium on Networked Systems Design and Implementation (NSDI '05), Volume 2*. Berkeley: USENIX Association, pp. 15–28.
5. Verkaik, P., D. Pei, T. Scholl, A. Shaikh, A. C. Snoeren, J. E. van der Merwe. 2007. Wresting control from BGP: Scalable fine-grained route control. In *Proceedings of the 2007 USENIX Annual Technical Conference (ATC '07)*. Berkeley: USENIX Association, pp. 23:1–23:14.
6. Feamster, N., H. Balakrishnan, J. Rexford, A. Shaikh, J. van der Merwe. 2004. The case for separating routing from routers. In *Proceedings of the ACM SIGCOMM Workshop on Future Directions in Network Architecture (FDNA '04)*, New York: ACM, pp. 5–12.
7. McKeown, N., T. Anderson, H. Balakrishnan, G. Parulkar, L. Peterson, J. Rexford, S. Shenker et al. 2008. OpenFlow: Enabling innovation in campus networks. *SIGCOMM Comput. Commun. Rev.* 38(2):69–74.
8. Casado, M., M. Freedman, J. Pettit, J. Luo, N. Gude, N. McKeown, S. Shenker. 2009. Rethinking enterprise network control. *IEEE/ACM Transactions on Networking* 17(4):1270–1283.
9. Casado, M., M. J. Freedman, J. Pettit, J. Luo, N. McKeown, S. Shenker. 2007. Ethane: Taking control of the enterprise. In *Proceedings of the 2007 Conference on Applications, Technologies, Architectures, and Protocols for Computer Communications (SIGCOMM '07)*, New York: ACM, pp. 1–12.
10. Casado, M., T. Garfinkel, A. Akella, M. J. Freedman, D. Boneh, N. McKeown, S. Shenker. 2006. SANE: A protection architecture for enterprise networks. In *Proceedings of the 15th Conference on USENIX Security Symposium (USENIX-SS '06), Volume 15*. Berkeley: USENIX Association.
11. Gude, N., T. Koponen, J. Pettit, B. Pfaff, M. Casado, N. McKeown, S. Shenker. 2008. Nox: Towards an operating system for networks. *SIGCOMM Comput. Commun. Rev.* 38(3):105–110.

12. Koponen, T., M. Casado, N. Gude, J. Stribling, L. Poutievski, M. Zhu, R. Ramanathan et al. 2010. Onix: A distributed control platform for large-scale production networks. In *Proceedings of the 9th USENIX Conference on Operating Systems Design and Implementation (OSDI '10)*. Berkeley: USENIX Association, pp. 1–6.

13. Yu, M., J. Rexford, M. J. Freedman, J. Wang. 2010. Scalable flow-based networking with DIFANE. In *Proceedings of the 2010 ACM SIGCOMM Conference*. New York: ACM, pp. 351–362.

14. Curtis, A. R., J. C. Mogul, J. Tourrilhes, P. Yalagandula, P. Sharma, S. Banerjee. 2011. DevoFlow: Scaling flow management for high-performance networks. In *Proceedings of the 2011 ACM SIGCOMM Conference*. New York: ACM, pp. 254–265.

15. Das, S., G. Parulkar, N. McKeown, P. Singh, D. Getachew, L. Ong. 2010. Packet and circuit network convergence with OpenFlow. In *2010 Conference on Optical Fiber Communication, Collocated National Fiber Optic Engineers Conference (OFC/NFOEC)*, pp. 1–3.

16. Kind, M., F. Westphal, A. Gladisch, S. Topp. 2012. Split architecture: Applying the software-defined networking concept to carrier networks. In *World Telecommunications Congress (WTC)*, pp. 1–6.

17. Yap, K.-K., M. Kobayashi, R. Sherwood, T.-Y. Huang, M. Chan, N. Handigol, N. McKeown. 2010. OpenRoads: Empowering research in mobile networks. *SIGCOMM Comput. Commun. Rev.* 40(1):125–126.

# 6

# Controller Architecture and Performance in Software-Defined Networks

## TING ZHANG AND FEI HU

**Contents**

## Introduction

The control plane is usually embodied by a central controller, providing a global view of the underlying network situation to the upper application layer. In this chapter, we look into the architecture and performance of the controller in software-defined networks (SDNs) [1] and also provide comparisons of their features. We first review the framework of SDNs, and then we describe the controller in SDNs. We summarize the new techniques and challenges in recent years for this area, and then we illustrate two areas of implementations: in Ethernet and in wireless sensor networks (WSNs).

The split architecture of SDN brings substantial benefits [2], such as consistent global policy enforcement and simplified network management. While the standardization progresses, steered OpenFlow has been envisioned as a promising approach to next-generation programmable and easy-to-manage networks [3]. However, the inherent heavy switch-controller communications in OpenFlow may throttle controller performance and, ultimately, network scalability. In this chapter, we identify that a key cause of this problem lies in flow setup and propose a control-message quenching (CMQ) scheme to address it. CMQ requires minimal changes to OpenFlow; imposes no overhead on the central controller, which is often the performance bottleneck; and is lightweight and simple to implement. Recognition and support have been from a large number of industry giants, such as Microsoft, Google, Facebook, HP, Deutsche Telekom, Verizon, Cisco, IBM, and Samsung. It has also attracted substantial attention from academia, where research has recently heated up on a variety of topics related to OpenFlow. Figure 6.1 shows a summarized picture of the control plane.

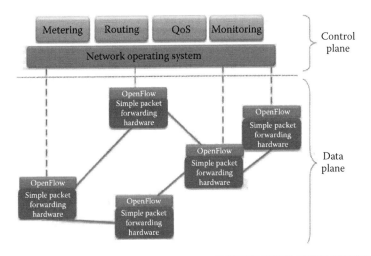

**Figure 6.1**   Summarized figure of a control plane. QoS, quality of service.

Prior studies on the scalability issue focus on designing a distributed control plane [4,5], devolving control to switches [6–8] or end hosts [9], or using multithreaded controllers [10,11]. These approaches either entail significant changes to OpenFlow (while a wide range of commercial OpenFlow products and prototypes have been off the shelf) or are complex to implement.

OpenFlow has been envisioned as a promising approach to the next-generation programmable and easy-to-manage networks. Based on the working procedures of the OpenFlow protocol, particularly the flow setup process, the major traffic that occurred in switch-controller communications is composed of two types of control messages: (1) packet-in message and (2) packet-out or flow-mod message. A packet-in message is sent by a switch to the controller to seek instructions on how to handle a packet upon table-miss, an event where an incoming packet fails to match any entry in the switch's flow table. The controller, upon receiving a packet-in message, will compute a route for the packet based on the packet header information encapsulated in the packet-in message and will then respond to the requesting switch by sending a packet-out or flow-mod message: (1) if the requesting switch has encapsulated the entire original packet in a packet-in message (usually because of no queue), the controller will use a packet-out message, which also encapsulates the entire packet; (2) if the packet-in message only contains the header of the original packet, a flow-mod message will be used instead, which also only includes the packet header, so that the switch can associate with the original packet (in its local queue). Either way, the switch can be instructed on how to handle the packet (e.g., which port to send to) and will record the instruction by installing a new entry in its flow table to avoid reconsulting the controller for subsequent packets that should be handled the same way.

This mechanism works fine when the switch-controller connection, referred to as the OpenFlow channel, has ample bandwidth and when the controller is super fast or the demand of the flow setup is low to moderate. However, chances are that neither is met. On one hand, some analysis and measurement show that the round-trip time (RTT)—the interval between the instance when a packet-in message is sent out and the instance when a packet-out or flow-mod message is received—can be considerably long. For instance, RTT is 3.67 to

4.06 ms even in low-load conditions, which is in line with what others reported. On the other hand, the flow inter-arrival time (reciprocal of the flow setup request rate) can reach 0.1 to 10 μs, which is several orders shorter than the RTT.

## Controllers in SDNs

### Types of Controllers in SDNs

Network architectures, in which the control plane is decoupled from the data plane, have been gaining popularity. Among the main arguments for this approach is that they provide a more structured software environment for developing network-wide abstractions while potentially simplifying the data plane. As has been adopted elsewhere [3], we refer to this split architecture as SDN.

While it has been argued that SDN is suitable for some deployment environments (such as homes [12,13], data centers [14], and the enterprise [15]), delegating control to a remote system has raised several questions on the control-plane scaling implications of such an approach. Two of the most often voiced concerns are (1) how fast the controller can respond to datapath requests and (2) how many datapath requests it can handle per second.

There are some references to the performance of SDN systems in the literature [15–17]. For example, a study shows that a popular network controller (Nox) handles approximately 30,000 flow initiation events per second while maintaining a 10-ms flow install time. For software controllers, there are four publicly available OpenFlow controllers: Nox, Nox-MT, Beacon, and Maestro [18].

*Examples of OpenFlow Controllers*   One example of OpenFlow controllers is Nox-MT. Nox—whose measured performance motivated several recent proposals on improving control-plane efficiency—has a very low flow setup throughput and a large flow setup latency. Fortunately, this is not an intrinsic limitation of the SDN control plane: Nox is not optimized for performance and is single threaded.

Nox-MT is a slightly modified multithreaded successor of Nox. With simple tweaks, we significantly improve the throughput and response time of Nox. The techniques used to optimize Nox are quite well known: batching I/O to minimize its overhead, porting the I/O handling harness to Boost asynchronous I/O (ASIO) library (which simplifies multithreaded operation), and using a fast multiprocessor-aware malloc implementation that scales well in a multicore machine.

Despite these modifications, Nox-MT is far from perfect. It does not address many of the performance deficiencies of Nox, including, but not limited to, the heavy use of dynamic memory allocation and redundant memory copies on a per-request basis, and the use of locking while robust wait-free alternatives exist. Addressing these issues would significantly improve the performance of Nox. However, they require fundamental changes to the Nox code base. Nox-MT was the first effort in enhancing controller performance and motivated other controllers to improve. The SDN controllers can be optimized to be very fast.

*Methods to Increase Controller Performance in SDN*

OpenFlow scalability's performance can be increased by enhancing controller responsiveness. This was motivated by the imbalance between the flow setup rate offered by OpenFlow and the demand of real production networks. The flow setup rate, as measured by Curtis et al. [19], on an HP ProCurve 5406zl switch, is 275 flows per second. On the other hand, a data center with 1500 servers, as reported by Kandula et al. [20], has a median flow arrival rate of 100,000 flows per second, and a 100-switch network, according to Benson et al. [21], can have a spike of 10 million flow arrivals per second in the worst case. This indicates a clear gap. The inability of OpenFlow to meet the demanding requirements stems from its inherent design, which leads to frequent switch-controller communications: to keep the data plane simple, OpenFlow delegates the task of routing (as required by flow setup) from switches to the central controller. As a result, switches have to consult the controller frequently for instructions on how to handle incoming packets, which taxes the controller's processing power and tends to congest switch-controller connections, thereby imposing a serious bottleneck to the scalability of OpenFlow.

## New Techniques in the Controller of SDNs

*Drawbacks on Current Available Controllers*

Unfortunately, recent measurements of some deployment environments suggest that these numbers are far from sufficient. This causes relatively poor controller performance and high network demands to address perceived architectural inefficiencies. However, there really has been no in-depth study on the performance of a traditional SDN controller. Instead, most published results were gathered from systems that were not optimized for performance. To underscore this point, researchers improved the performance of Nox, an open source controller for OpenFlow networks, by more than 30 times (see Figure 6.2) [19].

SDN introduces centralized controllers to dramatically increase network programmability. The simplicity of a logical centralized controller, however, can come at the cost of control-plane scalability. McNettle, an extensible SDN control system whose event processing throughput scales with the number of system CPU cores, supports control algorithms and requires globally visible state changes occurring at flow arrival rates. Programmers extend McNettle by writing event handlers and background programs in a high-level functional

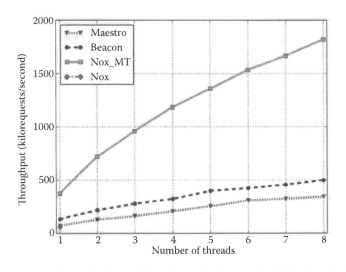

**Figure 6.2**   Average maximum throughput achieved with different number of threads.

programming language extended with shared state and memory transactions.

A study from Andreas et al. [11] implements a framework in Haskell and leverages the multicore facilities of the Glasgow Haskell Compiler (GHC) and run-time system. The implementation schedules event handlers, allocates memory, optimizes message parsing and serialization, and reduces system calls to optimize cache usage, OS processing, and run-time system overhead. Experiments show that McNettle can serve up to 5000 switches using a single controller with 46 cores, achieving a throughput of more than 14 million flows per second, a near-linear scaling of up to 46 cores, and a latency of less than 200 μs for light loads and 10 ms for loads consisting of up to 5000 switches.

**Controller in an Ethernet-Defined SDN**

*Overview of the Controller in an Ethernet-Defined SDN*

One of the prevalent trends in emerging large-scale multitenant data centers is network virtualization using overlays. Here, we investigate application performance degradation in such an overlay applied to commodity 10-GB Ethernet networks. In Ref. [17], they have adopted partition/aggregate as a representative commercial workload that is deployed on bare metal servers and is notoriously sensitive to latency and transmission control protocol (TCP) congestion. Using query completion time as the primary metric, they evaluate the degree to which an SDN overlay impacts this application's behavior and the performance bounds of partition/aggregate with an SDN overlay, and whether active queue management (AQM), such as random early detection (RED), can benefit this environment. They introduce a generic SDN overlay framework, which we can measure in hardware and simulate using a real TCP stack extracted from FreeBSD v9, running over a detailed layer 2 commodity 10-G Ethernet fabric network simulator.

*Partition/Aggregate in Commodity 10-G Ethernet—Network Visualization*

In recent years, there have been many fundamental and profound changes in the architecture of data center networks. Traditional data centers consist of lightly used servers running a bare metal OS or a hypervisor with a small number of virtual machines (VMs). Applications reside on a physical server, and each workload has an associated network state that resides on a switch. These networks currently are static and are often manually configured and managed, which is a costly and unsustainable mode of operation. Modern data centers are transitioning toward a dynamic infrastructure, including highly used servers running many VMs. Although server virtualization has existed since the 1960s, it has only become widely available on commodity ×86 servers within the last decade or so. Modern data centers contain tens of thousands of servers, each potentially hosting tens, and soon hundreds, of VMs. This places unique new requirements on the data center network, which must now cope with multitenancy and the automated creation, deletion, and migration of VMs.

### Controller in WSNs

*Motivation for WSN Controller Architecture*

WSNs have been conceived to be application specific [22], which has probably formed a belief thus far. However, we deem it worth an afterthought, with the upsurge of sensor applications nowadays and the advancement of sensor technologies, such as multimodal and high-end mote platforms (e.g., Imote2, Stargate, and NetBridge). These new changes render application-specific WSNs prone to (1) resource underutilization, where multiple WSNs for respective applications are deployed in the same or overlapping terrain while a single versatile WSN could achieve the same objectives; and (2) counter-productivity, where different vendors develop WSNs in isolation without adequately reusing common functionalities that would considerably expedite prototyping and production.

Another problem with WSNs is that they are rigid to policy changes. Policies are rules related to network-exogenous factors, such as business operation and user access, as opposed to network-endogenous factors, such as node and link properties that have been extensively studied over the years and handled algorithmically. Policy changes, engendered by the ever-changing business needs, are hard to cope with by algorithms and often dictate manual reconfiguration or reprogramming of WSNs, which is difficult because of the vendors' complicated and often proprietary implementations. As a result, it is usually inevitable to involve vendors and, hence, incur delay in policy enforcement, and considerable financial and opportunity cost.

A third problem is that WSNs are hard to manage. This is because developing a network management system (NMS) for distributed WSNs is a demanding task in the first place. Furthermore, in practice, this task is often scheduled as phase 2 in project planning and, hence, entails hacking existing codes on sensor nodes, which is a big headache to developers and is highly error prone.

The aforementioned problems are not superficial symptoms but are inherent to WSNs and deep-rooted in the architecture: each node is fully fledged with all physical up to application layer functionalities collectively behaving like an autonomous system that performs various networking functions, such as data forwarding and network control. Although this architecture works well most of the time because of many well-designed algorithms, it lacks good abstraction and carries too much complexity, making WSNs unwieldy, inelastic to change, and hard to manage.

While it has been a belief for more than a decade that WSNs are application specific, we argue that it can lead to resource underutilization and counter-productivity. We also identify two other main problems with WSNs: rigid to policy changes and difficult to manage.

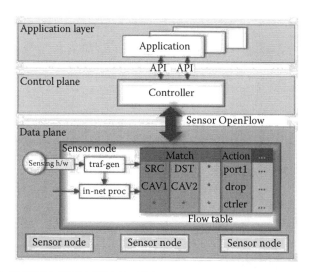

**Figure 6.3** Software-defined sensor networks. API, application programming interface; h/w, hardware.

In this chapter, we take a radical, yet peer–compatible, approach to tackle these problems inherent to WSNs. A software-defined WSN architecture (see Figure 6.3) can be used to address key technical challenges for its core component, Sensor OpenFlow.

### Discussion and Conclusion

We discussed the architecture and performance of the controller in SDNs and provided a comparison and a summary. In detail, we first introduced the framework of SDNs, and then we described the controller in SDNs. We summarized the new techniques and challenges in recent years about this area, and then we illustrated two areas of implementations: in Ethernet and in WSNs.

## References

1. Open Networking Foundation. 2012. Software-defined networking: The new norm for networks. ONF (open networking foundation) white paper, April 2012, https://www.opennetworking.org/images/stories/downloads/sdn-resources/white-papers/wp-sdn-newnorm.pdf.
2. McKeown, N., T. Anderson, H. Balakrishnan, G. Parulkar, L. Peterson, J. Rexford, S. Shenker et al. 2008. OpenFlow: Enabling innovation in campus networks. *SIGCOMM Comput. Commun. Rev.* 38(2), pp. 69–74.
3. Open Networking Foundation. http://www.opennetworking.org.
4. Koponen, T., M. Casado, N. Gude, J. Stribling, L. Poutievski, M. Zhu, R. Ramanathan et al. 2010. Onix: A distributed control platform for large-scale production networks. In *The 9th USENIX Conference on Operating Systems Design and Implementation (OSDI)*, pp. 1–6.
5. Tootoonchian, A. and Y. Ganjali. 2010. HyperFlow: A distributed control plane for OpenFlow. In *INM/WREN*. USENIX Association. Berkeley, CA, USA, Page 3.
6. Curtis, A. R., J. C. Mogul, J. Tourrilhes, P. Yalagandula, P. Sharma, and S. Banerjee. 2011. DevoFlow: Scaling flow management for high-performance networks. In *ACM SIGCOMM*. ACM, New York, NY, USA, 254–265.
7. Yu, M., J. Rexford, M. J. Freedman, and J. Wang. 2010. Scalable flow-based networking with DIFANE. In *ACM SIGCOMM*. 40, 4, 351–362.
8. Tam, A. S.-W., K. Xi, and H. J. Chao. 2011. Use of devolved controllers in data center networks. In *IEEE INFOCOM Workshop on Cloud Computing*. Shanghai, China. pp. 596–601, 10–15 April 2011.
9. Curtis, A. R., W. Kim, and P. Yalagandula. 2011. Mahout: Low overhead data center traffic management using end-host–based elephant detection. In *IEEE INFOCOM*. Shanghai, China. pp. 1629–1637, 10–15 April 2011.

10. Cai, Z., A. L. Cox, and T. S. E. Ng. 2010. Maestro: A system for scalable OpenFlow control. *Rice University Technical Report TR10-08*.

11. Tootoonchian, A., S. Gorbunov, Y. Ganjali, and M. Casado. 2012. On controller performance in software-defined networks. In *USENIX Workshop on Hot Topics in Management of Internet, Cloud, and Enterprise Networks and Services (Hot-ICE)*. Berkeley, CA, USA, Page 10.

12. Yiakoumis, Y., K.-K. Yap, S. Katti, G. Parulkar, and N. McKeown. 2011. Slicing home networks, In *Proceedings of the 2nd ACM SIGCOMM Workshop on Home Networks*. ACM, New York, NY, USA, pages 1–6.

13. Sundaresan, S., W. de Donato, N. Feamster, R. Teixeira, S. Crawford, and A. Pescap'e. 2011. Broadband Internet performance: A view from the gateway, in *Proceedings of the ACM SIGCOMM Conference*. ACM, New York, NY, USA, 134–145.

14. Al-Fares, M., A. Loukissas, and A. Vahdat. 2008. A scalable, commodity data center network architecture, in *Proceedings of the 2008 ACM SIGCOMM Conference on Data Communication*. ACM, New York, NY, USA, 63–74.

15. Casado, M., M. J. Freedman, J. Pettit, J. Luo, N. McKeown, and S. Shenker. 2007. Ethane: Taking control of the enterprise. *SIGCOMM CCR* 37(4):1–12.

16. Yan, H., D. A. Maltz, T. S. E. Ng, H. Gogineni, H. Zhang, and Z. Cai. 2007. Tesseract: A 4-D network control plane, in *Proceedings of the 2007 USENIX NSDI Conference*. Berkeley, CA, USA, 27.

17. Caesar, M., D. Caldwell, N. Feamster, J. Rexford, A. Shaikh, and J. van der Merwe. 2005. Design and implementation of a routing control platform, in *Proceedings of the 2005 USENIX NSDI Conference*. USENIX Association. Vol. 2. USENIX Association, Berkeley, CA, USA, 15–28.

18. Cai, Z., A. L. Cox, and T. S. E. Ng. 2010. Maestro: A system for scalable OpenFlow control. *Technical Report TR10-11*, Department of Computer Science, Rice University.

19. Curtis, A. R., J. C. Mogul, J. Tourrilhes, P. Yalagandula, P. Sharma, and S. Banerjee. 2011. DevoFlow: Scaling flow management for high-performance networks, in *ACM SIGCOMM*. ACM, New York, NY, USA, 254–265.

20. Kandula, S., S. Sengupta, A. Greenberg, P. Patel, and R. Chaiken. 2009. The nature of data center traffic: Measurements and analysis, in *ACM Internet Measurement Conference (IMC)*. ACM, New York, NY, USA, 202–208.

21. Benson, T., A. Akella, and D. A. Maltz. 2010. Network traffic characteristics of data centers in the wild, in *ACM Internet Measurement Conference (IMC)*. ACM, New York, NY, USA, 267–280.

22. Heinzelman, W. 2000. Application-specific protocol architectures for wireless networks. Ph.D. dissertation. Cambridge, MA: Massachusetts Institute of Technology.

# 7

# Mobile Applications on Global Clouds Using OpenFlow and Software-Defined Networking

## SUBHARTHI PAUL, RAJ JAIN, JAY IYER, AND DAVE ORAN

## Contents

## Introduction

In recent years, there has been an explosive growth in mobile applications (apps), most of which need to serve global audiences. This increasing trend of service access from mobile computing devices necessitates more dynamic application deployment strategies. Cloud computing provides unique opportunities for these application service providers (ASPs) to manage and optimize their distributed computing resources. For this trend to be successful, similar facilities need to be developed for the on-demand optimization of connectivity resources to enhance user experience.

Application delivery, on the surface, is simply connecting a user to a server. The original host-centric Internet architecture was well designed for this end-to-end two-host communication using source and destination IP addresses or names. However, today's Internet is mostly service centric, where users are interested in connecting to a service instead of a particular host. Delivering services over a host-centric design has led to complex application deployment environments, as discussed below, for the case of a single private data center (Figure 7.1), a single cloud environment, and a multicloud environment.

### Private Data Center

*Service Replication*   To scale a service, the service needs to be replicated over multiple servers. Network-layer load balancers were introduced into the network datapath to dynamically map the requests to

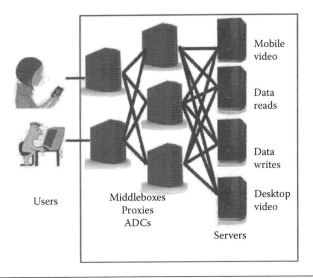

**Figure 7.1** Application delivery in a private data center.

the least loaded server. Pure control plane mechanisms, such as rotating domain name system (DNS), provide static load balancing.

*Service Partitioning* To improve performance, services often need to be partitioned, and each partition is hosted on a separate server. Each partition may be further replicated over multiple servers. A service may be partitioned based on content and/or context.

*Content-Based Partitioning* Even for the same service (e.g., xyz. com), accounting messages, recommendation requests, and video requests are all sent to different server groups. This can be done either by giving a different DNS name to each content type server or, as is common, by putting a proxy that classifies messages based on content.

*Context-Based Partitioning* User context, network context, or service context may require the application messages to be routed differently. An example of user context is a mobile smart phone user versus a desktop user. An example of network context is the geographical location of the user and the state of network links. An example of service context is that database reads and writes may be sent to different servers. Context-based partitioning is supported by proxies that classify messages based on the context.

*Service Composition* A service may represent a composed context where accessing the service actually requires going through a sequence of devices providing security (e.g., firewalls, intrusion detection systems [IDSs]), transformation/translation (e.g., transcoders, data compression), and performance enhancement (e.g., secure socket layer [SSL] offloaders, wide area network [WAN] optimizers) functions to the service deployment. These services, either provided by separate devices, generically called middleboxes, or by monolithic, integrated platforms, generically called application delivery controllers (ADCs), are now very common. In fact, the number of middleboxes in a data center is comparable to the number of routers [1].

*Multisegments* In many of the aforementioned examples, the proxy/middlebox/ADC terminates the transmission control protocol (TCP) and starts a new TCP connection. In general, a user-to-server connection is no longer end to end; it consists of many segments. Each of these segments can be served by multiple destinations (based on the replication of middleboxes). The ASPs implement complex application policy routing (APR) mechanisms inside the data centers.

*TCP Multiplexing* A special case of multisegments is TCP multiplexing. The goal is to improve the latency over long TCP connections by minimizing the effect of TCP's self-regulating window adjustment algorithm. ASP may place several proxy servers in different geographical locations so that the TCP connection between the user and the proxy has a short round trip and the window increases fast. Although the connection between the proxy and the server is long, it is permanently connected and, therefore, operates at optimally large windows. This technique is commonly used for short data transfers, such as Web services.

### Single Cloud Environment

In a private data center, as discussed above, high-capacity monolithic proxies are used for APR. When the applications move to a public cloud environment, the situation becomes more complex because private monolithic hardware boxes cannot be deployed. Moreover, the

ASP may not allow the cloud service provider (CSP) to look at its data for APR, and so software-based virtual machines are used for such services, and the amount of replication and dynamics (caused by Virtual Machine [VM] mobility and hardware failures) is much more common than a private data center.

### Multicloud Environment

Mobile apps that need to serve global audiences can easily get computing and storage facilities using cloud services from multiple cloud providers distributed throughout the world, for example, Amazon, Google, Rackspace, Microsoft, and so on. However, the problem of routing using ASP's policies in a very dynamic multicloud environment is not possible because Internet service providers (ISPs) offer no service to dynamically route messages to a different server using an ASP's policies.

Enterprises that operate multiple data centers already have this problem. For example, Google operates multiple data centers across different geographical locations and has installed a WAN-like infrastructure [2] that intercepts most of the traffic for Google-owned services at edge-network points of presence (POPs) and sends them over its private high-speed WAN infrastructure. At these POPs, Google (probably) operates application layer (layers 5–7) proxies to intelligently route service requests to appropriate data centers. However, for smaller ASPs, it is prohibitively expensive to operate such global networking infrastructures. Moreover, Google proxies have a complete view of the application data, which may not be desirable if such forwarding decisions were to be made by ISPs.

Our vision is to design a new session-layer abstraction called open application delivery networking (OpenADN) that will allow ISPs to offer services similar with Google WAN to smaller ASPs. ASPs can express and enforce application-traffic management policies and application delivery constraints to ISPs. It allows ASPs to achieve all the application delivery services that they use today in private data centers (mentioned above) in the global multicloud environment. As shown in Figure 7.2, using OpenADN aware data plane entities, any new ASP can quickly set up its service by using ADN services provided by ISPs.

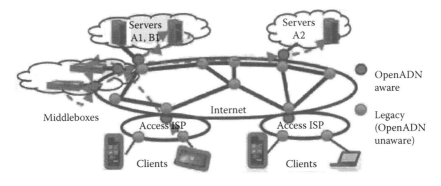

**Figure 7.2**   Open application delivery network.

To achieve this, we combine the following innovations:

1. OpenFlow
2. Software-defined networking (SDN)
3. Session splicing
4. Cross-layer communication
5. Multistage late binding
6. Identifiers (IDs)/locator split
7. MPLS-like application flow labels
8. Rule-based delegation

The rest of this chapter is organized as follows: "SDN and Open Flow" briefly explains the features of OpenFlow and SDN, which are helpful in our goal. "OpenADN Concepts" explains the several new concepts that we need to explain OpenADN. "OpenADN Extensions" discusses the aforementioned extensions that make OpenADN possible. "Related Work" provides a brief survey of the background literature, followed by the "Summary."

## SDN and OpenFlow

SDN is an approach toward taming the configuration and management complexities of large-scale network infrastructures through the design of proper abstractions. It proposes a separation between the network control and data planes. This would allow the control plane to be logically centralized, making it easier to manage and configure the distributed data plane components (switches and routers)

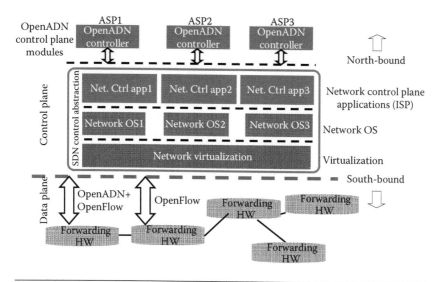

**Figure 7.3**  OpenADN uses meta-tags in the headers, to be used by forwarding elements as an extension of OpenFlow, and uses a north-bound interface from the controller for policy communication. HW, hardware; OS, operating system; Net. Ctrl, network controller.

involved in the actual packet forwarding. To implement this separation, SDN needs to design a basic abstraction layer interposed between the control plane and the data plane. This is explained further in Figure 7.3.

The evolving OpenFlow [3] standard is the protocol used between the controller and data plane forwarding entities (see Figure 7.3). The data plane entities are significantly simplified compared with today's switches in the sense that they do not do the usual control plane activities of preparing forwarding tables. Instead, they simply classify the packets based on their headers and use the forwarding table prepared by the central controller. This simplification of the data plane is expected to result in a significant cost savings in data centers where a large number of switches are used. This is one of the reasons why the industry is having an interest in this technology.

A flow in OpenFlow refers to a set of packets that are classified into the same policy class based on a combination of layer 2, layer 2.5 (MPLS), layer 3, and layer 4 header fields. We call such flows network-level flows. In OpenFlow, all packets are classified into separate network flow classes, and all packets belonging to the same network flow class are applied with the same control plane policies. Examples of control plane policies include different forwarding mechanisms (e.g., multicast, unicast) and different traffic engineering policies optimizing different parameters, such as energy efficiency, congestion, latency, etc.

## OpenADN Concepts

In this section, we explain some of the concepts required to understand the OpenADN extensions that are described in the next section.

### Waypoints

As previously discussed, the path between the user and the server may consist of multiple segments connecting intermediate middleboxes, proxies, or ADCs. Some of the middleboxes terminate TCP, and some do not. We use the term waypoint to indicate all such intermediate nodes. If a service is composed of multiple services, the packets may be forwarded to multiple servers, with intermediate nodes between those servers. The waypoints include these intermediate servers as well.

### Streams

Each waypoint may be replicated for fault tolerance or performance. We call a single stage connection between two specific nodes a stream. Thus, each segment may have multiple available streams. Once a session is assigned to a particular stream (a particular destination waypoint), all packets of that session will follow that stream. This is called session affinity.

### Application-Level Policies

Application-level policies specify the rules for forwarding the application traffic to help ASPs manage their distributed and dynamic application deployment environments.

The ASPs may design policies for setting up various session segments based on replication, content-based partitioning, and context-based partitioning as previously explained. These policies, when implemented in the network, will provide optimal user experience.

*Application Flows*

Enforcing the application-level policies in the Internet requires ISP to classify application traffic into separate application-specific application flow classes. This is currently not possible because this information is not available in the headers.

Note that application flows are different from network flows because the packets with the same L1 to L4 headers used to determine the network flow may contain different data types that need to be routed differently (a.k.a. content-based routing).

OpenFlow has a very limited context for expressing application-level policies through the transport layer port number and transport protocol ID header fields. This is inadequate for designing control applications for managing application-traffic flows.

OpenADN solves this problem by putting meta-tags in the message header, which help classify the packets appropriately. Thus, the data can be kept private from the switches and the routers. Of course, the waypoints that need to operate on such data (virtual machines or virtual appliances) will be under the control of the ASP.

In OpenADN, after the application-level flow classification is done at the network edge, the application flow class is included as a meta-tag in the OpenADN header, as discussed later in this section.

*Affinity*

OpenADN is designed for a very dynamic global environment in which each intermediate point is replicated and may move inside or among clouds while the virtual machine on which it is running moves. We need some rules on how often to change forwarding decisions. This is called affinity. OpenADN offers both message affinity and session affinity.

*Message Affinity*   All packets that are part of an application-layer message need to be classified into the same application flow class and applied the same application-level policy.

*Session Affinity* All segments of a session are initially bound to appropriate physical segments. Each segment is bound to a particular stream in that segment. This binding remains until the end of the session, where the definition of the end of the session is application specific (explicit or implicit). Setting up such multisegment sessions is called session splicing.

### Sender and Receiver Policies

OpenADN does not distinguish between the sender and the receiver, and allows both the sender and the receiver to express their policies in an end-to-end communication. Note that the user may also be a service (instead of a human), as is the case with Web 2.0 mash-up applications and service oriented architecture (SOA).

### OpenADN Extensions

As previously mentioned, OpenADN extends OpenFlow and SDN concepts and combines them with several recent networking paradigms to provide application delivery. These extensions are discussed in this section.

### Cross-Layer Communication

OpenADN provides a session-layer abstraction to applications. It is a cross-layer design where the application layer (layer 7) places the meta-tag (the result of the application-level classification that indicates the application flow class) in the OpenADN header. The header is split across layers 4.5 and 3.5 of the TCP/IP protocol stack. Layer 4.5 implements the OpenADN data plane slow path (dynamic policy-based binding), whereas layer 3.5 implements the data plane fast path (static switching transport). The L4.5 meta-tag is used for setting up various session segments. It has the information required to enforce the session-forwarding policies. The L3.5 header is used by OpenADN aware switches to forward packets in the data plane.

*Application Label Switching*

Layer 3.5 meta-tag processing mechanism uses techniques similar to MPLS label processing, with semantic differences. We, therefore, call this layer application label switching (APLS). APLS uses a mechanism similar to label stacking (label pushing and popping) for enforcing sender and receiver policies on an application-traffic flow. Moreover, APLS uses a mechanism similar to label switching for switching a packet through multiple application-level waypoints. Space constraints do not permit us to include all the details of the label processing.

*Multistage Late Binding*

OpenADN is designed to support dynamic application deployment and access. It uses indirection as the key primitive to support dynamicity. The first indirection mechanism that OpenADN uses is multistage late binding. This mechanism is the basis of the session splicing primitive. Each session consists of many session segments. The endpoint of each segment is determined at the time that the segment is set up based on the application state and the meta-tags in the L4.5 header.

*ID/Locator Split*

This is the second indirection mechanism in OpenADN. All client, waypoints, and servers in OpenADN are assigned with fixed IDs, which are separate from their locators (IP addresses). The indirection layer mapping the ID to a locator adds intrinsic support for mobility to the architecture. It also provides other benefits to construct policy and security frameworks [4].

*SDN Control Application (Control Plane)*

As shown in Figure 7.3, SDN consists of three abstraction layers: virtualization, network operating systems, and network control applications. ASPs can implement OpenADN-based control applications and place them at the top of the SDN stack.

The ASP's control application computes session-forwarding tables implementing the ASP's deployment/delivery/management policies. The ISP's SDN abstraction provides the ASP controller with a virtual view of the APLS data plane as if it was implemented over a single, centralized APLS switch. The ASP controller passes the session-forwarding tables to the ISP's SDN controller that is then responsible for deploying it (preferably distributedly) over the OpenADN-enabled OpenFlow switches in the data plane. Note that, now, ASPs can also invoke the network-level services provided by the ISPs (as proposed by the application-layer traffic optimization [ALTO] [5] framework).

*OpenADN Aware OpenFlow Switches (Data Plane)*

OpenFlow switches classify packets based on L2, L3, and L4 headers. OpenADN aware OpenFlow switches also use the L3.5 and L4.5 headers for packet classification and forwarding. As shown in Figure 7.4, explicitly chained virtual tables specified in the OpenFlow data plane specification 1.1 [3] can be used for this. Incoming packets are first passed through a generic flow-identification table, which then redirects the packet through a virtual table pipeline for a more specific flow processing context. Using this virtual table support, the OpenADN data plane may interpose application-traffic flow processing before handing off the flow for network-level flow processing.

We propose a three-level naming hierarchy for virtual tables. The first level identifies whether it is performing application-level or

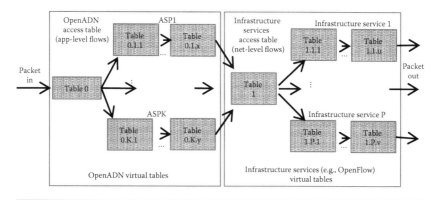

**Figure 7.4**   OpenADN and OpenFlow processing.

network-level flow processing. The second level identifies the SDN control module that configures the virtual table (e.g., ASP IDs for OpenADN, infrastructure service IDs for OpenFlow). The third level identifies the specific flow-processing context within an SDN control module.

Different types of data, e.g., Voice over IP (VoIP), video, Web, etc., require different quality of service at Internet switches and routers. The meta-tags in OpenADN headers easily allow this to be accomplished.

### Rule-Based Delegation

The rule-based delegation mechanism is one of the key innovations of OpenADN. It allows ASPs to create and optimize their specific networking environment over the common infrastructure. Unlike a private WAN (e.g., that of Google), in OpenADN, the ASPs and ISPs are different organizations, and so they do not completely trust each other and do not want to give full control to each other. Rule-based delegation solves this by allowing ASPs to securely communicate with the control plane of the ISP, which then arranges the data plane to satisfy the ASP's requirements, as shown in Figure 7.5. How the ISP distributes these rules to the data plane APLS entities is completely

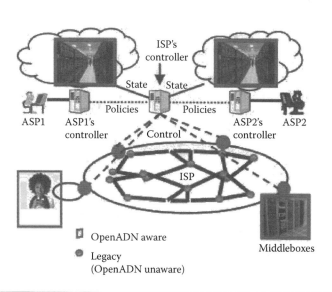

**Figure 7.5**   In OpenADN, ASP conveys its policies to ISP in the control plane.

up to the ISP. Moreover, the ISP does not need to look into the application data fields in the packets as required by Content Distribution Networks (CDNs). Furthermore, rule-based delegation creates a network-wide distributed intelligence that dynamically adapts to the dynamic changes of the applications and network conditions.

### Prototype Implementation

We have implemented a proof-of-concept prototype of an OpenADN switch using the click modular router [6]. To validate the functionality, we simulated a simple use-case scenario (Figure 7.6) consisting of three application servers (AppServer1$_A$, AppServer1$_B$, and AppServer2), two waypoints (IDS A and IDS B), a user (simulating multiple traffic sources), an OpenADN controller, and an OpenADN switch.

The use-case scenario is derived from the example of real-time services from smart cell towers. In this simulated scenario, the OpenADN controller belongs to the ISP that programs its OpenADN switch. Moreover, for this prototype, we assume a scenario where all traffic belongs to a single ASP. The application in this example deploys two different types of session segments: (1) SS1 (IDS, AppServer1) and

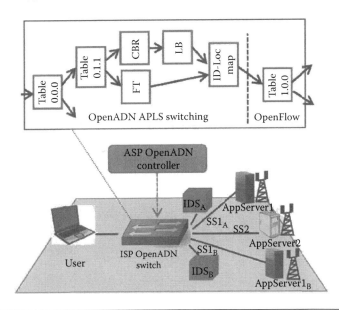

**Figure 7.6**   Prototype implementation.

(2) SS2 (AppServer2). SS1 has two streams (<$IDS_A$, $AppServer1_A$> and <$IDS_B$, $AppServer1_B$>).

This use-case scenario shows a centralized OpenADN switch–based implementation where all the entities are connected to the switch. The switch is responsible for indirecting the traffic for SS1 through an IDS to the application server. The IDS also implements an OpenADN stub that is statically configured to return the traffic to the interface over which it received it. When our final implementation of OpenADN (as a general session layer for the networking stack) is completed, it will be possible for the OpenADN controller to configure the OpenADN layer in the IDS to directly forward the traffic to the application server in the session segment. Moreover, in this example, we specifically named the tables FT, CBR, ID-Loc Map, LB, etc., to explain their functions. In fact, no such naming semantics is used, and tables are only numbered using the three-level numbering system previously discussed.

### Related Work

Application-specific packet processing has eluded network researchers for long. However, the full generality of in-network application-specific packet processing proposed by active networks research [7] failed to motivate real deployments. The active networks approach required applications to be allowed to run custom application processing codes on network nodes creating policy and security concerns for the network infrastructure providers.

Delegation-oriented architecture (DOA) [8] was proposed for off-path middlebox deployment to avoid the need to interpose middleboxes directly in the datapath. However, DOA was not designed for dynamic application delivery environments made available through cloud computing today. OpenADN borrows the principles of delegation from DOA and applies it to modern application delivery contexts. More recently, a flexible forwarding plane design has been proposed by the rule-based forwarding (RBF) architecture [9]. RBF proposes that packets must be forwarded to a rule rather than to a destination address. The rule would encode the specific processing required by a packet at a network node. However, rules bind early a packet to a set of processing nodes. Moreover, rules only allow enforcing receiver-centric policies. In OpenADN, packets carry application context, and it is bound late to a rule in the network. Moreover, OpenADN provides a standardized data plane abstraction for application-traffic flow processing and is thus more suitable for being deployed on high-performance network switches as compared with the (more) general purpose rule processing required by RBF.

Serval [10] is another recent approach for service-centric networking. However, Serval implements a separate control plane mechanism to support service replication across multicloud environments. OpenADN, on the other hand, is a data plane mechanism that allows dynamic service partitioning and service composition, in addition to service replication, for multicloud environments.

CloudNaaS [11] proposed an OpenFlow-based data plane to allow ASPs to deploy off-path middleboxes in cloud data centers. OpenADN is different from CloudNaaS in that it provides a session-layer abstraction to applications preserving session-affinity properties in an environment where the application servers and virtual appliances may need to scale dynamically. OpenADN allows both user and ASP policies, unlike CloudNaaS that only allows ASP policies. Moreover, CloudNaaS does not allow application-level flow processing such as OpenADN and, thus, does not provide support for dynamic service partitioning and replication, routing on user context, and context-based service composition. Some other works, such as CoMB [1] and APLOMB [12], have proposed delegating middlebox services to third-party providers. However, they directly conflict with the design principle of OpenADN that third-party providers should not have access to the ASPs' application-level data for reasons of privacy. Therefore, OpenADN provides a platform, where, instead of third parties, ASPs themselves can manage their middleboxes distributed across different cloud sites more easily.

## Summary

A recent explosion of mobile apps serving a global audience requires smart networking facilities that can help ASPs to replicate, partition, and compose their services on cloud computing facilities around the world on demand to dynamically optimize routing of their traffic.

The key features of OpenADN that may be of interest to the industry are as follows:

1. OpenADN is an open networking platform that allows ISPs to offer routing services for application delivery.
2. OpenADN uses OpenFlow concepts of data and control plane separation, extends headers to move application flow classification to the edge, and uses SDN concepts of a virtual centralized view of a control plane.
3. OpenADN offers ISPs and ASPs benefits of easy and consistent management and cost efficiencies that result from SDN and OpenFlow.
4. OpenADN takes network virtualization to the extremes of making the global Internet look like a virtual single data center.
5. Proxies can be located anywhere on the global Internet. Of course, they should be located in proximity to users and servers for optimal performance.
6. Proxies can be owned and operated by ASPs, ISPs, or CSPs.
7. Backward compatibility: Legacy traffic can pass through OpenADN boxes, and OpenADN traffic can pass through legacy boxes.
8. No changes to the core Internet.
9. Only some edge devices need to be OpenADN/SDN/Open Flow aware. The rest of the devices and routers can remain legacy.
10. Incremental deployment: It can start with just a few Open ADN aware OpenFlow switches.
11. Economic incentives for first adopters: Those ISPs that deploy few of these switches and those ASPs that use OpenADN will benefit immediately from the technology.
12. Full resource control: ISPs keep complete control over their network resources, whereas ASPs keep complete control over their application data that may be confidential and encrypted.
13. All this can be done now while the SDN technology is still evolving. This will also help in the development of northbound APIs for SDN.
14. CSPs, such as Google, Amazon, Rackspace, etc., can also add these features to their offerings for networks inside their clouds and also for networks connecting their multiple cloud sites.

## Acknowledgment

This work was sponsored in part by a grant from Cisco University Research Program and NSF grant number 1249681.

## References

1. Sekar, V., N. Egi, S. Ratnasamy, M. K. Reiter, and G. Shi. 2012. Design and implementation of a consolidated middlebox architecture. *Proceedings of the 9th USENIX Conference on System Design and Implementation (NSDI'12)*, pp. 24–37.
2. Gill, P., M. Arlitt, Z. Li, and A. Mahanti. 2008. The flattening Internet topology: Natural evolution, unsightly barnacles, or contrived collapse. *9th International Conference on Passive and Active Network Measurement*, pp. 1–10.
3. OpenFlow Switch Specification 1.3.1. 2012. http://www.opennetworking. org/images/stories/downloads/specification/openflow-spec-v1.3.1.pdf.
4. Paul, S., R. Jain, J. Pan, and M. Bowman. 2008. A vision of the next-generation Internet: A policy-oriented perspective. *Proceedings of the British Computer Society (BCS) International Conference on Visions of Computer Science*. pp. 1–14.
5. Seedorf, J. and E. Burger. 2009. Application-layer traffic optimization (ALTO) problem statement. *RFC 5963*.
6. Kohler, E., R. Morris, B. Chen, J. Jannotti, and M. F. Kaashoek. 2000. The Click modular router. *ACM Transactions on Computer Systems* 18(3), pp. 263–297.
7. Tennenhouse, D. L., J. M. Smith, W. D. Sincoskie, D. J. Wetherall, and G. J. Minden. 1997. A survey of active network research. *IEEE Comm.* (35)1, pp. 80–86.
8. Walfish, M., J. Stribling, M. Krohn, H. Balakrishnan, R. Morris, and S. Shenker. 2004. Middleboxes no longer considered harmful. *Proceedings of the 6th Conference on Symposium on Operating Systems Design & Implementation (OSDI'04) - Volume 6*.
9. Popa, L., N. Egi, S. Ratnasamy, and I. Stoica. 2010. Building extensible networks with rule-based forwarding (RBF). *Proceedings of the 9th USENIX Conference on Operating Systems Design and Implementation (OSDI'10)*.
10. Nordstrom, E., D. Shue, P. Gopalan, R. Kiefer, M. Arye, S. Y. Ko, J. Rexford, and M. J. Freedman. 2012. Serval: An end-host stack for service-centric networking. *Proceedings of the 9th USENIX Conference on Networked Systems Design and Implementation (NSDI'12)*.
11. Benson, T., A. Akella, A. Shaikh, and S. Sahu. 2011. CloudNaaS: A cloud networking platform for enterprise applications. *2011 Symposium on Cloud Computing (SOCC)*.

12. Sherry, J., S. Hasan, C. Scott, A. Krishnamurthy, S. Ratnasamy, and V. Sekar. 2012. Making middleboxes someone else's problem: Network processing as a cloud service. *Proceedings of the ACM SIGCOMM 2012 Conference on Applications, Technologies, Architectures, and Protocols for Computer Communication (SIGCOMM'12)*, pp. 13–24.

# 8

# HYBRID NETWORKING TOWARD A SOFTWARE-DEFINED ERA

CHRISTIAN ESTEVE ROTHENBERG,
ALLAN VIDAL, MARCOS ROGERIO
SALVADOR, CARLOS N. A. CORRÊA,
SIDNEY LUCENA, FERNANDO FARIAS,
JOÃO SALVATTI, EDUARDO CERQUEIRA,
AND ANTÔNIO ABELÉM

## Contents

## Introduction

While software-defined networking (SDN) has been an increasing buzz word with diverse interpretations, the first SDN products [1–3] based on the OpenFlow protocol—the original trigger of the SDN term [4]—have seen the light in 2012, targeting virtual data center networks for the enterprise and cloud service providers. In addition, the wide area network (WAN) has been targeted by Google's inter–data center network design based on OpenFlow to program (in-house built) 100-G equipment, with the forwarding information base (FIB) generated from open-source routing stacks (Quagga intermediate system to intermediate system [IS-IS]/border gateway protocol [BGP]) augmented with a centralized traffic-engineering application [5]. The result is a cost-effective, predictable, and highly automated and tested network that reaches unprecedented levels of use.

The first SDN deployments have been mainly greenfield scenarios and/or tightly controlled single administrative domains. Initial roll-out strategies are mainly based on virtual switch overlay models or OpenFlow-only network-wide controls. However, a broader adoption of SDN beyond data center silos—and between themselves—requires considering the interaction and integration with legacy control planes providing traditional switching; routing; and operation, administration, and management (OAM) functions. Certainly, rip-and-replace is not a viable strategy for the broad adoption of new networking technologies.

This chapter introduces the motivation and problem statement of hybrid networking models in OpenFlow/SDN and discusses different approaches to combine traditional control plane functions with OpenFlow control ("Hybrid Networking in OpenFlow/SDN"). Hybrid networking in an SDN scenario should allow deploying OpenFlow for a subset of all flows only, enable OpenFlow on a subset of devices and/or ports only, and, overall, provide options to interact with existing OAM protocols, legacy devices, and neighboring domains. As in any technology transition period where fork-lift upgrades may not be a choice for many, allowing for migration paths is critical for adoption. We introduce the design and implementation of two SDN control platforms that aim at addressing the requirements of hybrid networking deployments.

While the LegacyFlow project ("LegacyFlow: Leveraging the Legacy Switching Infrastructure") aims at turning legacy devices into OpenFlow-capable nodes in the topology available to the controller, the RouteFlow project ("RouteFlow: When IP Routing Meets OpenFlow") bridges the gap with traditional IP routing stacks. These proposals allow for transitioning opportunities from existing networks to OpenFlow/SDN. We will discuss how OpenFlow direct FIB manipulation can help IP routing control applications and enable cost-effective routing architectures by reducing the number of L3 control units and removing them from the devices themselves to centralized clusters. Details on currently implemented prototypes will be provided in addition to the results from the experimental evaluation work ("Prototype Implementations and Experimental Evaluation"). We will conclude with final remarks and a glimpse on our future work ("Conclusions and Future Work").

## Hybrid Networking in OpenFlow/SDN

The concept carried by the word hybrid spans several levels. The Hybrid Working Group of the Open Networking Foundation (ONF) [6] especially focuses on hybrid switch architectures. The goal is to have an equipment that can be configured to behave as a legacy switch or as an

OpenFlow switch and, in some cases, as both at the same time. This duality can be achieved, for example, by partitioning the set of ports of a switch, where one subset is devoted to OpenFlow-controlled networks, and the other subset, to legacy networks. For these subsets to be active at the same time, each one having its own data plane, multitable support at the forwarding engine (e.g., via ternary content-addressable memory [TCAM] partitioning) should be a requisite.

Besides port-based partitioning, it is also possible to have virtual local area network (VLAN)-based (prior to entering the OpenFlow pipeline) or flow-based partitioning using OpenFlow matching and the LOCAL and/or NORMAL actions to redirect packets to the legacy pipeline or the switch's local networking stack and its management stack. Flow-based partitioning is the most flexible option, and it allows each packet entering a switch to be classified by an OpenFlow flow description and treated by the appropriate data plane (OpenFlow controlled or legacy controlled).

The ONF Hybrid Working Group (WG) is currently discussing the different models, requirements, and use cases for a hybrid programmable forwarding plane (HPFP). One architectural approach for hybrid switches is called ships in the night (SIN). As its name suggests and, similarly, how the term has been used in the traditional IP networking domain [7], in the SIN architecture, there is no interaction of OpenFlow and legacy control planes, and the respective data planes are also isolated. Thus, there is no need to synchronize states between control or management planes. It can be considered a conservative and practical strategy for a migration path of network switch architectures, where new OpenFlow functionalities are incorporated to existing legacy equipments. The clear decoupling of OpenFlow and legacy control planes simplifies operation and makes each control plane be unaware of the other, like old SIN. The approach is not free of issues, and some use cases deserve special attention, for example, when using tunnel interfaces, the logical interface of the tunnel and the physical one used as an ingress or egress port may not be managed by the same control plane.

The ONF Hybrid WG is also discussing the integration of OpenFlow and the legacy control planes of a hybrid switch. Several aspects must be considered to enable this kind of integration, one of them is how to deal with shared resources between control planes, such as ports, tables, or meters. Safe operations of this integrated hybrid model require local resource arbitration, or an external orchestration, to manage the conflicts and integrate control states and resources. Notifications from the legacy data plane to the OpenFlow controller is another important and useful aspect to be considered, as well as communication between both control planes.

The best-known public implementation example of a software-based OpenFlow switch is OpenVSwitch (OVS). Generally deployed as a software switch running on top of a Linux kernel, it can be configured to act like an OpenFlow-only or a hybrid switch, supporting the SIN architecture with per-VLAN and per-port segregation. OVS has been also ported to hardware-based switches. HP Procurve has SIN implementation based on OVS, with per-port, per-VLAN, and per-flow segregation. Pica8 hardware switches also use OVS, with their latest product versions supporting OpenFlow version 1.2. Nippon Express Co. Programmable Flow Switch (NEC PFS) has a cascade model of integration, where packets are treated by the OpenFlow control plane then directed to the legacy control plane, without any other integration of both control planes. The NEC PFS has several shared resources, such as TCAM, buffers, and queues.

Among use cases involving hybrid switches, one is related with the need to complement OpenFlow switch capabilities to make this (still evolving) technology fully compatible with current networks in operation. One example is quality of service (QoS) policing, which is not completely defined on the earliest OpenFlow protocol versions and implementations. The opposite rationale for hybrid use cases also apply: it may be useful to have an OpenFlow switch complementing the capabilities of a legacy switch, for example, adding flexibility for rules matching. Another example related to hybrid switches with shared resources is the use of legacy mechanisms to set up tunnels and the OpenFlow control plane to decide which packets are directed to which tunnel.

When talking about hybrid networks, as already discussed, the focus is on how to coexist OpenFlow-enabled network devices with legacy ones. In such a concept, a hybrid switch can be very useful if, for instance, it can hold both data planes active at the same time and provide ways to intercommunicate both control planes. So, there are some different approaches that can be used to interconnect OpenFlow and the legacy amount of a network. On a hybrid network, a flow can traverse distinct unsynchronized data planes, each of them governed by OpenFlow or legacy distinct control planes. This transition between data planes can occur within a single switch or through different switches of a network.

From this perspective, there are several ways to define the meaning of hybrid network, where hybrid is the coexistence of the traditional environments of closed vendor's switches (and routers) with new OpenFlow-enabled devices. However, for very clear decoupling as well as for avoiding any upgrade of existing devices, the hybrid approach here focused refers to the interconnection of both control and data planes of legacy and new OpenFlow network elements (NEs).

*The Emergence of Hybrid Networks*

In a general sense, in any network of any kind, there is a need to establish a communication between different nodes to achieve a path (hopefully, the best one) between an origin-destination pair. Recall that, in current networks, there are basically two ways to establish a path: by the use of a distributed protocol running at each node or by directly configuring the data plane at each node along a chosen path. With respect to distributed protocols, for example, we have routing protocols, such as open shortest path first (OSPF), IS-IS, BGP, etc., or multiprotocol label switching (MPLS) signaling. With respect to the direct configuration of the data plane, it is possible to manually stitch VLANs through the network or use a generalized multiprotocol label switching (GMPLS)-based approach [8].

From an SDN perspective, hybrid networks need to be discussed based on the major characteristic of OpenFlow-enabled networks, which is the inherent external, typically centralized control of network nodes. This perspective induces at least three major approaches to be used for hybrid networking:

1. Make each OpenFlow node behave as a router, so it does not matter if such node is connected to another OpenFlow node or to a legacy router, supposing that the legacy nodes understand the respective L3 protocol.
2. Make each OpenFlow node understand circuit technology signaling to establish label switching paths (LSPs), supposing that other legacy nodes along the path also understand the same signaling protocol.

3. Use an OpenFlow-based control plane for the whole hybrid network and adopt an intermediate layer to translate respective OpenFlow forwarding rules to the configuration syntax of each legacy non-OpenFlow switch.

With respect to OpenFlow controllers, the following considerations arise:

1. For each approach above, OpenFlow domains may be controlled by one or more OpenFlow controllers, so subsets of the whole OpenFlow domain are permitted.
2. For the number 3 approach, consider OpenFlow as the uniform way to control the whole hybrid network and allow multiple OpenFlow controllers.
3. It is possible to plug solution numbers 1 or 2 on top of solution number 3; thus, the complete hybrid network will act like a pure OpenFlow network doing L3 routing or MPLS signaling.

The advantage of the latter compound approach (numbers 1 or 2 on top of number 3) relies at the specificities of the possible implementations of solution numbers 1 and 2. Because both are OpenFlow based, decisions on how to establish IP routes or LSPs can be centralized, which is interesting for a network operation/management point of view. Another interesting issue is related to services that can be offered on top of these solutions, such as the routing as a service (RaaS) concept [9,10]. For instance, RaaS can be engineered based on an OpenFlow architecture and applied to a hybrid network in which legacy devices are supported by solution number 3. Therefore, this approach offers a high-level interface for the description of routing services. Related to this scenario, the Internet Engineering Task Force (IETF) has recently started a discussion list on an interface to the Internet routing system (I2RS) [11], which aims to improve interactions between the underlying routing system and network control or management applications. Another IETF activity that allows for control split architectures is forwarding and control element separation (ForCES) [12], a framework that started long before OpenFlow but focused on the specifics of L3 devices, whereas OpenFlow devoted its initial pragmatic efforts to reuse TCAM capabilities in existing devices to offer a flat (L1–L4) flow abstraction programmable via a small instruction set.

Previously described approach numbers 1 and 3 are implemented by two OpenFlow-based architectures, that is, RouteFlow and LegacyFlow, which will be detailed further in this chapter. RouteFlow implements an IP level control plane on top of an OpenFlow network, making the underlying nodes act like routers, with different arrangement possibilities. LegacyFlow extends the OpenFlow-based controlled network to embrace non-OpenFlow nodes.

The common grounds of these pieces of work are (1) considering hybrid as the coexistence of traditional environments of closed vendor's routers and switches with new OpenFlow-enabled devices; (2) targeting the interconnection of both control and data planes of legacy and new NEs; and (3) taking a controller-centric approach, drawing the hybrid line outside of any device itself, but into the controller application space. Note that this approach is fundamentally different than an SIN strategy, where legacy and new control planes interconnect, and coexistence is device centric and not controller centric.

As an example of this deployment scenario for the case of RouteFlow, Figure 8.1 illustrates the new control plane added in the form of a BGP controller, which acts as a gateway between existing route reflectors (RRs) and OpenFlow controllers programming the datapath devices. Legacy neighboring routers close their (external BGP [eBGP]) control

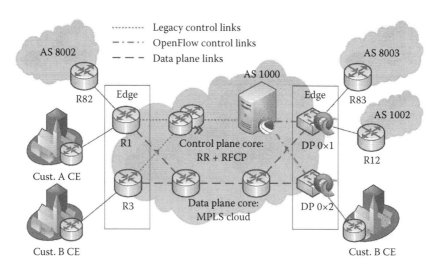

**Figure 8.1**    Routing architectures: current (left) vs. OpenFlow-based controller-centric hybrid SDN approach (right). AS, autonomous system; Cust, customer; CE, customer equipment; MPLS, multiprotocol label switching; DP, datapath; RFCP, RouteFlow control platform.

plane sessions with the routing engines running in centralized servers but effectively act as if they were at the edge devices themselves, which are reduced to plain packet forwarding as instructed via OpenFlow, the protocol used to download the FIB and to serve as the indirection point and encapsulation vehicle for the control plane traffic.

### LegacyFlow: Leveraging the Legacy Switching Infrastructure

The OpenFlow protocol allows production networking, such as campus networks, metropolitan networks, or research and development (R&D) networks, to experiment with so-called future Internet software, protocols, and architectures in parallel with production traffic. However, during roll-out, there are practical problems that arise when dealing with legacy elements or complete networks that do not support OpenFlow. Equipment replacement or upgrades involve high costs of reengineering and are often not feasible at all. In practice, the successful deployment of such a cutting-edge networking is far from seamless [13].

The overarching problem that we aim to address in the LegacyFlow project revolves around the question of how to integrate OpenFlow with legacy networks.

#### *Challenges between OpenFlow and Legacy Networks*

We start by conceptualizing legacy networks as networks that are composed of either non-OpenFlow equipment (e.g., the current Internet) or technologies that are currently not supported by the OpenFlow protocol (e.g., L1–L0 circuit technologies).

Legacy networks will always be a levee in the integration or connection among these new infrastructures (R&D experimental facilities or OpenFlow networks). Hence, it is important to understand some of the underlying problems between these two worlds—legacy and OpenFlow networks—including (1) the integration of OpenFlow and legacy networks, and (2) the convergence of transportation technologies not supported by OpenFlow into a unified control and management of legacy equipment by OpenFlow. The former challenge is related to the compatibility requirements of legacy network protocols. The latter issue is OpenFlow being unable to handle or manage legacy equipment using the same OpenFlow protocol in a unifying approach.

As a result of these two main challenges, the connection of multiple OpenFlow environments through legacy networks is very difficult.

Another important issue to the success of OpenFlow is its extensibility to support the relevant layers 1 and 0 circuit technologies of legacy networks, such as MPLS, GMPLS, synchronous optical networking/synchronous digital hierarchy (SONET/SDH), or wavelength division multiplexing (WDM). On the other hand, the new OpenFlow version 1.2 allows the inclusion of new match conditions and actions, but it is still unclear how the control capabilities negotiation between transport elements and the controller should happen in addition to a multiversion OpenFlow operation.

*Integrating OpenFlow and Legacy Networks*    In the current state-of-the-art, the integration of different OpenFlow islands requires different engineering and handcrafted strategies on legacy network devices and their protocols. A transparent and automatic manner to integrate legacy network is still an unresolved issue, because of the current absence of legacy protocol emulation applications available on OpenFlow controllers (e.g., address resolution protocol [ARP] replier, link layer discovery protocol [LLDP]) that are truly compatible with the distributed protocols running in the legacy equipments. Figure 8.2 illustrates an example of such a mismatch in the case of link layer topology discovery. The openflow discovery protocol (OFDP) [14] is an artificial term* of how LLDP is being leveraged with subtle modifications to perform topology discovery in an OpenFlow network. OFDP provides a controller-driven means of finding out neighboring OpenFlow switches.

Scenario A shows two OpenFlow domains using the OpenFlow LLDP implementation to find links between domains. In this case, the discovery works well because all OpenFlow equipment implement the same protocol code (e.g., Ethertype set to 0x88cc). However, in Scenario B, there is a legacy domain between the OpenFlow domains, and the OpenFlow discovery process fails because the legacy switches do not correctly support the OpenFlow LLDP code. Thus, the LLDP packets are dropped, and the link between the OpenFlow domains is not detected.

*Support of Legacy Circuit Technologies*    Another common problem when dealing with legacy networks is the support of popular circuit

---

* http://groups.geni.net/geni/wiki/OpenFlowDiscoveryProtocol.

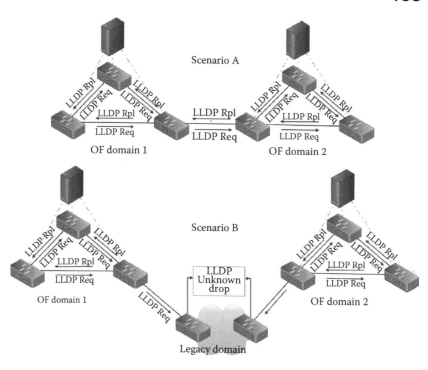

**Figure 8.2** An OpenFlow and legacy network scenario. OF, OpenFlow; Rpl, reply; Req, request.

technologies. A practical example of the legacy network is the Internet backbone, which consists of a wide range of transport technologies. Current OpenFlow specifications do not offer consistent support of circuit-based technologies, such as MPLS, GMPLS, SONET/SDH, or WDM. Because these technologies are essential elements on operational networks, they should be somehow embraced because an equipment rip-off may not be wise or even possible. Therefore, the question is how to absorb these technologies into a unifying architecture.

The problem is that most of these technologies do not offer essential features for architectural evolution or virtualization based on slicing and programmability. For example, GMPLS is considered a conservative protocol suite, which leaves little room for innovation because of the complexity of integrating new features in the distributed control plane. On the other hand, OpenFlow can be easily virtualized along different dimensions (switches, flow spaces, controllers, etc.) while allowing the network control layer to be sliced and the system performance to be improved [15].

*Management of Legacy Vendor Equipment*   We now turn our attention to examining legacy equipment, such as switches, routers, and optical devices. Issues about whether (and where) to use OpenFlow are common concerns among network and system administrators these days.

Despite being a costly proposition, the ecosystem is not ready to replace every networking gear to turn the network 100% OpenFlow enabled. Some devices are paramount in the production network, and currently, there is no equivalent OpenFlow-enabled equipment for every network device, for instance, 100-G core packet switches, expensive reconfigurable optical add-drop multiplexer (ROADM) optical switches, or very low-latency legacy switches. Hence, it becomes necessary to find a means of using the OpenFlow without losing compatibility with the legacy equipment.

*Design and Architecture*

LegacyFlow proposes a hybrid model that focuses on the three main issues previously presented: (1) the integration of OpenFlow with legacy networks, (2) the (re)use of legacy technologies, and (3) the management of legacy equipment. Some advantages of our approach include reusing existing equipment in operation, reducing the cost of OpenFlow deployment, and offering ways for gradual deployment while contributing to the capabilities of a unifying OpenFlow control plane.

The first step in the project was to look for a merge point between the two networks (legacy and OpenFlow) without weakening the key capabilities of programmability and network virtualization. One requirement is being able to manage the legacy equipment via OpenFlow without the chance of developing an OpenFlow agent running inside the control plane of the switch.

The adopted model defined in the LegacyFlow framework introduces a new virtual datapath component in the form of a software switch that acts as a control glue between OpenFlow and the Legacy network. Acting as a proxy to OpenFlow actions, it translates OpenFlow rules to the equivalent operations available in the underlying legacy equipment using interfaces, such as command-line interface (CLI), WebService, or simple network management protocol (SNMP). Of course, this operation mode imposes some limitations to the type of actions supported by the devices. On the other hand, it allows us to expose to

OpenFlow some switch features that are not supported in the standard protocol (e.g., Lambda switching).

One fundamental restriction of this approach is sacrificing the reactive mode of operation of OpenFlow, which packets without a matching rule are forwarded to the controller via packet-in events. At this stage, LegacyFlow does not support this type of send-to-controller actions in non-OpenFlow devices. Hence, these legacy devices will generally be used in network segments that interconnect OpenFlow switches located at the ingress or egress point of a flow. Figure 8.3 illustrates edge switches that are fully OpenFlow enabled and supported, thus capable of forwarding the packet to the controller, either as an explicit flow action or as an implicit no-match switch configuration.

For instance, when a new flow enters a LegacyFlow network, the first packet can be recognized by the edge OpenFlow device and sent to the controller via a packet-in. In turn, the controller installs the required flow rules on the datapaths along the path, for instance, creating a tunnel or a circuit based on VLAN identifier (ID) throughout the legacy and OpenFlow equipment. The circuit technology could not be limited to VLAN, but to the MPLS/GMPLS LSP or the WDM light path.

*Architecture* The LegacyFlow architecture shown in Figure 8.4 is divided into two layers: (1) datapath and (2) switch controller.

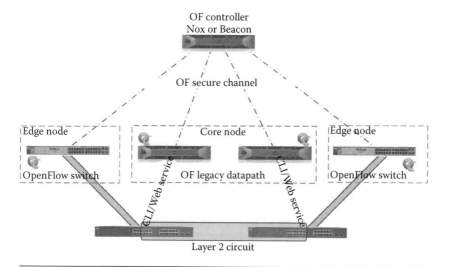

**Figure 8.3** A view of the LegacyFlow solution.

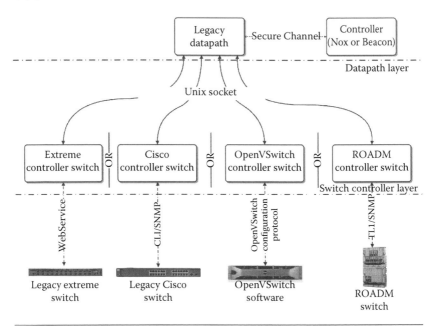

**Figure 8.4**  The LegacyFlow architecture. ROADM, reconfigurable optical add-drop multiplexer; CLI, command-line interface; SNMP, simple network management protocol; TL1, transaction language 1.

In the datapath layer (DL), there is a new software switch (LegacyFlow datapath [LD]) that (1) communicates with the controller via the OpenFlow protocol and (2) provides a bridge of access of different features from the legacy equipment to the controller.

The switch controller layer (SCL) provides the configuration module for each legacy equipment (e.g., switch, router, or optical switch) along with the information related to the available access method depending on each vendor, commonly WebService, Telnet/CLI, and SNMP. The communication between DL and SCL is based on a client/server interprocess communication (IPC). This modular approach keeps independent the development on the legacy datapath and the controller switch, allowing a simpler integration of new modules to the LegacyFlow architecture.

The link between the legacy datapath and the controller switch is point to point—each legacy datapath is responsible for a kind of controller switch. Currently, LegacyFlow supports the legacy equipment of the following vendors: Extreme and Cisco. At this point of writing, there is an ongoing development for Juniper routers and ROADM optical equipment.

*LegacyFlow Datapath* The LD acts as a proxy receiving OpenFlow commands from the controller and applying the corresponding actions in the real switch. Each legacy switch (part of the hybrid OpenFlow network) is assigned to an LD that runs in a guest machine (real or virtual) with Gnu's Not Unix (GNU)/Linux OS, as shown in Figure 8.5.

From an outside point, the LD represents some information about the features of the legacy switches, such as port numbers, link speed, and sent-and-received packets rate, among other capabilities specified in OpenFlow. At the same time, it translates the proactive installation of OpenFlow entries into the corresponding switch commands whenever possible, returning an error if the match field or action is not controllable via the vendor-specific application programming interfaces (APIs).

To initialize an OpenFlow datapath, it is necessary to pass through the parameters of the network interface (NIC). Likewise, the LD must be started with information of the switch interfaces. As shown in Figure 8.6, the LegacyFlow virtual interface (LVI) is a Linux virtual interface module that represents a switch port of the LD. The LD is initialized by creating virtual NICs according to the number of switch ports of their features, mirroring from the real switch characteristics, such as link speed and maximum transmission unit (MTU) size, among other values. Depending on how the switch offers this information, translation from the switch to the virtual interfaces via SNMP or WebService may be required.

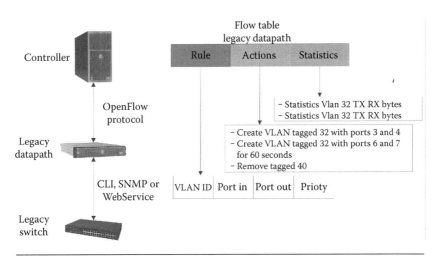

**Figure 8.5** LegacyFlow datapath. CLI, command-line interface; SNMP, simple network management protocol; VLAN ID, virtual local area network identifier; TX, transmitter; RX, receiver.

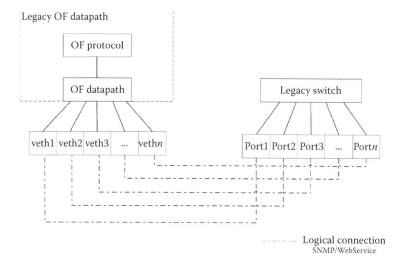

**Figure 8.6**   Connection between LVI and legacy switch. OF, OpenFlow; SNMP, simple network management protocol.

The LD module allows the controller to acquire a view of the legacy equipment and features in a way similar with the OpenFlow equipment, managing the state and collecting statistics in a unified manner. Currently, the LVI can represent the layer 2 interface, such as the MAC address, speed, link status, name, and modes. In the future, it will represent layer 1 port information (WDM), such as name, wavelength and bandwidth.

*LegacyFlow Controller Switch*   The Legacy controller switch (LCS) is a set of applications used to configure the legacy equipment. The communication between the LD and the LCS is based on an IPC, where the LD is a client and the LCS is a server. This approach facilitates the integration of a new LCS independent from the programming language or the configuration interface (i.e., SNMP, NetConf, WebService, or Telnet/CLI).

An out-of-band control channel between the LCS and the legacy equipment is used to send configuration commands. The port management equipment is generally used to establish this communication. Each LCS is limited to handling just one piece of legacy equipment. The information about the equipment being managed is obtained from the configuration file during the initialization process of the LD.

*LegacyFlow Controller Components (LCC)*    The LCC refers to the set of applications that integrate to the OpenFlow controller. These components are necessary to keep a transparent combination of the OpenFlow and the legacy network. Currently, the components are developed using Nox API and can be divided into four parts: LegacyFlow API, LegacyFlow application, LegacyFlow manager (LM), and LegacyFlow database (LDB). Figure 8.7 illustrates a high-level view of the LCC.

The LegacyFlow API is responsible for offering access interfaces to the LegacyFlow actions. These new actions are only recognized and executed on the appropriate LD. The LM is a component developed to manage communication with other elements, such as Web graphic

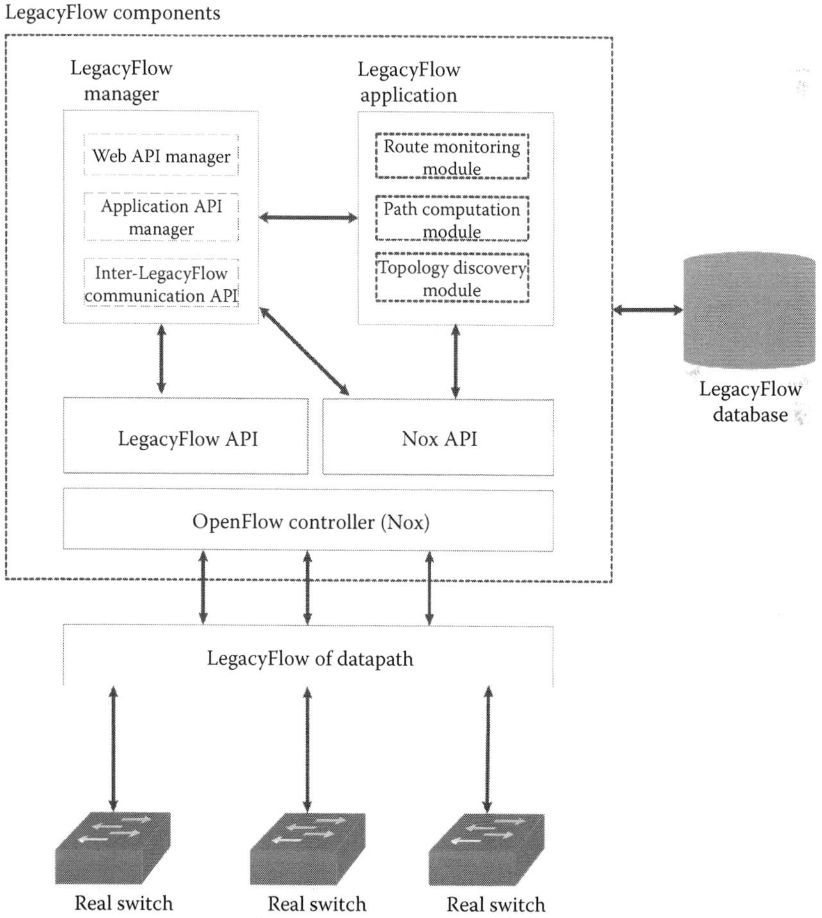

LegacyFlow components

**Figure 8.7**    Connection between LVI and legacy switch. OF, OpenFlow, API, application programming interface.

user interface (GUI), helper applications, or other LegacyFlow controller instances. The three main subcomponents are as follows:

1. **Web manager API:** It interacts with the WEB GUI based on hypertext transfer protocol (HTTP)+JavaScript object notation (JSON).
2. **Application manager API:** It provides interfaces to helper applications, such as route calculation, topology management, monitoring, etc.
3. **Inter-LegacyFlow communication API:** It manages the communication between LCCs located in other OpenFlow domains and offers resource allocation, topology exchange, and resource updates.

The LegacyFlow application assists the LCC on decisions, such as policies, routes, or topology, which are common applications developed by the network administrator. Currently, the following applications are available: route monitoring, to track created circuits or paths; path calculation, to compute routes between source and destination OpenFlow switches; and topology discovery, to maintain the topology formed by legacy switches.

Finally, the LDB stores every information created by LCC, including circuits, interface statistics, or monitoring. Today, we use a Not Only Structured Query Language (NoSQL) database (DB) that provides JSON APIs.

*Use-Case Operations*

We now illustrate the LegacyFlow in operation in the case of a circuit path among legacy switches based on VLAN ID.

The activity diagram is presented in Figure 8.8. Initially, the LD is started and receives as a mandatory parameter the LD switch model describing the available communication protocols. The LVI is initiated by creating virtual interfaces on the operational system according to the ones available in the switch. Using an outer and dedicated out-of-band channel between the switch and the LD, SNMP connections are created to collect information of the switch interfaces and apply them to the virtual interfaces. Currently, this state is updated every 3 s to keep information relatively accurate.

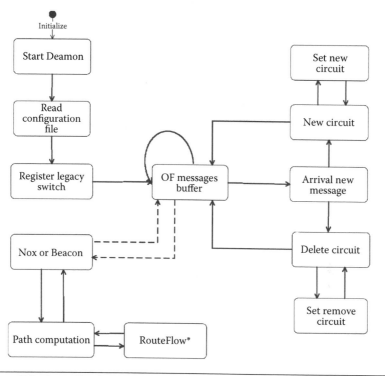

**Figure 8.8** LegacyFlow datapath. OF, OpenFlow.

After the switch is registered with the LD, one vendor-specific switch controller is initialized to manage the real switch using an interprocess pipeline. Following the setup phase, the interfaces are connected to the LD, which can start receiving flow commands from the controller. OpenFlow messages received in the LD are interpreted and checked for compatibility with actions supported by the legacy switches. The OF message buffer is responsible for receiving all OpenFlow protocol messages and handling them to the corresponding actions, such as delete or create circuit. When the action is called, a specific interprocess call is executed. In this example, these actions are the configuration counterparts to remove or create VLANs on the real switch, using parameters such as ingress port, egress port, and VLAN ID.

When a flow packet gets into a qualified OpenFlow switch, the controller identifies the source and destination of the flow and uses the path computation module to calculate one path among the Legacy switches. When the circuit is established, the flow is forwarded to the destination.

*Concluding Remarks and Work Ahead*

The LegacyFlow efforts continue toward improving gradual deployment options for OpenFlow in production environments. The architectural proposal prioritizes reusing the legacy equipment without changing the entire infrastructure. The approach introduces OpenFlow datapath abstractions per legacy device, following a proxy model that translates OpenFlow commands into vendor-specific control and configuration equivalents. Dynamic circuits across legacy switches are used to bridge the connectivity among OpenFlow-enabled devices.

Similar work on dynamic circuits is being done in the global environment for network innovations (GENI) experimental network and the stitching architecture based on VLANs. The ongoing activities toward a Slice Federation Architecture include investigations into the interactions with the controller proxy FlowVisor that allows sharing the programmable infrastructure among multiple controllers to realize isolated virtual networks.

As for the future, there are multiple lines of work that we would like to pursue. One front seeks to adapt LegacyFlow to interwork with the dynamic circuit network (DCN) similar with interdomain controller (IDC)/on-demand secure circuits and advance reservation system (OSCARS) [16], dynamic resource allocation via GMPLS optical networks (DRAGON) [17], or bandwidth on demand (AUTOBAHN) [18], the shared goal being on-demand dynamic circuit management between OpenFlow domains and/or testbed facilities. We will also investigate the applicability of LegacyFlow as a tool to create circuits using multilayer technologies and handle expected issues of a unified control plane, such as protection and restoration, standardization, and so on. Further integration work is likely to be required in addition to the extensibility of the components to embrace more legacy technologies and additional features of the OpenFlow framework. To strengthen the architecture performance, measurements need to be undertaken to stress test the different components. Finally, experimental evaluation work on the feasibility and effectiveness of the resulting artifact will be pursued within the facilities of the future Internet testbeds/experimentation between Brazil and Europe (FIBRE)-federated testbed facilities [19].

### RouteFlow: When IP Routing Meets OpenFlow

RouteFlow was born as a *Gedankenexperiment* (thought experiment) on whether the Linux-based control plane embedded in a 1U merchant silicon Ethernet switch prototype could be run out of the box in a commodity server, with OpenFlow being the sole communication channel between data and control planes. Firstly baptized as QuagFlow [20] (Quagga + OpenFlow), the experiment turned out to be viable in terms of convergence and performance when compared with a traditional laboratory setup [21]. With increasing interest from the community, the RouteFlow project emerged and went public to serve the goal of gluing together open-source routing stacks and OpenFlow infrastructures.

Fundamentally, RouteFlow is based on three main modules: the RouteFlow client (RFClient), the RouteFlow server (RFServer), and the RouteFlow proxy (RFProxy).* Figure 8.9 depicts a simplified view of a typical RouteFlow scenario: routing engines in a virtualized environment generate the FIB according to the configured routing protocols (e.g., OSPF, BGP) and ARP processes. In turn, the routing and ARP tables are collected by the RFClient daemons and then translated into OpenFlow tuples that are sent to the RFServer, which adapts this FIB to the specified routing control logic and finally instructs the RFProxy, a controller application, to configure the switches using OpenFlow commands.

This approach leads to a flexible, high-performance, and commercially competitive solution, at this time, built on available resources, such as (1) programmable low cost switches and small-footprint embedded software (i.e., OpenFlow), (2) a stack of open-source routing protocols (e.g., Quagga), and (3) commodity ×86 server technology.

The project counts with a growing user base worldwide (more than 1000 downloads and more than 10,000 unique visitors since the project started in April 2010). External contributions range from bug

---

* As a historical note, the first QuagFlow prototype implemented RFServer and RFProxy as a single Nox application. After the separation (in the first RouteFlow versions) RFProxy was named RouteFlow controller. This caused some confusion because it actually is an application on top of an OpenFlow controller, so we renamed it. Its purpose and general design remain the same.

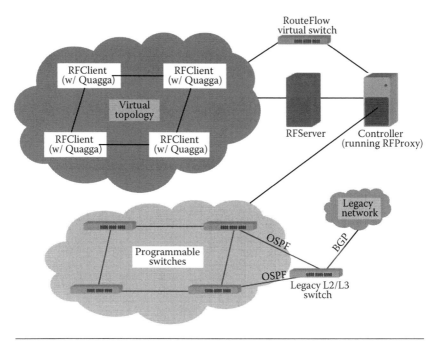

**Figure 8.9**  A typical, simplified RouteFlow scenario. OSPF, open shortest path first; BGP, border gateway protocol.

reporting to actual code submissions via the community-oriented GitHub repository. To cite a few examples, Google has contributed with an SNMP plug-in and is currently working on MPLS support and the new APIs of the Quagga routing engine maintained by the Internet systems consortium (ISC) Open Source Routing initiative. Indiana University has added an advanced GUI and run pilots with hardware (HW) switches in the U.S.-wide network development and deployment initiative (NDDI) testbed. Unirio has prototyped a single node abstraction with a domain-wide eBGP controller. Unicamp has done a port to the Ryu OpenFlow 1.2 controller and is experimenting with new data center designs. While some users look at RouteFlow as Quagga on steroids to achieve an HW-accelerated open-source routing solution, others are looking at cost-effective BGP-free edge designs in hybrid IP-SDN networking scenarios, where RouteFlow offers a migration path to OpenFlow/SDN [22]. These are ongoing examples of the power of innovation resulting from the blend of open interfaces to commercial HW and open-source community-driven software (SW) development.

*Architectural Details*

The current RouteFlow architecture is an evolution from previous prototype designs to a better layered, distributed system, flexible enough to accommodate different virtualization use cases (*m:n* mapping of routing engine virtual interfaces to physical OpenFlow-enabled ports) and ease the development of advanced routing-oriented applications. Past rearchitecting efforts have attempted to solve problems revealed during the first year of the public release, including feedback from third-party users and lessons learned from demonstrations using commercial OpenFlow switches.* The main issues addressed include configurability, component flexibility, resilience, easy management interfaces, and collection of statistics.

Anticipating the need for updating (or even replacing) parts of the architecture while facilitating multicontroller support, the implementation is segregated into the following three components (cf. Figure 8.10):

1. **RFClient:** collects routing and forwarding information generated by the routing engine (e.g., Quagga [23]) of the Linux system,† where it runs as a user-space daemon. Optionally, to extract additional routing information (e.g., all BGP paths in the routing information base [RIB]-in), it hooks into or peers (e.g., interior BGP [iBGP]) with the routing engine(s).

2. **RFServer:** a stand-alone application responsible for the system's core logic (e.g., event processing, virtual machine [VM]-to-datapath [DP] mapping, etc.). RouteFlow control platform (RFCP) services are implemented as operator-tailored modules that use the knowledge information base (KIB) to deliver arbitrary, high-level routing logics (e.g., load balancing, preferred exit points, etc.).

3. **RFProxy:** Simple proxy application on top of an OpenFlow controller (e.g., Nox, Pox), which serves the RFCP with switch interaction and state collected from topology discovery and monitoring applications.

---

* Open Networking Summit I (October 2011) and II (April 2012), Super Computing Research Sandbox (November 2011), OFELIA/CHANGE Summer School (November 2011), and Internet 2 NDDI (January 2012), 7th API on SDN (June 2012). See details on http://www.sites.google.com/site/routeflow/updates.

† Typically, it is a lightweight virtual container like LXC [26].

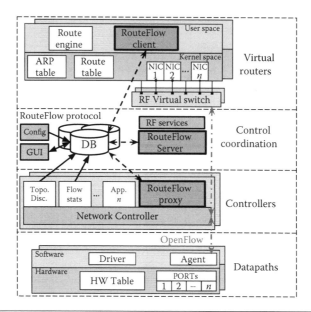

**Figure 8.10** Architecture design of the RFCP. ARP, address resolution protocol; RF, RouteFlow; Config, configuration; GUI, graphic user interface; HW, hardware; Stats, statistics.

In line with the best design practices of cloud applications, we opted for a scalable, fault-tolerant data store that centralizes (1) the RFCP core state (e.g., resource associations), (2) the network view (logical, physical, and protocol specific), and (3) any information base (e.g., traffic histogram/forecasts, flow monitoring, and administrative policies) used to develop routing applications. Hence, the data store embodies the network information base (NIB) [24] and the KIB [25].

In addition to acting as the back-end storage, the distributed NoSQL DB of choice (e.g., MongoDB, Redis, Cassandra) is used as the pub-sub–like message queuing IPC that loosely couples the modules via an extensible JSON-based implementation of the RouteFlow protocol. Without sacrificing performance, the data store–based IPC eases fault management, debugging, and monitoring by natively keeping a history of the RouteFlow workflow, allowing for replay or catch-up operations. Noteworthy, the IPC implementation (e.g., message factory) is completely agnostic to the DB of choice, should we change this decision.*

---

* While some may call database-as-an-IPC an antipattern we debate this belief when considering NoSQL solutions, such as MongoDB, acting as a messaging and transport layer (e.g., http://www.shtylman.com/post/the-tail-of-mongodb/).

Altogether, the design tries to follow principles that allow architectural evolvability [26]: layers of indirection, system modularity, and interface extensibility.

*Messaging Workflow* To allow for legacy networks integration, RouteFlow use flow entries to match known routing messages received by the physical infrastructure, handling them into the virtual topology. Conversely, pertinent routing messages that originated in the virtual topology are sent through the physical infrastructure. For example, adjacent Open Shortest Path First (OSPF) routers instantiated in the VMs will have their HELLO messages sent down to the OpenFlow switches and forwarded out via the corresponding physical ports. In turn, the OSPF messages are received via the physical ports of the adjacent OpenFlow switch and will match a flow entry based on the OSPF protocol information. The messages will then be sent up via the RouteFlow middleware into the corresponding virtual interface.

After the OpenFlow physical topology is discovered and registered as an available resource, the RFServer maps available VMs (interfaces) to OpenFlow switch (ports). The routing service will decide on how this mapping shall be done and allows for different modes of operation (later detailed in "OpenFlow Concept"). For instance, in case of an aggregated routing model, a single control plane VM may span multiple switches that are effectively programmed to act as a single router. Each VM runs a routing protocol stack and is instantiated with as many NICs as there are active ports in the corresponding OpenFlow domain. The NICs are bound to software switches through which physical connectivity can be mapped. Routing protocols in the VMs execute as usual and compute their individual FIBs. For each FIB update, the RFClient captures the event and sends an update message to the RFServer. In turn, based on the implemented routing services, the RFServer transforms each FIB update (e.g., route our ARP entry add/remove) into a (number of) flow modification commands that the RFProxies receive and execute by issuing the corresponding OpenFlow protocol flow-mod commands.

In the datapath devices, matching packets on routing protocol and control traffic (e.g., ARP, BGP, routing information protocol [RIP], OSPF) are directed by the RFProxy to the corresponding

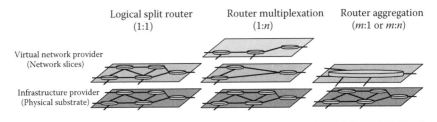

**Figure 8.11**   Different modes of operation and usage scenarios of router virtualization.

VM interfaces via a software switch [27]. The behavior of this virtual switch* is also controlled by the RFProxy and allows for a direct channel between the physical and virtual environments, eliminating the need to pass through the RFServer and the RFClient, reducing the delay in routing protocol messages, and allowing for distributed virtual switches and additional programmability. Note that only control plane traffic is sent via OpenFlow to the RouteFlow domain. User traffic is solely forwarded in the data plane via the OpenFlow-compiled FIB state installed in the datapath devices.

*Modes of Operation*

Separating the control plane from the forwarding substrate allows for a flexible mapping and operation between the virtual elements and their physical counterparts. Figure 8.11 shows three main modes of operation that RouteFlow aims to support.

1. **Logical split:** This 1:1 mapping between hardware switches and virtualized routing engines basically mirrors the physical substrate (number of switch ports, connectivity) into the virtual control plane. Two submodes of operation can be defined depending on whether the routing protocol messages are sent through the physical infrastructure (i.e., traditional routing) or are kept in the virtual plane. The latter allows to separate the problems of physical topology discovery and maintenance from the routing state distribution. This enables an optimization of the routing protocols, where fast connectivity maintenance can be run in the data plane (e.g., bidirectional

---

* We use OpenVSwitch for this task: http://www.openvswitch.org/.

forwarding detection [BFD]-like), whereas route updates (e.g., link state advertisement [LSA]) can be flooded inside the virtual domain.

2. **Multiplexing:** This 1:$n$ mapping of physical to virtual substrate represents the common approach to router virtualization, where multiple control planes run simultaneously and install their independent FIBs on the same hardware. Multitenant virtual networks can be defined by letting control protocol messages flow through the virtual plane and stitching the data plane connectivity accordingly.

3. **Aggregation:** This $m$:1 mapping of hardware resources to virtual instances allows to simplify the network protocol engineering by bundling a group of switches, such that neighboring devices or domains can treat the aggregated as if it were a single element. This way, intradomain routing can be independently defined per software, whereas legacy interdomain or interzone routing (e.g., BGP) can be converged into a single control unit for signaling scalability and simplified, centralized management purposes.

*Protocols and Abstractions*

The RouteFlow protocol glues together the different modules with a simple command/response syntax north bound to the RFClients and a subset of the OpenFlow messages south bound to the RFProxy application. As one approach to the OpenFlow version polyglotism, the RouteFlow protocol layer abstracts most of the details of the differences from the controller implementing OpenFlow versions 1.x and the slightly higher-level RouteFlow APIs. There is no conceptional barrier to support flexible $m$:$n$ mapping as pursued within the IETF ForCES WG [12] and the definition of control elements (CE) and forwarding elements (FE), which compound forms an NE that logically constitutes a router.

Besides being the API to the datapaths, OpenFlow is used as the vehicle to deliver control plane messages to/from the interfaces of the VMs running the routing engine and the RFClient. Programming the virtual switches (OVS in Figure 8.10), we can select a mode of operation where routing protocol messages are sent down to the physical devices or are kept up in the virtual network plane, which can be a reproduction of the discovered physical connectivity or a simplified/arbitrary version (e.g., single router abstractions [22,28]) of the HW resources. In the former, messages follow the physical path, so no additional fault-detection mechanism are required at the cost of extra propagation delays. In the latter, signaling is kept in the virtual domain, benefiting from low latencies and contributing to better scalability. This option provides a more natural support for virtual networks but requires extra programmability and fault-detection extensions to update VMs' logical associations caused by physical link changes.

All in all, the resulting software artifact is capable of instructing the SDN data plane with regard to IP forwarding and addresses the following premises:

- The ability to define a desired routing criteria (i.e., business policies) in an abstraction level, which does not require individual configuration of multiple routing processes;
- A unified, programmer-friendly DB that conveys knowledge and network information, allowing for provisioning and traffic-engineered route control services; and
- Scalable virtual control plane elements, which provide strong automation and flexibility for in-memory representation of intended control scenarios.

*Use Cases*

Pioneering work on routing control platforms (RCP) [29,30] has shown many of the potential benefits in more flexible routing [31], security [32], and connectivity management [33]. Replacing iBGP control with the direct FIB manipulation and the extended action set of OpenFlow allows for a wider range of rich IP routing application. We briefly present some use-case scenarios that we want to pursue with the RouteFlow platform.

*Engineered Path Selection*  Leveraging the knowledge of all available paths in a domain, RouteFlow offers advanced load-balancing and cost/performance path selection [31,34,35] per application or customer using OpenFlow flexible matching to push traffic into traffic-engineered paths or virtual routing and forwardings (VRFs).

Policy-based routing becomes untangled from flow classification restrictions or technology requirements. Moreover, OpenFlow allows direct preparation for optical switching in the core as well as edge to edge. Recent research [36] has shown that network-wide visibility and control enable joint traffic engineering and failure recovery by efficiently balancing the load while potentially simplifying router design and reducing operation costs for Internet service providers (ISPs).

*Path Protection and Prefix-Independent Convergence*  The complete view of route state allows for novel IP route protection schemes like per-edge computation of primary and secondary paths for each prefix, considering shared risk group inferences [25]. Group and multitables introduced in version 1.1 allow for an implementation of hierarchical FIB organization [37] to quickly recover from edge failures. Network-wide, loop-free alternate next hops can be precalculated and maintained in the group tables, removing the burden of one flow-mod per prefix update and allowing a fast local repair, provided that an OAM extension is available, for instance, triggered by a BFD state change. The indirection points introduced by OpenFlow group tables allow to flexibly point and update next hops, a feature known as prefix-independent convergence capabilities. The common design principle is moving away from dynamic route computation by decoupling failure recovery from path computation [38].

*Data Plane Security*  One generic OpenFlow/SDN application is to distribute firewall and intrusion prevention system/intrusion detection system functionality along the path, potentially on a customer basis, instead of forcing all traffic to either pass through or be detoured to a specialized box. Once detected, a selective distributed denial of service blackholing strategy (akin Ref. [33]) can be deployed by inserting higher priority flow entries with drop action only on the inbound ports under attack. OpenFlow fine granular rules (inport, src, dst, tp − port) allow to blackhole traffic without taking out of service the target IPs (as with today's common filtering and BGP-based mechanisms). Similar filtering capabilities can be used to duplicate traffic for legal intercept purposes or advanced monitoring applications in spirit of SPAN/test access point (TAP) ports but pin out by OpenFlow flexible matching.

*Secure Interdomain Routing*   Secure origin BGP (soBGP) or secure BGP (SBGP) deployment gets easier because of the readily available CPU power when consolidating BGP into commodity server technology. Benefits of the centralization approach include (1) the elimination of duplicate processing, signing once per neighbor rather than once per neighbor peering location; (2) avoidance of code upgrades of all existing autonomous system (AS) border routers; (3) maintenance of keys in one place rather than on every router; and (4) duplicate processing plus the nontrivial SW and HW upgrades of AS border routers. Deployment gets easier because of the plenty CPU and crypto assistance on the controller, which centralizes key management and reduces the signing overheads to once per neighbor rather than per peering location. Moreover, route selection can prioritize those trustable AS participating in address registries, public key infrastructure (PKI), domain name system security extensions (DNSSEC), or new security techniques beyond today's concepts. For instance, Morpheus [32] proposes algorithms (e.g., history-based anomaly detection) to achieve pretty good BGP security without requiring massive S*BGP adoption [39].

*Simplifying Customer Multihoming and IPv6 Migration*   Enterprise networks lack customer-friendly and effective multihoming solutions, especially in the absence of PI addresses or dual-stack IPv4/v6. RouteFlow-based policy routing could be used as an upstream affinity mechanism, sending only packets sourced from ISP1 networks via ISP1 (and ISP2 networks via ISP2) by flow matching based on the source IP subnet. Effectively, it delivers a VRF per WAN link, with source address–based VRF selection for egress traffic. The OpenFlow API allows to update flows based on performance and/or cost [35]. Simultaneously, edge migration to IPv6 could be accelerated and cheaply rolled out with OpenFlow match plus encapsulate actions. IPv6/v4 migration/translation (e.g., network address translation IPv6-IPv4 [NAT64], dual-stack lite [DS-Lite]) services could be flexibly inserted into the datapath, potentially enabling operators to offer a hosted service for the IPv6 control plane.

*Internet Exchange Point (IX or IXP)*   IX or IXP is the physical infrastructure through which ISP networks (autonomous systems) exchange

Internet traffic. High-capacity L2/L3 switches serve as the main interconnection of these networks, where BGP is used at the protocol level to implement the business-driven traffic policies. The growing number and size of IXPs in addition to the increase and evolution of peering relationships (flatter Internet effect) make IXPs an attractive ecosystem, with a fundamental role in overall Internet traffic, especially in these days of distributed cloud and content delivery networks [40]. We believe that RouteFlow-like architectures leveraging route servers and traffic monitoring may contribute to fasten the pace of IXP operations and enable new service models beneficial to all parties involved. Initial activities along this application space have been recently reported in New Zealand.*

*Concluding Remarks*

While the trigger for this piece of work was looking at a transition scenario from traditional IP to SDN via a hybrid OF controller, the work on centralized IP routing has shown several advantages of the proposed IP split (hybrid) architecture—some applicable in general and inherent from OpenFlow/SDN and some routing protocol-specific–like BGP routing applications.

The horizon is, however, not without challenges. Maybe the most pressing ones from an architectural point of view is reaching a solution with high availability in all component layers, from the OpenFlow switches connecting to master and slave controllers, through the RouteFlow middleware, to the virtualized routing stacks. Other open challenges that may be more transient because of the maturity of OpenFlow switch products are (1) the processing limitations of the OpenFlow agent in the switch CPU (limiting the number of OpenFlow events, for example, flow-mod/sec, data-to-control capacity and delay); and (2) the currently available hardware-based flow table sizes in the range of a few thousands.

### Prototype Implementations and Experimental Evaluation

Throughout the life of the project, many experiments and demonstrations have to be conducted to prove the concept of the RouteFlow

---

* http://www.list.waikato.ac.nz/pipermail/nznog/2012-December/019635.html.

architecture. In this section, we will present the latest collected data to evaluate the current RouteFlow prototype and exemplify some of the recent use cases, such as multicontroller deployments. This section tries to answer commonly asked questions about the performance and scalability of the architecture.

*How Much Latency Is Introduced between the Data and Control Planes?*

To answer this question, we conducted two types of experiments to measure the different paths between the data and control planes. The first one measures the latency of events (e.g., ARP) that result in flow-mods carried through the RouteFlow IPC/messaging layer. The second one measures the latency of control traffic (e.g., interenet control messaging protocol [ICMP]) carried via the two OpenFlow channels that glue together VMs and physical devices.

*Route Update Events*    Performance measurements were made using the cbench tool [41], which simulates several OpenFlow switches generating requests and listening for flow installations. We adapted cbench to fake ARP requests (inside 60-B packet-in OpenFlow messages). These requests are handled by a modified version of the RFClient so that it ignores the routing engine. This way, we could effectively eliminate the influences of both the HW and the SW, which are not under our control, measuring more closely the specific delay introduced by RouteFlow.

The code and tools used to run these tests are openly available.* Except otherwise specified, the benchmarks were made on a Dell Latitude e6520 with an Intel Core i7 2620M processor and 3-GB RAM.

In latency mode, cbench sends an ARP request and waits for the flow-mod message before sending the next request. The results for RouteFlow running in latency mode on Pox and Nox are shown in Figure 8.12.

Each test is composed of several rounds of 1 s in duration, in which fake packets are sent to the controller and then handled by RFProxy that redirects them to the corresponding RFClient. For every test

---

* http://www.github.com/CPqD/RouteFlow/tree/benchmark.

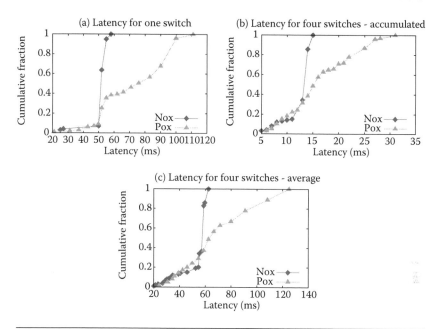

**Figure 8.12** Latency CDF graphs for Nox and Pox controlling a single network with one and four switches (taken from 100 rounds).

packet, the RFClient is configured to send a flow installation message. By doing this, we are testing a worst-case scenario in which every control packet results in a change in the network. These tests are intended to measure the performance and behavior of the new IPC mechanism.*

Figure 8.12 illustrates the cumulative distribution of latency values in three tests. Figure 8.12a shows the latency distribution for a network of only one switch. In this case, the IPC polling mechanism is not used to its full extent because just one message will be queued every time. Therefore, the latency for most of the rounds is around the polling timeout. Figure 8.12b shows the accumulated latency, calculated considering all four switches as one. When compared with Figure 8.12c, which shows the average latency for all the four

---

* The IPC mechanism uses a 50-ms polling time to check for unread messages in MongoDB. This value was chosen because it optimizes the ratio of DB access to message rate when running in the latency mode. Whenever a polling timeout occurs, the IPC will read all available messages.

switches, the scales differ, but the behavior is similar. The accumulated latency shows that the IPC performs better in relation to the case in Figure 8.12a, mostly because the IPC will read all messages as they become available; when running with more than one switch, it is more likely that more than one message will be queued at any given time, keeping the IPC busy in a working cycle, not waiting for the next poll timeout.

Another comparison based on Figure 8.12 reveals that RouteFlow running on top of Nox (RFProxy implemented in C++) is more consistent in its performance, with most cycles lasting less than 60 ms. The results for Pox (RFProxy implemented in Python) are less consistent, with more cycles lasting almost twice the worst case for Nox.

*OpenFlow Control Channel Performance*   We now turn our attention to the latency of the data to control plane channel formed by two OpenFlow segments: one south bound from the controller to the datapath switches and another north bound from the controller to the OVS, where the virtualized routing engines are attached. Recall that this control to data plane channel is used to exchange control plane traffic entering the switch and allowing neighboring legacy routers to interact with the RouteFlow control plane.

These experiments were conducted in the testing setup presented in Figure 8.13. The two OpenFlow control channels are depicted as red slashed lines. The tests consisted in PING requests originate from the Linux name space (LXC)-virtualized control plane, with the destination being the IP addresses of Host1, Host2, and each of the eight interfaces of the test equipment plugged to the physical ports of the hardware-based OpenFlow switch under test (Pronto 3290). In a sequential manner, five PINGs were sent to each IP, and the experiment was repeated for 100 runs, obtaining altogether 5000 round trip time (RTT) values (5 PINGs times 10 IPs times 100 repetitions).

Figure 8.14 presents the cumulative distributed function obtained for the RTT values. The RTT of each first PING request also accounts for the ARP resolution process that takes place prior to sending the ICMP packet. The large fraction of the first PING takes less than 15 ms. The next PINGs travel back and forth through both OpenFlow control segments in less than 3 ms. PING (and ARP) packet exchanges match first the OVS in the control plane and are first carried in

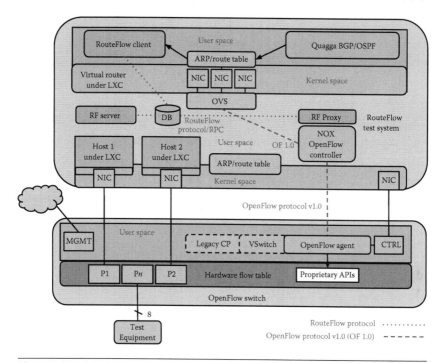

**Figure 8.13** RouteFlow test system built by router analysis [43]; the reference test controller design is based on a SuperMicro X9SAE-V with a single Intel I7-3770 CPU, 32-GB ECC RAM, 240-GB Intel 520 Series SSD, and four Intel I340-T2 GbE NICs. Equipm, equipment; API, application programming interfaces; CTRL, control; CP, control plane; MGMT, management; RPC, remote procedure call; RF, RouteFlow; LXC, linux name space; OSPF, open shortest path forwarding; BGP, border gateway protocol.

OpenFlow packet-in messages to the RFProxy instance and then sent out via OpenFlow packet-out messages to the datapath physical ports. The reverse path uses the same OpenFlow message sequence, but in a different order. The physical switch originates packet-in messages carrying packets that match the control traffic match rules installed,

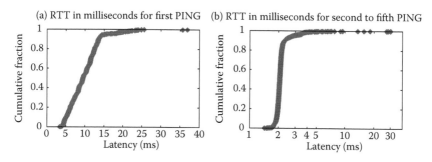

**Figure 8.14** CDF of the RTT values for the first PING (a) and the second to fifth PING request/reply (b). RTT, round trip time.

and the RFProxy decapsulates the payload and sends it via packet-out to the corresponding interfaces of the control plane OVS.

Note that these values are obtained in positive load conditions because neither the hosts nor the OpenFlow switch is busy with other tasks. CPU consumption of the OpenFlow agent resulting from additional (serial) messaging processing load is expected to increase the latencies to the control plane. This effect has been observed and reported by others [42,43] and should be further investigated in future work on stress testing under realistic workload conditions.

*How Many Control Plane Events Can Be Handled?*

In the throughput mode, cbench keeps sending as much ARP requests as possible to measure how much flow installations are made by the application (Table 8.1). The throughput test stresses RouteFlow and the controller, showing how many flows can be installed in a single round lasting for 1 s. The results in Table 8.2 show how many flows can be installed in all of the switches in the network during a test with 100 rounds lasting for 1 s each. The results show that the number of switches influence the number of flows installed per second, more than the choice of the controller.

**Table 8.1**   Total Number of Flows Installed per Second When Testing in Throughput Mode

|  | ONE SWITCH | | FOUR SWITCHES | |
| --- | --- | --- | --- | --- |
| CONTROLLER | Flows (average) | No. flows (90%) | No. flows (average) | No. flow (90%) |
| Pox | 915.05 ± 62.11 | 1013.0 | 573.87 ± 64.96 | 672.0 |
| Nox | 967.68 ± 54.85 | 1040.0 | 542.26 ± 44.96 | 597.0 |

*Note:*  Average, standard deviation, and 90% percentile taken from 100 rounds.

**Table 8.2**   RTT (ms) from a Host to the Virtual Routers in RouteFlow

| CONTROLLER | HOST AT CPQD | HOST AT USP |
| --- | --- | --- |
| Pox | 22.31 ± 16.08 | 24.53 ± 16.18 |
| Nox | 1.37 ± 0.37 | 3.52 ± 0.59 |

*Note:*  Average and standard deviation taken from 1000 rounds.

*What Is the Actual Performance in a Real Network?*

Tests on a real network infrastructure were performed using the control framework of the FIBRE project,* with resources in two islands separated by 100 km (the distance between the CPqD laboratory in Campinas and the LARC laboratory at University of Sao Paulo in Sao Paulo). To evaluate the delay introduced by the virtualized RouteFlow control plane, we measured the round-trip time from end hosts when sending ICMP (PING) messages to the interfaces of the virtual routers (an LXC container in the RouteFlow host). This way, we are effectively measuring the compound delay introduced by the controller, the RouteFlow virtual switch, and the underlying network, but not the IPC mechanism. The results are illustrated in Table 8.2 for the case where the RouteFlow instance runs in the CPqD laboratory with one end host connected in a local area network and the second end-host located at USP. The CPqD-USP connectivity goes through the GIGA network and involves approximately 10 L2 devices. The end-to-end delay observed between the hosts connected through this network for PING exchanges exhibited line-rate performance, with a constant RTT of approximately 2 ms. The results in Table 8.2 also highlight the performance gap between the controllers. The Nox version of RFProxy introduces little delay in the RTT and is more suited for real applications. These results are in line with the previous OpenFlow control channel performance tests in the section "How Much Latency Is Introduced between the Data and Control Planes?"

*How Do We Split the Control over Multiple OpenFlow Domains?*

To validate the new configuration scheme, a simple proof-of-concept test was conducted to show the feasibility of more than one network being controlled by RouteFlow. This network setup is illustrated in Figure 8.15 and makes use of the flexibility of the new configuration system. A central RFServer controls two networks: one contains four OpenFlow switches acting as routers being controlled by a Pox instance and another contains a single OpenFlow switch acting as a learning switch being controlled by a Nox instance. In this test,

---

* http://www.fibre-ict.eu/.

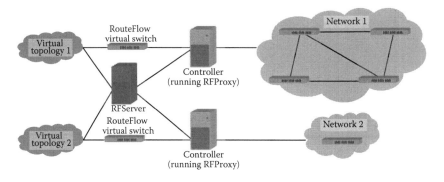

**Figure 8.15**   Test environment showing several controllers and networks.

RouteFlow properly isolated the routing domains belonging to each network while still having a centralized view of the networks.

*How Does the Virtualized Control Plane Scale?*

To evaluate the virtualization elements considered for the RouteFlow control plane architecture, experiments were done under different connectivity scenarios and scaling factors. The control planes tested were composed of virtual routers based on the Quagga routing stack v0.99.10 [23], virtualized with the LXC containers integrated into mainline kernel 2.6.32 [44], and attached to a single OVS [27] instance (v1.1.0).

Two types of topologies were explored corresponding to different scalability requirements:

1. **Grid-based topologies:** Two network topologies were defined with nodes arranged in a 3×3 grid and a 4×4 grid. This arrangement was adopted to evaluate the scaling behavior of the resource consumption by increasing the number of nodes and virtual network connections.

2. **Full-mesh topologies:** Three fully meshed network topologies of 15, 25, and 35 nodes were evaluated under the operation of the OSPF protocol. All subnets were attached to the OSPF area 0, and none of the interfaces had a summarization activated. While this configuration would not be recommended for real-world deployments because of the LSA N-squared problem [45], this setup is valid as a stress test for the scalability of the virtualized control plane.

The low-scale set of experiments (i.e., grid-based topologies) was run on a Intel Core 2 Duo 2.93-GHz platform with 3-GB RAM, and the second set of experiments (i.e., full-mesh topologies) was executed on a system with an Intel Xeon X5660 CPU and 48-GB RAM, both running Debian GNU/Linux 5.0 distribution.

Three different stages in the operation of the virtual control plane were introduced: (1) initialization, (2) regular operation (10 min after the initialization of each virtual topology), and (3) topology changes (i.e., link down and link up events). The following metrics and data were collected:

- **Boot time of the nodes:** the time interval needed for nodes to be fully operational after the request for its instantiation
- **CPU and memory usage:** allows to infer the degree of overhead in moments of intense demand
- **Convergence time:** measures the convergence time of the OSPF protocol at different stages. During initialization, the measured time is the interval between the topology boot-up and the first convergence of the OSPF protocol, that is, the time when the last OSPF announce message (LSAs) is exchanged. During topology changes, the convergence time interval is measured between the first and the last link state update (LSU) messages, after link disconnections and reestablishment.

To confer statistical relevance to the obtained results, each metric had its data collected through 30 distinct executions of each network topology. Table 8.3 presents a consolidated view of the results. After a 77% growth in network (from 9 to 16 nodes), initialization time grew to 96%. The result is almost linear, considering that the increase in the number of nodes also leads to a higher count of virtual links (from 12 to 24) and NICs. During the boot time of the virtual routers, CPU usage was in the range of 99% to 100%. Immediately after,

**Table 8.3** Control Plane Operation Metrics

| TOPOLOGY | INITIALIZATION (S) | RAM UTILIZATION (MB) | | |
| --- | --- | --- | --- | --- |
| | | Initialization | Regular Operation | Connectivity Events |
| 3×3 | 32 | 175 | 185 | 228 |
| 4×4 | 63 | 269.75 | 286 | 369 |

it dropped below 3% and has never gone up this level again for the duration of each experiment execution. Because none of the resource management features of LXC were used, this result is consistent with container-based virtualization premises. The evolution of memory consumption is also approximately linear with the increasing scale of topology. RAM usage peeks on connectivity modification events, and the 3×3 grid uses 23% more memory during OSPF reconvergence, whereas the larger 4×4 network increases by 29%. Again, considering the increased count of links and interfaces, this is not unexpected.

Figure 8.16a covers how long OSPF convergence takes. Figure 8.16a shows that greater number of links and increased network diameter have low impact on the first convergence when opposing 3×3 and 4×4 topologies. Fully meshed 15- and 25-node topologies were also evaluated and performed better, on average, than grids. This can

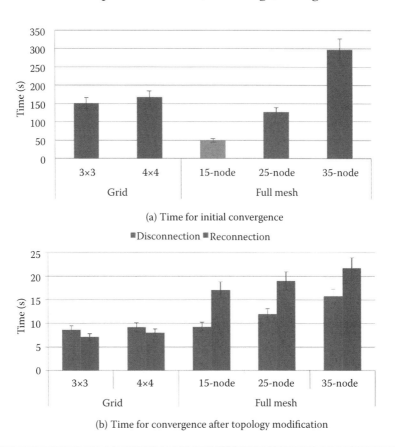

(a) Time for initial convergence

(b) Time for convergence after topology modification

**Figure 8.16**    OSPF convergence in grid and full-meshed network topologies.

be explained by the topology itself: because all nodes are connected, few OSPF discoveries offer good enough alternatives to be disseminated between multiple nodes (only those happened before all nodes' connections are up). For 35-node networks, however, the exponential growth in link numbers shows a much worse result, compatible with what the LSA N-squared problem suggests.

The timings shown in Figure 8.16b indicate that connectivity events take longer to converge on all full-meshed topologies, when related to 3×3 and 4×4 grids. This could be explained by the fact that connectivity changes always propagate through the entire network on full-meshed topologies (as all nodes report a dead neighbor), whereas this does not always hold true for grids. The same logic can be applied to explain why host reconnections have slower convergence in full meshes, whereas in grids, disconnections take longer to propagate—a reconnected node always brings up new best routes to all nodes in a full mesh, whereas this does not hold true in grids.

The stable growth in a grid's convergence time, allied to stable and predictable results in relation to full meshes, suggests that LXC and OpenVSwitch allow for a scalable control plane. Moreover, the OpenVSwitch's approach to attach and connect the LXC routers allows to implement a distributed control plane environment to scale CPU and memory requirements over a cluster of servers. Such a divide and conquer approach has been previously shown in this section when splitting control over multiple OpenFlow domains.

### Conclusions and Future Work

In this chapter, we have tried to introduce the motivation and problem statement of hybrid networking models in OpenFlow/SDN. More specifically, we have presented two controller-centric approaches to combine traditional control plane functions with OpenFlow control. Hybrid networking will certainly be a concern for SDN adoption because it will touch both economic and technical issues. The RouteFlow and LegacyFlow approaches present point solutions that allow for different migration strategies and deployment scenarios.

Further than enabling a gradual adoption path for OpenFlow technology, we have discussed how centralized IP routing engines coupled with OpenFlow-based installation of IP-oriented flow rules leads to a new degree of control that enables several applications that are a real challenge (if not impossible) to realize in classic multivendor networks.

As for the future, we are devoting efforts to increase the stability and scalability of the software components. This is especially critical because we start deploying pilots with operational traffic. Extensions to support MPLS and IPv6 are also underway. To support new protocols and realize use cases that involve QoS, load balancing, and fast failover, we will upgrade the prototypes with the newest OpenFlow protocol version.

We hope that the presented pieces of work are only examples of disruption on the horizon enabled by OpenFlow (in particular) and SDN (in general). Fueled by industry modularization and the forces of community-driven open-source developments, the SDN era promises unprecedented levels of innovation in networking and computing technologies.

## Acknowledgments

This work is partially supported by the GIGA Project (FINEP/FUNTTEL) and was done while the first author was employed by CPqD. Carlos N. A. Corrêa was supported by Unimed Federação Rio de Janeiro. We also thank colleagues and coauthors that contributed to previous publications related to the RouteFlow (CPqD*) and LegacyFlow (GERCOM-UFPA†) projects. We thank Steven Noble from Router Analysis, who contributed with the experiments described in "How Much Latency Is Introduced between the Data and Control Planes?" and Figure 8.13, and further detailed on an extended report and analysis on OpenFlow testing [43].

## References

1. Big Switch Networks. Open Software-Defined Networking (SDN) Product Suite. http://www.bigswitch.com/.
2. NEC. Programmable Flow Networking. http://www.necam.com/pflow/.
3. Nicira. Network Virtualization Platform. http://www.nicira.com/en/network-virtualization-platform.

---

\* http://www.cpqd.com.br.
† http://www.gercom.ufpa.br.

4. Greene, K. 2009. TR10: Software-defined networking. *MIT Technology Review.* http://www2.technologyreview.com/article/412194/tr10-software-defined-networking/.

5. Google Inc. Inter–data center WAN with centralized TE using SDN and OpenFlow. http://www.opennetworking.org/images/stories/downloads/misc/googlesdn.pdf.

6. ONF. Open Networking Foundation. http://www.opennetworking.org.

7. P. Gross. 1992. *Choosing a Common IGP for the IP Internet.* RFC 1371 (Informational). http://tools.ietf.org/html/rfc1371.

8. Mannie, E. 2004. *Generalized Multi-Protocol Label Switching (GMPLS) Architecture.* RFC 3945 (Proposed Standard).

9. Chen, C.-C., L. Yuan, A. Greenberg, C.-N. Chuah, and P. Mohapatra. 2011. Routing-as-a-service (RAAS): A framework for tenant-directed route control in data center. In *Proceedings of the 2011 IEEE INFOCOM (INFOCOM, '11)*, pp. 1386–1394, China: IEEE.

10. Lakshminarayanan, K. K., I. Stoica, S. Shenker, and J. Rexford. Routing as a service. Technical Report. Electrical Engineering and Computer Sciences, University of California at Berkeley, Berkeley, CA. http://www.cs.princeton.edu/~jrex/papers/ras.pdf.

11. I2RS Discussion List. 2012. Interface to the Internet routing system. http://www.ietf.org/mailman/listinfo/i2rs.

12. Yang, L., R. Dantu, T. Anderson, and R. Gopal. 2004. *Forwarding and Control Element Separation (ForCES) Framework.* RFC 3746 (Informational). http://tools.ietf.org/html/rfc3746.

13. Davy, M., G. Parulkar, J. van Reijendam, D. Schmiedt, R. Clark, C. Tengi, I. Seskar, P. Christian, I. Cote, and C. George. 2011. A case for expanding OpenFlow/SDN deployments on university campuses. GENI Workshop White Paper Technical Report. http://archive.openflow.org/wp/wp-content/uploads/2011/07/GENI-Workshop-Whitepaper.pdf.

14. Dudhwala, T. 2012. Discovery in OpenFlow networks: A report from the ONF PlugFest. http://www.blogs.ixiacom.com/ixia-blog/discovery-in-openflow-networks-at-onf-plu gfest/.

15. Das, S., Y. Yiakoumis, G. Parulkar, N. McKeown, P. Singh, D. Getachew, and P. D. Desai. 2011. Application-aware aggregation and traffic engineering in a converged packet-circuit network. In *National Fiber Optic Engineers Conference*, Los Angeles: Optical Society of America.

16. OSCARS. 2012. Oscars Project. http://www.wiki.internet2.edu/confluence/display/DCNSS/OSCARS.

17. Dragon. 2012. Dragon Project. http://www.dragon.maxgigapop.net/.

18. Autobahn. Autobahn Project. 2012. http://www.geant2.net/server/show/nav.756.

19. FIBRE. 2012. Future Internet testbeds experimentation between Brazil and Europe. http://www.fibre-ict.eu/.

20. Nascimento, M. R., C. E. Rothenberg, M. R. Salvador, and M. F. Magalhães. 2010. QuagFlow: Partnering Quagga with OpenFlow. In *Proceedings of the ACM SIGCOMM 2010 conference (SIGCOMM '10)*, pp. 441–442, New York: ACM.

21. Nascimento, M. R., C. E. Rothenberg, M. R. Salvador, C. N. A. Corrêa, S. C. de Lucena, and M. F. Magalhães. 2011. Virtual routers as a service: The RouteFlow approach leveraging software-defined networks. In *Proceedings of the 6th International Conference on Future Internet Technologies (CFI '11)*, pp. 34–37, New York: ACM.

22. Rothenberg, C. E., M. R. Nascimento, M. R. Salvador, C. N. A. Corrêa, S. C. de Lucena, and R. Raszuk. 2012. Revisiting routing control platforms with the eyes and muscles of software-defined networking. In *Proceedings of the 1st Workshop on Hot Topics in Software-Defined Networks (HotSDN '12)*, pp. 13–18, New York: ACM.

23. GNU Project Quagga. 2010. Quagga Software Routing Suite. http://www.quagga.net/.

24. Koponen, T., M. Casado, N. Gude, J. Stribling, L. Poutievski, M. Zhu, R. Ramanathan et al. Onix: A distributed control platform for large-scale production networks. In *Proceedings of the 9th USENIX Conference on Operating Systems Design and Implementation (OSDI '10)*, pp. 1–6, Berkeley, CA: USENIX Association.

25. Saucez, D. et al. 2011. Low-level design specification of the machine learning engine. FP7 ECODE Deliverable D2.3. http://www.ecode-project.eu/pmwiki/uploads/Ecode/ECODE%20D23-v1.0.pdf.

26. Ghodsi, A., S. Shenker, T. Koponen, A. Singla, B. Raghavan, and J. Wilcox. 2011. Intelligent design enables architectural evolution. In *Proceedings of the 10th ACM Workshop on Hot Topics in Networks (HotNets '11)*, pp. 3:1–3:6, New York: ACM.

27. Pfaff, B., J. Pettit, T. Koponen, K. Amidon, M. Casado, and S. Shenker. 2009. Extending networking into the virtualization layer. In *Proceedings of the 2009 Workshop on Hot Topics in Networks (HotNets-8)*, New York: ACM.

28. Keller, E. and J. Rexford. 2010. The "Platform as a Service" model for networking. In *Proceedings of the 2010 Internet Network Management Conference on Research on Enterprise Networking (INM/WREN '10)*, p. 4–4, Berkeley, CA: USENIX Association.

29. Caesar, M., D. Caldwell, N. Feamster, J. Rexford, A. Shaikh, and J. van der Merwe. 2005. Design and implementation of a routing control platform. In *Proceedings of the 2nd conference on Symposium on Networked Systems Design and Implementation (NSDI'05)*, vol. 2. pp. 15–28, Berkeley, CA: USENIX Association.

30. Feamster, N., H. Balakrishnan, J. Rexford, A. Shaikh, and J. van der Merwe. 2004. The case for separating routing from routers. In *Proceedings of the ACM SIGCOMM Workshop on Future Directions in Network Architecture (FDNA '04)*, pp. 5–12, New York: ACM.

31. Wang, Y., I. Avramopoulos, and J. Rexford. 2009. Design for configurability: Rethinking interdomain routing policies from the ground up. *IEEE J. Sel. A. Commun.* 27(3):336–348.

32. Rexford, J. and J. Feigenbaum. 2009. Incrementally deployable security for interdomain routing. In *Proceedings of the 2009 Cybersecurity Applications and Technology Conference for Homeland Security*, pp. 130–134, Washington, DC: IEEE Computer Society.

33. van der Merwe, J., A. Cepleanu, K. D'Souza, B. Freeman, A. Greenberg, D. Knight, R. McMillan et al. 2006. Dynamic connectivity management with an intelligent route service control point. In *Proceedings of the 2006 SIGCOMM Workshop on Internet Network Management (INM '06)*, pp. 29–34, New York: ACM.

34. Duffield, N., K. Gopalan, M. R. Hines, A. Shaikh, and J. van der Merwe. 2007. Measurement-informed route selection. In *Proceedings of the 8th International Conference on Passive and Active Network Measurement (PAM '07)*, pp. 250–254, Berlin/Heidelberg, Germany: Springer-Verlag.

35. Motiwala, M., A. Dhamdhere, N. Feamster, and A. Lakhina. Toward a cost model for network traffic. *SIGCOMM Comput. Commun. Rev.* 42(1):54–60.

36. Suchara, M., D. Xu, R. Doverspike, D. Johnson, and J. Rexford. 2011. Network architecture for joint failure recovery and traffic engineering. In *Proceedings of the ACM SIGMETRICS Joint International Conference on Measurement and Modeling of Computer Systems (SIGMETRICS '11)*, pp. 97–108, New York: ACM.

37. Filsfils, C. et al. 2011. BGP prefix independent convergence (PIC). Technical Report, Cisco. http://inl.info.ucl.ac.be/publications/bgp-prefix-independent-convergence-pic-technical-report.

38. Caesar, M., M. Casado, T. Koponen, J. Rexford, and S. Shenker. 2010. Dynamic route recomputation considered harmful. *SIGCOMM CCR* 40(2):66–71.

39. Gill, P., M. Schapira, and S. Goldberg. 2011. Let the market drive deployment: A strategy for transitioning to BGP security. In *Proceedings of the 2011 ACM SIGCOMM Conference (SIGCOMM '11)*, pp. 14–25, New York: ACM.

40. Ager, B., N. Chatzis, A. Feldmann, N. Sarrar, S. Uhlig, and W. Willinger. 2012. Anatomy of a large European IXP. In *Proceedings of the 2012 ACM SIGCOMM Conference on Applications, Technologies, Architectures, and Protocols for Computer Communication (SIGCOMM '12)*, pp. 163–174, New York: ACM.

41. Tootoonchian, A., S. Gorbunov, Y. Ganjali, M. Casado, and R. Sherwood. 2012. On controller performance in software-defined networks. In *Proceedings of the 2nd USENIX Conference on Hot Topics in Management of Internet, Cloud, and Enterprise Networks and Services (Hot-ICE '12)*, p. 10, Berkeley, CA: USENIX Association.

42. Mogul, J. C., J. Tourrilhes, P. Yalagandula, P. Sharma, A. R. Curtis, and S. Banerjee. DevoFlow: Cost-effective flow management for high-performance enterprise networks. In *Proceedings of the 9th ACM SIGCOMM Workshop on Hot Topics in Networks (HotNets '10)*, pp. 1:1–1:6, New York: ACM.

43. Router Analysis, Inc. 2013. *The State of OpenFlow 2012: Report and Analysis*. RA-011513-01. http://www.routeranalysis.com/the-state-of-openflow-2012-report-and-analysis.

44. LXC. 2011. LXC Linux containers. http://www.lxc.sourceforge.net/.

45. Aho, A. V. and D. Lee. Abstract hierarchical networks and the LSA N-squared problem in OSPF routing. In *Proceeding of Global Telecommunications Conference, 2000 (GLOBECOM '00)*, vol. 1, IEEE.

# 9

# NETWORK VIRTUALIZATION FOR OPENFLOW

## RUI MA AND FEI HU

### Contents

### Introduction

As technology develops, the modern network becomes larger and more capable of providing all kinds of new services. Cloud computing, and some frameworks, such as Global Environment for Network Innovations (GENI), French National Lab on Future Internet (F-Lab), etc., uses the large-scale experimental facilities from networks. However, resources are always limited, and users' demands keep increasing as well. The sharing of network hardware resources among users becomes necessary because it could use the existing infrastructure more efficiently and satisfy users' demands.

Network virtualization is a good way to provide different users with infrastructure-sharing capabilities, and software-defined networking (SDN) offers a powerful network control and management. The term OpenFlow emerges with network virtualization these years.

OpenFlow is a flow-oriented technology, which defines a flow as a set of header fields, counters, and associated actions. It runs on Nox or some other platforms and provides people the software to control the switches. FlowVisor is a special type of OpenFlow controller, which is a middleware between OpenFlow switches and multiple OpenFlow controllers. It decomposes the given network into virtual slices and delegates the control of each slice to a different controller.

Both OpenFlow and FlowVisor have their limitations in terms of network management, flexibility, isolation, and quality of service (QoS). OpenFlow offers the OpenFlow switches and common instructions but lacks standard management tools. FlowVisor only has access to the data plane, so the control plane and network controllers have to be managed by the users of the infrastructure. On the other hand, it can ensure a logical traffic isolation but with a constant level, which means that it lacks flexibility. Facing these challenges, researchers try to establish their own frame or architecture based on OpenFlow or FlowVisor for an improved network virtualization.

The remainder of this chapter is organized as follows: In "Background," the network virtualization technology based on OpenFlow is summarized. In "Architecture of OpenFlow-Based Network Virtualization," different approaches to improve the OpenFlow-based network virtualization are discussed. In "Discussion," the network virtualization schemes are provided for the corresponding challenges. This chapter ends with the "Conclusion."

## Background

### OpenFlow Review

There are many ways for network abstraction, and they could be classified by the level of details that they expose to the tenants. They could map all the physical components of the network to virtual ones, or they could just provide a one big switch or a single router as the network abstraction [1]. The first way provides virtual nodes, virtual routers, virtual switches, virtual ports, virtual links, and virtual topology. It gives the tenant more control to the network but needs more complex management. The second way is simpler and needs less management, but it cannot offer the required special functionality.

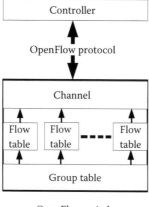

**Figure 9.1**    The architecture of the OpenFlow switch.

OpenFlow is a flow-oriented system and has switches and ports abstraction to control the flow (see Figure 9.1) [2]. In SDN, there is a software named controller, which manages the collection of switches for traffic control. The controller communicates with the OpenFlow switch and manages the switch via the OpenFlow protocol. An OpenFlow switch consists of flow tables, a group table, and an OpenFlow channel. The flow tables contain flow entries and communicate with the controller, and the group table can configure the flow entries.

OpenFlow ports are virtual ports at the network interface. They pass packets between OpenFlow processing and the rest of the network. OpenFlow switches connect with each other via OpenFlow ports.

*FlowVisor*

FlowVisor is a network virtualization platform based on OpenFlow, which can divide the physical network into multiple virtualization networks and thus achieve SDN. It can be preinstalled on the commercial hardware and provide the network administrator with comprehensive rules to manage the network rather than adjust the physical routers and switches.

FlowVisor creates slices of network resources and acts as the controlling proxy of each slice to different controllers, as shown in

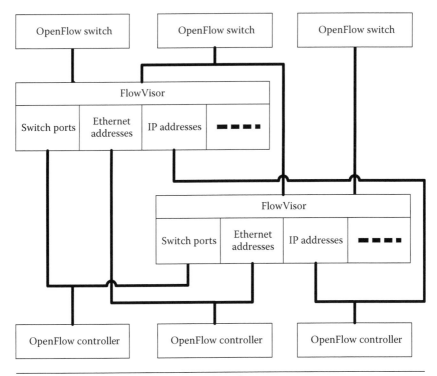

**Figure 9.2**   FlowVisor acts as proxy and provides slices.

Figure 9.2. The slices may be switch ports, Ethernet addresses, IP addresses, etc., and they are isolated and cannot control other traffic. It can dynamically manage these slices and distribute them to different OpenFlow controllers, and enables different virtual networks to share the same physical network resources.

*Virtualization Models*

In the context of OpenFlow, there are different virtualization models in the view of translation [3] (Figure 9.3). Translation maps identifiers between the physical network and the virtual ones. The translation unit is located between the application layer and the physical hardware. According to their placements, we could classify them into five models:

1. FlowVisor is the translation unit that delegates a protocol and controls multiple switches and multiple controllers.

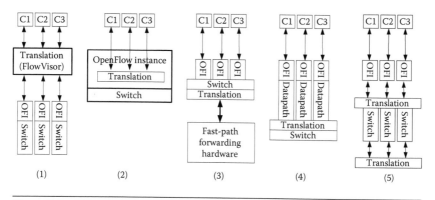

**Figure 9.3** Different placements of the translating functions in the system. C1, C2, and C3, different OpenFlow controllers; OFI, OpenFlow instance.

2. The translation unit is in the OpenFlow instance on the switch, and it performs translation among different controllers at the protocol level.
3. Multiple OpenFlow instances running on one switch are connected to one controller each. Translation is located between each OpenFlow instance and the fast-path forwarding hardware.
4. Multiple OpenFlow instances running on one switch, and the switch's datapath, are partitioned into several parallel datapaths, one per instance. It translates by adjusting the ports connected to the different parallel datapaths.
5. Multiple translation units—one for virtualization on the switch level and another for connecting multiple virtual switches and creating a virtual network representation.

### Architecture of OpenFlow-Based Network Virtualization

Several systems have been proposed to address the limitations of the OpenFlow-based network virtualization. These methods can be classified into three kinds:

1. **Improve the OpenFlow controller.** OpenFlow controller is a software, and it can be modified by users to satisfy their special demands.

2. **Improve the FlowVisor.** The FlowVisor itself already has basic management function, and it can be improved to overcome some limitations.

3. **Add new abstraction layer on the OpenFlow switch.** Researchers add new layers or new components to manage the virtual network. It actually creates new components, such as another FlowVisor.

We look into these approaches, focusing on the virtualization challenges that they have addressed.

*Flexibility*

The flexibility of the network virtualization denotes the scalability and the control level to the network. It is usually in conflict with the isolation demand.

In Ref. [4], a FlowN system is presented, which extends the Nox version 1.0 OpenFlow controller and embeds a MySQL version 14.14–based database with the virtual-to-physical mappings, as shown in Figure 9.4. This FlowN is a scalable virtual network and provides tenants full control of the virtual network: tenants can write their own controller application and define the arbitrary network topology. With the container-based architecture, the controller software that interacts with the physical switches is shared among tenant applications so that the resources could be saved when the controller becomes more and more complex these days.

This system is evaluated in two experiments: one measures the latency of the packets arriving at the controller when increasing the number of the network, and the other measures the fault time of the link used by multiple tenants when increasing the number of the network units. When the number of network units is large, the system has an equivalent latency to FlowVisor but is more flexible than FlowVisor, and its fault time could be small even if the number of the networks is large.

An integrated virtualization and management framework for Open Flow networks is presented in Ref. [5]. This system is capable of (1) running and managing multiple instances of OpenFlow switches with different OpenFlow versions, (2) running and configuring full controllers or network applications, and (3) configuring QoS in the network.

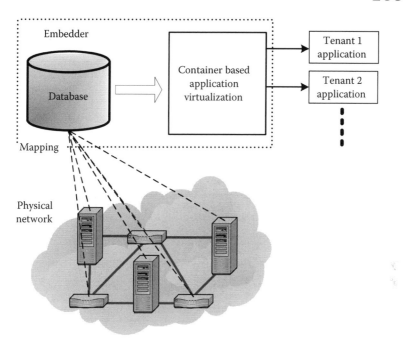

**Figure 9.4**   System design of FlowN.

The system configures the switches and maintains the physical topology of the network. It may reconfigure the physical-to-virtual port as a physical failure.

It uses Open Network Management System (OpenNMS) as the management framework because it is an open source. The virtual network management and the controller management modules are added, as shown in Figure 9.5. The prototype is successfully tested on the testbed consisting of six PCs, one switch, and one OpenFlow switch.

A novel layer 2 network virtualization approach based on a new medium access control (MAC) addressing scheme is presented in Ref. [6]. It is a centralized and locally administered MAC addressing scheme, and it overcomes scalability problems. This system efficiently supports cloud computing and sharing of the infrastructures, as shown in Figure 9.6.

The virtualization of the  local area networks (LANs) could be used to virtualize the network, but it has more complexity and overhead, and is not good at scalability. So, the virtualization of MAC is

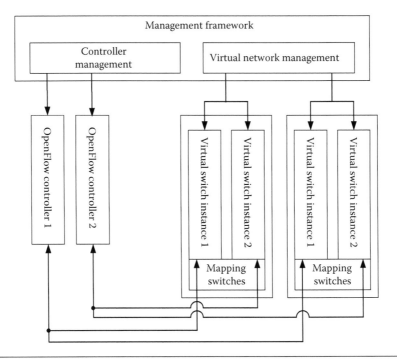

**Figure 9.5**    Integrated OpenFlow management framework.

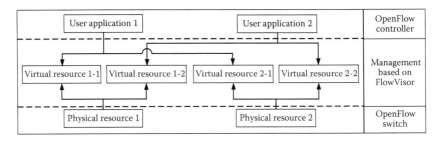

**Figure 9.6**    OpenFlow network virtualization for cloud computing.

proposed, and it is realized by reserving part of the remaining MAC address for the virtual nodes. This system reduces IP and control overhead, but the security issues need to be solved. Details of the system are provided, but the prototype is not tested in the experiment.

*Isolation*

To make sure that all the tenants of the virtual network can share the infrastructure without collision, the isolation problem must be addressed. The isolation may be in different levels or places, just like address space and bandwidth.

A research network named EHU OpenFLow enabled facility (EHU-OEF) is proposed in Ref. [7] (Figure 9.7). This network uses layer 2 prefix-based network virtualization (L2PNV) to separate different slices and allows users to have several virtual networks depending on the MAC source/destination. The system is built by devising some basic OpenFlow rules and developing some custom modules for the controller, as well as making some changes in the FlowVisor.

The system enforces slice isolation in the forwarding tables, and the flow traffic is distinguished on the value of the locally/globally administered address bits of the MAC header. Moreover, all the slices are handled by Nox controllers because of its rich module ecosystem. This solution has the benefit of enabling researchers to work with a larger number of header fields, including virtual local area networks (VLANs), and to test non-IP protocols by changing the MAC address.

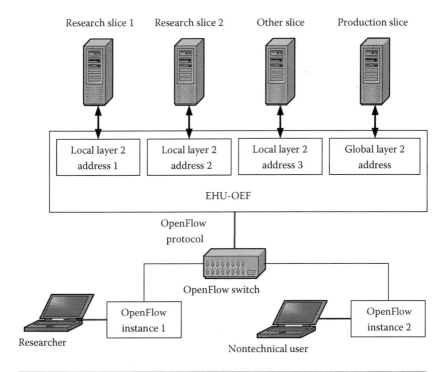

**Figure 9.7**   Integrated OpenFlow management framework.

The prototype is tested on the platform that is composed of seven NEC switches (IP8800/S3640), four Linksys WRT54GL, and two network field programmable gateway arrays (NetFPGAs). It is the first OpenFlow-enabled facility fully operational in Europe and allows production and research traffic in the same infrastructure without affecting each other.

It is proposed in Ref. [3] that a full virtualization system provides fair resource allocation both in the datapath and on the control channel, as shown in Figure 9.8. Different tenants of the network obtain the network resource by enforcing the resource allocation in the central controller, the datapath of the forwarding elements, and the control channel between the controller and the forwarding elements. The QoS tools of classification are applied to make a fair resource allocation.

It provides strict isolation between different virtual networks and more flexibility. The system makes improvements to the protocol, either through extensions or through restricting the usage of certain functionalities. There is no prototype tested.

Ref. [8] investigates the problem of isolation between the slices of a virtualized switch and presents an adapting isolation to the desired performance and flexibility while ensuring fairness between the slice. It developed a model that provides the choice between several levels of isolation, as shown in Figure 9.9. A slice isolator above the OpenFlow abstraction layer and the hardware switch is designed as an implementation of the model focusing on (1) interface isolation, (2) processing isolation; and (3) memory isolation. Thus, there are three isolators as well.

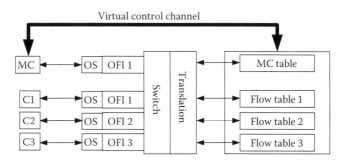

**Figure 9.8**   Full virtualization system. MC, master controller; C1, C2, C3, regular controllers; OS, operating system; OFI, OpenFlow instance.

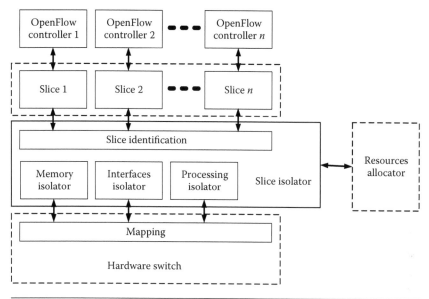

**Figure 9.9**  Network virtualization using slices isolator.

Evaluations of the system show that the isolation levels have a significant impact on performance and flexibility. The time for reconfiguring the hardware traffic manager increases fast when the isolation level goes up. High isolation level also leads to latency. So, the best isolation level can be determined based on the update time and the latency to achieve the required performance.

*Management*

Network virtualization management is involved with mapping, layer abstraction, or system design to make sure that the virtualized network can satisfy different demands. It is the integration of flexibility, isolation, and convenience.

A network virtualization architecture allowing management tools to be independent of the underlying technologies is presented in Ref. [1]. The article proposes an abstraction deployed as a library, with a unified interface toward the underlying network-specific drivers. The prototype is built on top of an OpenFlow-enabled network, as shown in Figure 9.10. It uses the single router abstraction to describe a network and has feasibility for creating isolated virtual networks in a programmatic and on-demand manner.

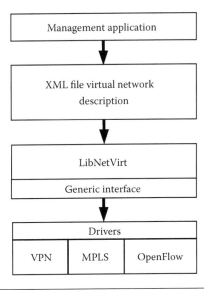

**Figure 9.10** LibNetVirt architecture. VPN, virtual personal network; XML, extensible markup language; MPLS, multiprotocol label switching.

In this system, the management tools can be independent of the working cloud platform so that different technologies can be integrated, and the system focuses on reducing the time of creating the virtual network. The prototype named LibNetVirt is separated into two different parts: generic interface and drivers. The generic interface is a set of functions that allow interaction with the virtual network and execution of the operations in the specific driver. A driver is an element that communicates to manipulate the virtual network (VN) in the physical equipment.

A solution based on OpenFlow technology is proposed in Ref. [9], as shown in Figure 9.11, which enables the creation of multiple isolated virtual experimental infrastructures all sharing the same physical infrastructure. This system implements a novel optical FlowVisor and has a cross-layer for management and high isolation for multiple users.

This architecture provides the following abstraction layers for the management:

1. The flexible infrastructure virtualization layer (FVL) is composed of the virtualized slicing and partitioning of the infrastructure.

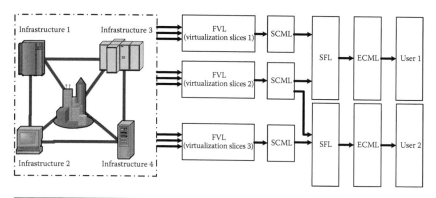

**Figure 9.11** Cross-layer experimental infrastructure virtualization.

2. The slice control and management layer (SCML) achieves the management, control, and monitoring functionality of the infrastructure.

3. The slice federation layer (SFL) federates infrastructure slices from different infrastructures together to create one virtual experimental infrastructure.

4. The experiment control and management layer (ECML) provides experimenters a set of management, control, and monitoring tools to manage their own virtual infrastructures. It is based on the extended OpenFlow controller.

The architecture is tested on the platform that is composed of eight NEC IP8800 OpenFlow-enabled switches and four OpenFlow-enabled Calient DiamondWave optical switches. The result shows that the setup time of establishing the flow path increases for a large number of hops.

There are other aspects of network virtualization designs, but the aforementioned three are the most important designs. Flexibility may include scalability, and isolation sometimes overlaps with the management. We compare previously discussed systems with respect to their focus points in Table 9.1.

FlowVisor becomes the standard scheme of network virtualization, so we compare these presented systems with FlowVisor. Most of the presented systems, whether it is improved based on FlowVisor or it is built totally in a new scheme, have abilities equivalent to that of FlowVisor but have one or more advantages over FlowVisor, such as flexibility, adjustable isolation levels, etc.

**Table 9.1** Comparison of the Reported Network Virtualization Systems Based on OpenFlow

| SYSTEM | FLEXIBILITY | ISOLATION | MANAGEMENT | FLOWVISOR |
|---|---|---|---|---|
| FlowN | √ | – | – | – |
| Integrated system | √ | – | √ | – |
| MAC addressing system | √ | – | – | √ |
| EHU-OEF | √ | √ | – | √ |
| Adaptable isolation system | √ | √ | – | – |
| LibNetVirt | – | – | √ | – |
| Cross-layer experimental infrastructure virtualization | – | – | √ | √ |

### Discussion

Network virtualization not only enables infrastructure sharing, but also provides new better ways to use the infrastructure or to reduce the cost.

Worldwide mobile network operators have to spend billions of dollars to upgrade their own networks to the latest standards for wireless communication of high-speed data for mobile phones. SDNs can overcome this problem. For example, a solution exploring OpenFlow as an architecture for mobile network virtualization has been proposed in Ref. [10]. The SDN-shared network has lower cost than the classical network and the SDN network. A case study for a German reference network is given. The considered capital expenditures can be reduced by 58.04% using the SDN-shared network instead of the classical one.

1. The technoeconomic analysis indicates that SDN and virtualization of the first and the second aggregation stage network infrastructure lead to substantial cost reductions of capital expenditures for the mobile network operator.
2. Qualitative cost evaluation of SDN shows that the continuous cost of infrastructure for the SDN scenario, maintenance cost, costs for repair, and cost of service provisioning will be lower.
3. The cost for the first-time installation of network equipment will alter significantly.

It is reported in Ref. [11] that the OpenFlow technology is deployed in wireless sensor networks (as shown in Figure 9.12) and can lead to

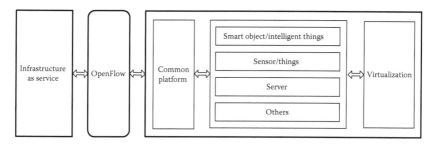

**Figure 9.12**    Abstraction layers of the virtual network.

a significant achievement in Internet of things and cloud computing arena through network virtualization.

In traditional sensor networks, some sensors away from the access point may not be reached. However, via virtualization, all the typical sensors are replaced by flow sensors, and the sensors of one network can use the sensors of other networks for data transfer. A flow sensor is just like a typical sensor associated with a control interface (software layer) and flow tables (hardware layer).

Figure 9.13 shows an example of a flow sensor network. Here, sensor networks 1 and 2 cannot communicate with each other without the access point, so node 4 is very far and is lost; for the flow sensor network, node 4 can talk to node 8, so that node 4 can be accessed.

The evaluation shows that all flow sensors achieved 39% more reachability than a typical sensor. Flow sensor generated more packets and took more time to simulate in comparison with a typical sensor in an ideal scenario. It is found to perform better in large-scale networks.

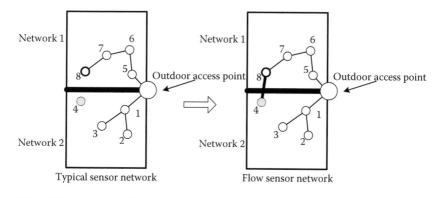

**Figure 9.13**    Typical sensor network and flow sensor network.

Conclusion

Network virtualization based on OpenFlow is a successful implementation of SDN, and its development offers people more and more convenient and amazing services. We have reviewed different architectures proposed recently, which focus on the improvement of network flexibility, isolation, and management. It can be seen that embedding an additional module or abstraction layer on the top of OpenFlow or FlowVisor provides solutions to these challenges. Using the database can also help simplify the creation of the abstraction layer.

# References

1. Turull, D., M. Hidell, and P. Sjodin. 2012. LibNetVirt: The network virtualization library. *2012 IEEE International Conference on Communications (ICC)*, pp. 5543–5547.
2. OpenFlow switch specification. Ottawa, ON, Canada. pp. 6622–6626.
3. Skoldstrom, P. and K. Yedavalli. 2012. Network virtualization and resource allocation in OpenFlow-based wide area networks. *2012 IEEE International Conference on Communications (ICC)*, pp. 6622–6626.
4. Drutskoy, D. *Software-Defined Network Virtualization*. Master's thesis, Princeton University, p. 28.
5. Sonkoly, B., A. Gulyas, F. Nemeth, J. Czentye, K. Kurucz, B. Novak, and G. Vaszkun. 2012. OpenFlow virtualization framework with advanced capabilities. *2012 European Workshop on Software Defined Networking (EWSDN)*, Darmstadt, Germany, pp. 18–23.
6. Matias, J., E. Jacob, D. Sanchez, and Y. Demchenko. 2011. An Open Flow-based network virtualization framework for the cloud. *2011 IEEE 3rd International Conference on Cloud Computing Technology and Science (CloudCom)*, pp. 672–678.
7. Matias, J., B. Tornero, A. Mendiola, E. Jacob, and N. Toledo. 2012. Implementing layer 2 network virtualization using OpenFlow: Challenges and solutions. *2012 European Workshop on Software Defined Networking (EWSDN)*, pp. 30–35.
8. El-Azzab, M., I. L. Bedhiaf, Y. Lemieux, and O. Cherkaoui. 2011. Slices isolator for a virtualized OpenFlow node. *2011 1st International Symposium on Network Cloud Computing and Applications (NCCA)*, pp. 121–126.
9. Nejbati, R., S. Azodolmolky, and D. Simeonidou. 2012. Role of network virtualization in future Internet innovation. *2012 17th European Conference on Networks and Optical Communications (NOC)*, pp. 1–4.

10. Naudts, B., M. Kind, F. Westphal, S. Verbrugge, D. Colle, and M. Pickavet. 2012. Technoeconomic analysis of software-defined networking as architecture for the virtualization of a mobile network. *2012 European Workshop on Software Defined Networking (EWSDN)*, pp. 67–72.

11. Mahmud, A., R. Rahmani, and T. Kanter. 2012. Deployment of flow sensors in Internet of things' virtualization via OpenFlow. *2012 3rd FTRA International Conference on Mobile, Ubiquitous, and Intelligent Computing (MUSIC)*, pp. 195–200.

# PART III
## QUALITY OF SERVICE

# 10
# Multimedia Over OpenFlow/SDN

COLBY DICKERSON, FEI HU,
AND SUNIL KUMAR

## Contents

## Introduction

Presently, streaming media requires a predifictable and steady network. It also requires a network with little delay and no packet loss. OpenFlow technology is a software-defined network (SDN) prototype that focuses mainly on the separation of the data plane and the control plane. Essentially, this is a network that directs traffic by decoupling the control and forwarding layers of routing. Currently, it can be extremely difficult and challenging not only to operate the network management, but also to maintain and ensure that the network is secure for communications. The intentions of improving and advancing to an SDN are to create switches in the data plane to enhance and simplify packet forwarding devices and to control the behavior of the whole network through the use of a centralized software program.

With our current Internet architecture, it is difficult to perform network routing on a per-flow basis, and the OpenFlow protocol provides a new hope in solving this problem. OpenFlow offers a new way to solve Internet architecture deficiencies by allowing users to define the routing rules associated with data flow. This would enable one to configure the network layout and modify the traffic control as in an SDN. The controller in an OpenFlow network is essential for determining routing changes. Controller algorithms associated with different data flows could lead to alternate routing choices. The controller uses forwarder interfaces to determine the network topology. It also uses an application interface to provide users with a way to make reservations of new data partitions and a means of defining new routing rules. Figure 10.1 outlines the architecture for OpenFlow protocol.

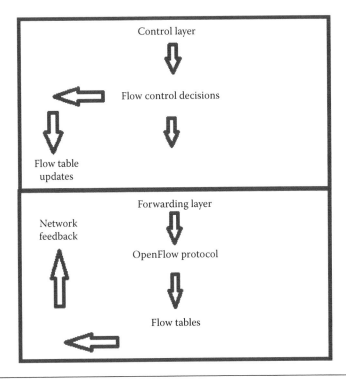

**Figure 10.1** Architecture for OpenFlow protocol.

### QoS (Quality of Service)-Enabled Adaptive Video Streaming Over SDN

An OpenFlow controller design has been presented in Ref. [1], with a means of optimizing the framework to provide QoS guarantees to certain video streams. This would allow for optimized routing with certain constraints to ensure the delivery of QoS flows. Multimedia applications encounter problems, such as packet loss, delay, and jitter. Obviously, for a better QoS, there should be minimum packet loss. Therefore, to solve the constrained shortest path problem, both the cost metrics and constraints need to be determined. With the Open Flow protocol, the controller is able to collect monitored per-flow statistics that are maintained in the forwarders. The solution provides the minimum cost route while satisfying a specific maximum delay variation from the source to the destination. The Dijkstra formula can help determine and calculate the shortest paths using link cost, link delays, and aggregated link cost. Moreover, the Lagrangian relaxation–based aggregated cost algorithm can help solve the maximization problem of the constrained shortest path problem. The Dijkstra formula would then be applied to solve the minimization problems in Ref. [1].

### Quality of Experience-Oriented Scheme

Multimedia delivery over a network has various limitations and user quality expectations depending on the various media types. In Ref. [2], different quality of experience (QoE) measurements are taken at both the service level and the network level. Currently, at the network level, all data flow along the same path from the source to the destination. This current type of technology enables media flow to be delivered over the best available path and the best service configuration. This, in return, should increase the total QoS for users. Although this is how our current network operates, SDNs have been proposed to combine the configuration of network elements and the end host. SDNs would enable vendors to generate their own rules and policies, thus enabling a more flexible control over network services, such as traffic, routing, and QOS. Logically, an SDN consists of three main layers: infrastructure, control, and application. OpenFlow networking would allow for a standardized boundary between the control plane and the infrastructure elements. Moreover, with this technology, applications would be allowed to use network services without knowing the network topology. Applications would essentially be able to generate requests that can be translated by the control layer to some device configuration. Figure 10.2 displays SDN functionalities.

Overall, the goal is to achieve optimized flow and maximize the network throughput. Although this does not ensure that the QoE would be increased, it would generate such path assignments at the infrastructure layer, which could lead to path optimization for multimedia flow and to an enhanced QoE.

The architecture of an SDN mainly consists of the QoS matching and optimization function and the path assignment function. The QoS matching and optimization function lies within the application layer. This function allows for optimized session delivery and session adaptation via end-to-end signal paths. Upon receiving a service request, the QoS matching and optimization function's role is to match processes with service configurations and to optimize the process to determine the best service configuration. At that point, the overall QoE is expressed as a combination of media flows. Moreover, the path assignment function is used to calculate the optimal path assignment. This application is executed on the OpenFlow controller. The path assignment function maintains a network topology database and a present view on the active flows in the network to essentially optimize path routes to maximize the session's QoE. With an OpenFlow protocol, the path assignment function can specify the flow and the minimum rate of any network device. This type of architecture proves

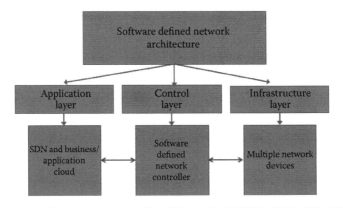

**Figure 10.2**   Basic SDN architecture.

to be very flexible because overlays are not required to be maintained, and path routing does not depend on the overlays.

### OpenFlow Controller Design for Multimedia Delivery

Internet is based on an end-to-end flow control. This allows for minimal network support, and the end host controls communication tasks. With this configuration, there are two basic advantages. It allows for a best-effort service for all data at the network layer. Moreover, it reduces the overhead and cost at the network layer while maintaining the reliability. This architecture proves to be valuable for data transmissions. However, with multimedia transmissions, users are more concerned with timely transmissions over reliability. With the best-effort Internet, multimedia delay requirements cannot be guaranteed. Therefore, it is important that the network infrastructure provide some type of QoS for multimedia traffic. Ultimately, the Internet lacks real-time configurations and adaptations.

The migration to SDNs provides for a logical and virtual entity of the network. OpenFlow is the first successful attempt to create an SDN. OpenFlow provides network visibility, monitoring, and virtualization to generate a refined network. OpenQoS [3] is a new OpenFlow controller design for multimedia delivery. To support this design, multimedia flows are placed on QoS-guaranteed routes, and they still maintain their shortest paths. OpenQoS is just an extension of the standard OpenFlow controller. OpenQoS provides a new routing prioritization scheme. To maintain the end-to-end QoS, a dynamic routing for multimedia traffic is introduced while the data keep its shortest path. The advantage of this design is minimized packet loss and latency.

Moreover, the Internet does not allow routing on a per-flow basis. When the router receives a packet, it checks the packet's source and destination address with the routing table, and the packet is forwarded. OpenFlow, on the other hand, is flexible in defining different types of flows based on a specified set of actions and rules, such as shortest path routing algorithms. OpenQoS exhibits OpenFlow's flow base forwarding, which allows for the differentiation of multimedia traffic. Multimedia flows are recognized by packet headers. Normally, lower layer packet headers define flows compared with the upper layers. Sometimes, upper layer header fields may be required to better determine a packet, and OpenFlow technology allows for this to be done easily because of its flexibility. To calculate the QoS routes, the up-to-date global network state information must be obtained. This information includes delay, bandwidth, and the packet loss rate in each link. The routing algorithm performance directly correlates with how well the network state information is. This is difficult, however, with the Internet. It is difficult because of the distributed architecture in the Internet, but OpenFlow technology allows for this task to be done easily by using a centralized controller. The controller obtains the forwarders' state information and determines the best routes according to the information that was obtained.

In Ref. [3], they have proven that, with QoS support, traffic has a minimal effect on the ultimate full video quality. On the other hand, a video without QoS support suffers a significant amount of quality loss. In Ref. [3], they show that, even with QoS support, videos suffer some loss, but it is quickly retained. The quality is retained in less than 1 s, and most viewers are not disturbed by the loss and never know that there was even a loss. It is hard for researchers to perform experiments on the Internet because the Internet is a closed environment. Currently, Internet router vendors provide a closed hardware and software that is not open to users. The great aspect about OpenFlow technology is that it removes the boundaries of the Internet.

### Network Management for Multimedia Delivery

Researchers are also working to improve the network management of SDNs for better multimedia delivery [4]. With the existing network management, there lie three main issues. First, operators are unable to make frequent changes to the network. Second, there is a lack of support for high-level network configurations. Finally, there is a lack of visibility and control when tests are being performed on the network.

Networks are very complex and consist of a large amount of switches, firewalls, routers, etc. Network operators are responsible for network events and for the configuration of the network with high-level languages. This is where an SDN can be implemented. By creating a network with just simple packet forwarding and controller logic, many benefits can be achieved. Some benefits include the ease of introducing software programs to the network and a strategic approach to a centralized network configuration versus a distributed managed network. Research is geared toward north-bound and south-bound SDN interfaces. North bound refers to the protocol between the controller and the programmable switch layer. South bound refers to the policy layer or the upper portion of the controller. Because of the listed problems and benefits that have been mentioned before, a north-bound SDN-based language and control framework called Procera is generated [4]. This is one effort to increase north-bound interface research and to implement reactive policies, whereas most other research activities pertain to OpenFlow or south-bound SDN interfaces.

Procera is based on an SDN prototype. It is an event-driven network control structure that is based on functional reactive programming. Operators have a way to express high-level policies by using Procera. Procera will also translate the policies into forwarding rules and network infrastructure policies using OpenFlow. This new infrastructure was first introduced and used for experiments at the Georgia Tech campus. Event-driven network policies can now be expressed as control domains using Procera. This means that, upon certain actions or conditions within the network, a resulting packet forwarding assignment will take place. Procera also uses various control domains to help ease the job of network operators. Currently, Procera only has four different control domains: time, data usage, status, and flow. Table 10.1 reveals the different control domains in Procera, with their specific policies and the reasoning for the actual control domain.

The time domain would help manage traffic based on peak traffic during certain times of the day, week, or year. The data usage domain would allow operators to specify policies based on the amount of data

**Table 10.1**   Procera Domains

| DOMAIN | POLICY | REASONING FOR THE CONTROL DOMAIN |
|---|---|---|
| Time | Peak traffic control | The operator can implement policies based on a certain time frame. |
| Data usage | Data usage amount | Operators can specify policies based on network behavior or the amount of data used. |
| Status | Identity and authentication | The operator can specify the privileges of a different status based on different users or different groups. |
| Flow | Network specification | Operators can specify different network behaviors based on the field values specified in a flow. |

usage over a certain time interval. The status domain would allow the operator to grant or deny certain user privileges. Finally, the flow domain would allow a network operator to state the different network behaviors caused by altering field values within layers. With the Procera interface, the controller is responsible for making traffic forwarding decisions. It also updates switch flow-table entities in the low-level network. The network controller is used to translate network policies into actual packet forwarding rules. It generates a connection with the OpenFlow switch through an OpenFlow protocol. This allows the insertion, deletion, or modification of packet forwarding rules. Figure 10.3 reveals the interactions between the controller, switch flow tables, and computers when using the OpenFlow protocol.

The use of Procera was first introduced in the campus environment at Georgia Tech, where network dynamics is very dynamic. The campus network on top of other enterprise networks is often complex and susceptible to errors. This is why Procera was deployed at Georgia Tech. At Georgia Tech, the network is based on a virtual local access network. This means that registered and unregistered devices are separated by different virtual local access networks. Constantly, up-to-date virtual local access network maps are being generated based on the authentication of registered and nonregistered devices. This allows for correct forwarding behavior. With Procera, such policies would be greatly simplified. Procera had also been used in home networks. One problem with home networks is that they only offer limited visibility into home broadband

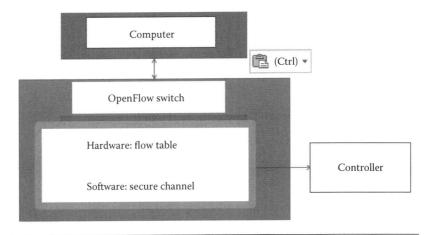

**Figure 10.3**   Controller components.

performance and status. The challenge that home networks face is that a new software cannot be installed to home gateways because of the lack of flexibility and the closed software that they currently have. An SDN was the true light for home network traffic management. As previously stated, it would allow a centralized controller to make trafficking decisions and enforce rule and policies on home gateways, which would drastically increase the flexibility of home network management. Currently, home users are unable to view their home network limitations and monthly data usage. They are also unable to monitor the usage of bandwidth on a daily basis. ISPs provide a way for users to monitor bandwidth usage on a household level, but not on an individual device level. Procera would allow users to implement new types of reactive network policies. SDNs work very well in performing such actions as these. Logic can be applied in the back-end server, and home network routers can act as packet forwarding devices.

With Procera, results have proved that the network configuration workload and management of the system were greatly reduced. The use of Procera in both a campus setting and a home network setting has generated reasonable and encouraging results. Still, there is some work to be done with the performance and scalability of the Procera interface. Procera is working to increase and expand the current four control domains. Ultimately, using the OpenFlow protocol to interact with the Procera controller and the network switches can lead to an automatic responding network, where state changes can be made more readily.

### From a Cloud Network Perspective

Cloud networking [5] is a feature that has increased and developed very rapidly in a short period. Both cloud service providers and cloud controller platforms have made rapid advances in offering multimedia data support. An SDN sheds a new light in integrating application provisions within the cloud. This can be achieved by programmable interfaces and automation in the network. For this cloud technology to be successful, the network environment needs to be highly responsive and flexible to support altering business requirements. The platform for cloud computing allows enterprises to use computing resources only when they are needed. Service providers have extended their contributions by including services that extend beyond virtual servers, storage volumes, and network connectivity because various applications and workloads moving to the cloud keep increasing. The modeling for cloud management systems aims to offer a virtualized form of components and functions similar with that of the network of a customer's data center.

An SDN offers cloud technology provisions through programmable interfaces and automation. An SDN is a perfect match for cloud applications because a greater demand over network control is increasing. Many SDN cloud solutions have been generated for producing virtual networks in multitenant clouds based on standards such as OpenFlow. Some advantages of these SDN cloud solutions include scalability, performance, and flexibility. OpenFlow provides for control through a programmable interface for things such as packet handling and forwarding of physical and virtual switches. This task remains challenging, however, because of issues such as hardware flow-table limitations and controller performance. With the new SDN approach, consistencies in virtual networking configuration for application in the cloud are made possible.

An SDN framework, called Meridian [5], was therefore created to support a service-level model for application networking, which would support various options for generating virtual networks on a physical network. This platform of cloud networking architecture was inspired by conceptual models of SDNs. The Meridian architecture is composed of three main logical layers: network model and application programming interfaces (APIs), network orchestration, and interfaces to network devices. The API layers' main goal is to present applications with a network model and APIs that reveal only necessary information for interaction with the network. The API enables cloud controllers to request policy-based connectivity between virtual servers and logical groups. The objective of the network orchestration layer lies within the SDN architecture. The orchestration layer provides key services to support cloud networking. This layer's primary components include planner and deployer modules that perform network operations. The first goal of this layer is to perform a logical-to-physical translation of commands that the abstraction layer sends. Moreover, this layer offers a set of network services to applications such as topology views and link or utilization availabilities. Finally, this layer provides coordination and mediation network requests and the mapping of the requests to the network.

Meridian can generate an integration point that connects cloud orchestration applications to virtual network services. Researchers have determined that Meridian cloud integration application requires a separate common module that offers identical functions to a high-level cloud platform. Meridian uses OpenFlow devices to generate virtual networks and services by implementing a logical driver. With such an OpenFlow-capable network, the driver could interface with the OpenFlow controller. This would allow for the addition of flow rules in virtual and physical switches. The Meridian prototype also uses a Floodlight controller platform. This is a Java-based OpenFlow controller that allows for OpenFlow protocols, link discoveries, and OpenFlow-enabled switches. The topology goal of Meridian services is to recognize cloud virtual networks by using various network resources. With all Meridian services, as well as all other SDN platforms, the ultimate goal is to achieve flexibility and ultimate handling of network devices.

The virtual network topology used by Meridian allows for the recognition of segments by using OpenFlow rules on switches. Different routing policies support various segments. The current Meridian implementation uses shortest path routing, middlebox waypoint routing, access control filters, and scoped broadcast based on OpenFlow policies. To implement certain segments, attachment endpoints are located using Floodlight's device manager and Meridian's topology services. At the specified endpoint, routing policies are defined to achieve flexibility of routing differing endpoints. One key problem that Meridian faces is the management of updates to virtual network topology. The problem occurs when segments are added, removed, or changed. For Meridian to obtain updates, a control block for each virtual network has to be maintained. The virtual network block records network operations for the virtual network. The control block creation for each virtual network and OpenFlow installations are taken and stored in the virtual network control block. Using this scheme, it is easy to manage virtual network updates by changing or altering the entire network. Although multiple instances of virtual networks can be supported, interactions and conflicts can arise between network configurations. New options and additional enhancement are currently being explored, such as a planner to recover failed plans and add a topology-discovering capability. Overall, this type of technology [5] has proven to be useful, but there is still some work to be done to further enhance its capabilities.

## Multimedia Service Composition Over a Programmable Network Substrate

Multimedia service composition [6] offers another flexible way to construct customized multimedia delivery. The Internet has devices such as routers and switches, which are involved in the service composition process. This essentially extends the level of control of data flow over the whole network and also provides reinforcement for network efficiency. An OpenFlow-based programmable substrate is used as the building block for enabling multimedia service composition over future networking services. Essentially, multimedia service composition is a way to construct multimedia application with composable multimedia constituents. The networking service section for multimedia service composition contains a media-related functionality that includes the creation, processing, and transmission of multimedia. This defines the services for both computing and networking functionalities because multimedia service composition is based on the assumption that it only has control over the computing devices in the network. Networking services provide a means to configure the basic network functionalities within a network device. This includes functionalities such as packet forwarding and the embedding of various functionalities.

Again, to obtain a programmable substrate, the network device is controlled by a central network controller. Each network device contains a simple networking service. The main prototype of networking services is the use of OpenFlow. This SDN allows for switching hardware to be controlled by software in the control plane. The control plane is able to make decisions, based on a flow table, of how each flow should be forwarded. OpenFlow is very beneficial because it offers programmability on a network substrate, where someone can configure the network as they please. However, it does not offer reconfiguration of the network substrate by directly putting in new functionalities. This can be viewed as a downside for the use of OpenFlow. If the spanning tree function is not enabled when the OpenFlow controller is started, it will be impossible to produce a spanning tree. The only way to implement new functionalities would be to restart the OpenFlow controller. Currently, experiments are being conducted to extend OpenFlow and create network services that would enable configurations of the network substrate. By enabling the OpenFlow controller, networking services can be configured and executed whenever the user pleases. Moreover, in Ref. [6], it provides network connectivity to end-host users that are part of two different OpenFlow islands. Advanced networking services can be solved by the use of tunneling services that maintain connectivity between OpenFlow-based networks and the OpenFlow islands.

### Multimedia Over Wireless Mesh Networks

Wireless mesh networks provide a low-cost broadband connectivity when cabled connectivity is not possible. Wireless mesh networks are beneficial to rural areas, and this technology allows the network to be expanded as a rural community grows. Two of the most well-known routing protocols for wireless mesh networks are the Better Approach to Mobile Adhoc Networking Advanced protocol (BATMAN Advanced) and the 802.11 standard protocol [7]. Although these protocols have proven to be very useful in rural communities, they have many restrictions as well. Wireless mesh networks have drawbacks, such as downgraded equipment, single channels and interfaces, and low performance quality.

OpenFlow technology sheds a new light on wireless mesh networking [7]. With OpenFlow, services such as access control, virtualization, management, and mobility would ultimately be enhanced. Research has proven that the BATMAN Advanced protocol is more stable and able to generate a higher quality of packet delivery, but the 802.11 protocol has proven to have the fastest recovery time. Using OpenFlow technology in wireless mesh networks has proven to obtain a more flexible and efficient productivity. Figure 10.4 reveals a view of the network infrastructure for a wireless mesh network.

The framework for this OpenFlow technology in wireless mesh networks [7] consists of a virtual wireless network that allows for control signaling and data transfer. SDNs have evolved into a well-suited way to manage a network using software from some external server. However, using SDNs in wireless mesh networks has several

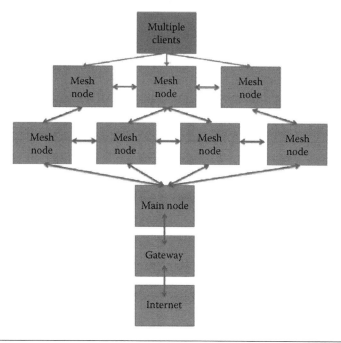

**Figure 10.4**  Wireless mesh network architecture.

challenges. First, if the wireless mesh routers are unable to communicate with the controller, then there is a total failure of new flows. Second, one controller is not capable of solving issues such as network splits. To solve these problems, OpenFlow technology can be implemented. Table 10.2 displays the advantages and disadvantages of OpenFlow technology over the commonly known wireless mesh network protocols—BATMAN Advanced and 802.11.

Ultimately, Table 10.2 proves that OpenFlow technology offers a viable solution to wireless mesh networks. Although it has some disadvantages, OpenFlow still maintains the highest level of security.

**Table 10.2**  Advantages of Using OpenFlow

| QOS METRICS | BATMAN ADVANCED | 802.11 | OPENFLOW |
| --- | --- | --- | --- |
| Jitter vs. hops | Highest | Average | Lowest |
| Throughput vs. hops | Highest | Average | Lowest |
| Loss of datagrams vs. hops | Highest | Average | Lowest |
| Maximum throughput with TCP | Highest | Lowest | Average |
| Packet delivery ratio | Highest | Lowest | Average |

While BATMAN appears to be a good solution and achieves a high transmission rate, much of the datagram information is lost in the process. For wireless mesh networks, OpenFlow definitely confirms to be more than acceptable when compared with the 802.11 standard protocol.

### Discussion

In summary, multimedia over OpenFlow/SDNs is a viable way to increase product quality, speed, and durability. Some multimedia over OpenFlow schemes have been presented. Each scheme has advantages and disadvantages as discussed. The ultimate goal of each scheme is to meet high-quality QoS in multimedia transmissions and to create switches in the data plane to enhance and simplify packet forwarding devices and to control the behavior of the whole network through the use of a centralized software program. This prototype will essentially supply product users with a new way to solve Internet architecture deficiencies by allowing the user to define routing rules associated with data flow. This step toward progression enables users to configure the network layout and modify traffic control as in an SDN. Although there are still some flaws with certain schemes, researchers are working diligently to solve the product issues to obtain a network that is user friendly, high speed, high quality, and durable.

## References

1. Egilmez, H. E., S. Civanlar, and A. M. Tekalp. 2013. An optimization framework for QoS-enabled adaptive video streaming over OpenFlow networks. *IEEE Transactions on Multimedia*, vol. 15, no. 3, pp. 710, 715.
2. Kassler, A., L. Skorin-Kapov, O. Dobrijevie, M. Matijasevic, and P. Dely. *Toward QoE-Driven Multimedia Service Negotiation and Path Optimization with Software-Defined Networking*. Software, Telecommunications and Computer Networks (SoftCOM), 2012 20th International Conference on, pp. 1, 5.
3. Egilmez, H. E., S. T. Dane, K. T. Bagci, and A. M. Tekalp. *OpenQoS: An OpenFlow Controller Design for Multimedia Delivery with End-to-End Quality of Service over Software-Defined Networks*. Signal & Information Processing Association Annual Summit and Conference (APSIPA ASC), 2012 Asia-Pacific, pp. 1, 8.

4. Kim, H. and N. Feamster. 2013. Improving network management with software-defined networking. *IEEE Communications Magazine*, vol. 51, no. 2, pp. 114–119.

5. Banikazemi, M., D. Olshefski, A. Shaikh, J. Tracey, and G. Wang. 2013. Meridian: An SDN platform for cloud network services. *IEEE Communications Magazine*, pp. 120–127.

6. Kim, N. and J. Kim. 2010. Designing networking service for multimedia service composition over programmable network substrate. *International Conference on P2P, Parallel, Grid, Cloud, and Internet Computing*, pp. 105–108.

7. Chung, J., G. Gonzalez, I. Armuelles, T. Robles, R. Alcarria, and A. Morales. 2012. Characterizing the multimedia service capacity of wireless mesh networks for rural communities. *1st International Workshop on Community Networks and Bottom-up Broadband*, pp. 628–635.

# 11

# QoS Issues in OpenFlow/SDN

## KUHELI L. HALDAR AND DHARMA P. AGRAWAL

### Contents

### Introduction

The architecture of a network's physical infrastructure significantly affects the overall performance of the network. In today's traditional network system, the control and the data plane are tightly coupled into the same hardware, as shown in Figure 11.1. However, each plane has separate tasks that provide the overall switching/routing functionality. The control plane is responsible for computing and programming the paths of the data flows. Once these paths have been determined, they are then sent and stored in the data plane. The typical function of the data plane is to make path decisions in the hardware based on the latest information provided by the control plane.

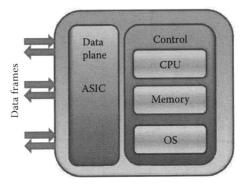

**Figure 11.1** Traditional hardware switches.

Managing a switch basically implies interacting with the control plane. When an IP packet reaches a given switch or local area network (LAN) interface, the primary job of the switch is to forward the packet to the next appropriate switch port or to the LAN interface. The decision making involved in this process is an inherent function of the control plane and is performed by accepting and processing the information frames such as the bridge protocol data unit (BPDU). Routing tables and spanning tree protocols (STP) are determined within the control plane. Once the path computation has been done, the routing table is pushed to the data plane's programmable application-specific integrated circuit (ASIC) chip. The basic responsibility of the data plane is then of sending the packets through a specified switch port or the LAN interface.

Another important function of the control plane involves obtaining the information needed for making those decisions. Routers use protocols to exchange routing information. Likewise, switches also use different proprietary and *de facto* approaches to determine the next outbound port for a packet. Traditionally, this has been a very effective method. The decision-making process in the hardware has been fast while the control plane manages the heavier processing and configuration functions. However, in the wake of server virtualization and cloud computing, these approaches pose some limitations to the service providers as well as the cloud customers. These challenges revolve around the limitations of the network equipment and, sometimes, the protocols themselves. In recent times, cloud customers want more control and flexibility in accessing applications, infrastructure, and other network resources, whereas server virtualization has greatly increased the number of hosts requiring network connectivity. Unfortunately, the traditional Ethernet routing and switching standards were developed for an era when the network configuration and topology were far more static, rendering them incapable of handling the high volume of changes inherent in today's networking environment.

**What Are the Shortcomings in the Perspective of QoS?**

The tightly coupled architecture of traditional networks is not well suited to meet the quality of service (QoS) requirements of today's enterprises, carriers, and end users. The first and the most important point to focus on is scalability [1]. During scheduling, QoS allows different data flows to be treated with different forwarding priority based on the characteristics of those flows. For example, delay-sensitive real-time traffic, such as voice and video, is given high priority, whereas non–real-time traffic, such as data, is treated with less priority to ensure best user experience. Traffic prioritization is performed based on the parameter specified in the class of service (CoS) field or the differentiated services code point (DSCP) of a packet. These parameters and the corresponding rules must be applied consistently to every packet entering the network. In the traditional network system, this becomes very cumbrous because the configuration must be replicated in every switching device.

Second, accurate classification and routing of traffic is a challenge in the network. Important questions, such as what traffic should be prioritized, who defines the priorities, etc., also arise when traffic prioritization comes into the picture. For example, voice traffic are delay sensitive, whereas data traffic profits from more bandwidth. So, different kinds of traffic have different requirements. Currently, very few tools exist in the network to differentiate traffic flows and route them through their appropriate paths, which would be beneficial for both the network and the end users.

Third, as the Internet evolves, Internet applications and associated traffic patterns have undergone a drastic change. This is most noticeable in the enterprise data centers. Most of the communications follow the classic north-south traffic pattern compared with the traditional client-server model. Today's Internet applications access a variety of databases and servers creating a huge amount of machine-to-machine traffic before returning the results to the end users [2]. In addition, current users contribute more to the changing pattern by accessing corporate content and applications from any type of devices, anywhere, and anytime. These applications with real-time and non–real-time traffic have very distinct QoS requirements. The traditional QoS framework is no longer efficient enough to support these mobile applications and provide a good experience to the end users.

Fourth, a rapid rise in users and applications, such as digital media streaming in the network, has given rise to an excessive demand for bandwidth. As cloud network infrastructure migrates toward mega-data centers, it is driving higher bandwidth to and between the cloud data centers. Managing these megadata sets requires a much larger network capacity because they require a massive amount of processing on thousands of servers with direct connections to each other. In today's network with limited bandwidth availability and increasing cost, maintaining any-to-any connectivity without going bankrupt is a daunting task.

Currently, each server that handles multiple virtual machine (VM) instances and traffic flows from multiple applications has to be merged over a single 10GbE physical interface in today's virtualization world. This is feasible because of an effective I/O QoS implementation that ensures required service levels as per the criticality of applications. In a traditional implementation, data center application servers take care of one application per physical server acting as a stove pipe. This forces server resources to be underutilized, and many physical servers are needed to execute an organization's applications. Virtualization allows multiple VMs to exist on a single host server that enables each VM to guarantee an appropriate service level for each application. In addition, by adopting virtualization, many enterprises today use an integrated IP network for voice, data, and video traffic. While existing networks can support differentiated QoS levels for various applications, the provisioning of those resources is typically done manually. Therefore, the network cannot dynamically adapt to changing traffic, application, and user demands.

Finally, with the emergence of cloud services, the computing infrastructures need to dynamically maintain computing platforms, made of cloud resources, and services. These applications will be defined by

QoS requirements, such as timeliness, scalability, high availability, trust, and security, as specified in the service level agreements (SLAs). An SLA is a legally binding contract of QoS to provide its applications. Current cloud technology is not designed to have a full SLA.

### What Is SDN and How Do We Address It?

Software-defined network (SDN) is an emerging paradigm in computer networking, which proposes to disjoin the components of traditional tightly coupled network switches and routers to improve network flexibility and architectural manageability, as illustrated in Figure 11.2. It is basically a conglomeration of multiple technologies, which addresses the need for an ever-increasing demand for speed, scalability, and resilience in today's networks. By opening up the control, data, and management planes of the network, SDN enhances the participation of operators and independent network developers through application programming interfaces (APIs). High-level control programs can be written, which specify the behavior of the entire network as compared with existing systems with specific functionalities of low-level device configuration.

A typical SDN architecture should allow the manipulation and programming of the network frame and the packet flows on a wider scale. This will allow the operators to gain a network-wide view and the ability to centrally manage the network. SDN accomplishes this by separating the control plane and the data plane and by having an interface to program the control plane. There are two reasons for this. First, the control plane forwards information to the data plane's ASIC hardware, making it more flexible. Second, it provides network-wide

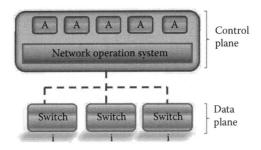

**Figure 11.2** SDN architecture.

visibility to the control plane, allowing it to have a comprehensive view of the entire network. Through such decoupling, SDN provides a platform for making advanced routing, security, and quality decisions efficiently and easily.

### The Three Key Capabilities of SDN

1. Capability to be programmed
2. Capability to manage the forwarding of frames/packets and apply policy
3. Capability to perform the aforementioned two functions at a large scale dynamically

### The Components of SDN

*Controller*   The controller is the main governing entity in the whole SDN ecosystem. Network intelligence is centralized in the controller, which maintains the holistic logical view of the entire network. Physically, the controller is a software that runs on a regular computer and connects through a dedicated logical interface to the routers or switches.

*OpenFlow*   OpenFlow is a programmable network protocol that provides an open standard-based interface to control and direct network traffic from multiple vendors [3]. It has the potential to dramatically increase data center services. The datapath of an OpenFlow Switch consists of a flow table and an action, such as outbound port, modify-field, or drop, for each flow entry. When an OpenFlow switch receives a packet that finds no match in the flow table, it queries the controller through the OpenFlow protocol. Then, the controller decides how the packet should be handled. OpenFlow provides several benefits compared with the traditional packet switching. First, it serves as a logically centralized controller that supervises all flow decisions, thereby eliminating the need for any global policies. Second, the central controller has full visibility of the network and its vendors, enabling the centralized management of the network traffic. Third, the switches can be kept simple and futuristic because all important decisions are made by the controller rather than the switches themselves.

OpenFlow is a relatively new protocol and is still undergoing heavy development. Despite the growing interest in the networking world, adequate research regarding the performance and scalability of the OpenFlow switches is yet to be achieved. For example, as the network grows, OpenFlow switches will need to handle more and more flow changes. The question that also arises is the number of flows that it can manage and the speed at which the flows can be installed. Again, the usefulness of a central controller is also questionable. First, although OpenFlow can provide globally supervised flow and admission control, the question is whether all flows ought to be mediated by a central controller. Certain flows can be preauthorized by creating wildcard rules and be delegated to the switches, whereas more security-sensitive flows can be authorized by the central controller. Second, all flows do not require QoS guarantee and can be treated as best-effort traffic. Only those flows that require such guarantees can be treated individually by the controller. Finally, there are some flows that are so brief that they become interesting only if they are aggregated. Hence, instead of confronting every flow at the initial setup, the controller can wait to distinguish the flows as they become significant and reroute them midconnection to avoid unnecessary overhead.

*QoS Issues in OpenFlow*   QoS is a very important consideration for OpenFlow. On one hand, it is essential for OpenFlow to have some basic support for QoS, whereas on the other hand, too much effort on its part will make OpenFlow too complex to handle. Therefore, it is essential to determine a threshold when a default set of QoS options can be built into the OpenFlow switches and routers. Additional flexibility should be available to enable vendors to have more QoS support and differentiate their product from others. At present, most switches and routers support some native QoS requirements. OpenFlow switches can be expected to provide provisions for mapping flows to a service class or may have its own service class.

QoS was not supported in the previous versions of OpenFlow. However, future versions are expected to have the ability to assign QoS to the packets. When a switch connects to its controller, it should report to the controller its QoS capabilities such as the supported classes. Each flow entry in the flow table can be assigned a certain level of service called the service class. A flow belonging to a service class will be placed in the respective service queue. Openflow. org has suggested the three QoS primitives to be incorporated into the switches.

- **Rate Promising or Minimum Rate:** An important feature, also called slicing, promises each flow a minimum bandwidth guarantee at each egress. If a switch is not able to provide a minimum egress rate on a per-flow basis, it could allow a flow to be part of a queue called the service class. A separate tag called Rate Promise ID could be added to a flow-entry to match it with a smaller set of queues that is less than one per flow. In reality, this feature is common in today's switches and has already been incorporated into OpenFlow switches version 1.0 onward. Under this QoS requirement, the controller is able to set up and configure the queues and then map each flow to a specific queue. The configuration of the queue regulates how a packet should be treated in the switch. The slicing mechanism is composed of two distinct parts:
  - **Configuration:** Queue configuration is a vendor extension, not the responsibility of the OpenFlow protocol. Within a switch, each queue may have different efficacies, leading to a per-port QoS characteristic rather than representing the entire switch. The characteristics of a port depend on the mechanism used.
  - **Flow Queue Mapping/Forwarding:** This involves unwrapping of the queues to the corresponding forwarding actions.
- **Rate Limiting or Maximum Rate:** Rate limiter controls the rate of the packet flow. In OpenFlow, the rate limiter is designed in such a way that it is orthogonal to the other QoS features to avoid any adverse effects on them. The main disadvantage of a rate limiter is its underutilization of the network capacity. However, rate limiter has many advantages too. First, the model is very simple, and their behavior is very easy to predict, making them widely applicable to many QoS frameworks. Second, their implementation in the hardware is very simple and efficient. Third, rate limiters are not bound to any port, making them very flexible to deploy and process any

set of packets. Rate limiters can be deployed both at the egress and the ingress ports of a switch.

- **Egress Rate Limit:** This feature enforces an optional upper bound on the speed of transmission of a flow at egress. If a switch is not able to control the outbound rate on a per-flow basis, it can be done by a shared queue. A separate tag called Rate Limit ID could be added to a flow entry to match it to a set of queues that is less than one per flow. However, it is important to define the maximum rate at the egress of a flow.
- **Ingress Rate Limit:** Also known as policing, ingress rate limit drops packets if they arrive faster than a predefined limit at the ingress. This characteristic is already present in several switches.
- **Strict Precedence:** This feature assigns priority levels to flows. For example, an HP ProCurve switch has eight priority queues per port. The OpenFlow flows are mapped to the priority queues using either the virtual local area network priority code point (VLAN PCP) or the the IP type of service/diffserve code point (ToS/DSCP), where the VLAN PCP offers direct mapping, and the IP ToS/DSCP leads to indirect mapping. In strict precedence, a queue is only served when the queues with higher priorities are empty.

### Resource Allocation

Resource allocation comes into play when there is a possibility of contention for bandwidth and buffer space in a switch if there is a scarcity of these resources. In a multitenant wide area network (WAN) or a fixed-mobile converged network environment, several resources of an OpenFlow switch must be separated from each other to prevent one VN from using more than its allocated share of resources. Although the packet classifier in the OpenFlow switch is quite simple, congestion may occur from the multicast traffic because a single incoming packet could be multiplied and could end up in several of the queues. Congestion caused by a high rate of incoming traffic can be limited by applying QoS tools, such as classification, metering, coloring, policing, shaping, and scheduling, both at the incoming and the outgoing ports. This will also limit the impact on the downstream switches. The same

QoS tools will guarantee reasonable resource usage and isolation in case of virtualization [4]. In the case of the control channel, the network connecting the controller to the switches must support rate limiters to prevent one VN from hogging the entire bandwidth of the control channel. In addition, control traffic must be given a higher priority than other traffic, such as data and configuration traffics, so that fairness in the use of control channel resources could be ascertained.

### Proposed QoS Solutions

#### QoS Framework

The current version of OpenFlow switches can only control DiffServ-like class-based traffic. To equip them with automated and fine-grained QoS control, a QoS controller can be introduced to the framework [5]. The controller will be responsible for allocating resources to QoS-sensitive flows in an automatic manner to guarantee a minimum bandwidth and to satisfy a maximum delay bound. To do so, the controller will create network slices and will automatically assign different application flows to different slices to ascertain performance requirements across all the applications. Basically, the main job of the controller is to reserve network resources so that SLA requirements can be guaranteed. The network administrators will only have to specify the slice specifications for the services. In this purview, the slice specifications can be applied to individual flows, aggregated traffic flows of a certain type, or a combination of both to have added flexibility. To automate the configuration and the management of QoS flows by the controller, QoS APIs are also added to OpenFlow. The QoS APIs attach the flows to the rate limiters and priority queues present in the hardware switch. The rate limiter APIs enable the controller to enforce total bandwidth usage limit on the flows, and the priority queue APIs enable the controller to map the flows to their respective priority queues in the outgoing ports of the switch. Thus, the QoS APIs manage bandwidth and latency allocation to the flows in the switches, and the controller manages the mapping across all the flows in a dynamic manner on a per-switch basis.

To avoid excessive computational and storage overhead for a controller in a large network, an additional adaptive flow aggregator (AFA) must be introduced. Instead of the controller computing resource allocation for each flow, the AFA will categorize flows into groups and will reserve resources based on each group. This is to be noted because all traffic does not require per-flow isolation. They only need resource allocation for the aggregated flow.

As shown in Refs. [6] and [7], QoS routing has a significant effect on the overall throughput of the network. Because the major portion of Internet traffic is composed of the best-effort traffic, an efficient routing algorithm that optimizes QoS must account for their existence and impact on the performance optimization. Moreover, the best-effort traffic allows for scalability in a network with a large number of nodes. Different routing protocols can be used within an OpenFlow controller to generate flow tables and manage QoS flows in the data plane [8]. The job of the controller should be to generate routes in the flow tables for QoS flows, keeping in mind that (1) QoS traffic should suffer minimum to zero packet loss; (2) in addition to the path length, the optimization function must consider other costs, such as delay; (3) QoS flows may have preemptive rights, causing more packet loss to the best-effort traffic on the shared path; and (4) optimal route calculation for the QoS traffic is done, considering the traffic pattern and the packet loss estimate of the best-effort flows. In addition to route calculation and route management, the QoS controller will also need to perform additional subfunctions, such as QoS contract management and QoS route management. To ensure that all switches conform to the SLA agreement, an additional supervisor node can be introduced in the datapath, which reports back to the controller with any violations. Each switch must track its current resource usage and report back to the controller for an efficient resource monitoring of the network. The report should also contain resource availability information in the switches for the controller to manage and allocate resources efficiently and provide load balancing in the network. Rather than the controller continuously polling the switches in the datapath of the QoS flows, improved efficiency in the network can be achieved if the switches can autonomously report back congestion condition in the network.

*Load Balancing Using Wildcards*

Load balancers act as a single point of entry in a network and have a drastic impact on performance and availability. It is expensive to deploy dedicated load balancers. OpenFlow introduces an alternative approach by distributing traffic to an array of replicated servers. Although the current OpenFlow architecture has a single controller, it has a provision for multiple controllers, and special controllers could be used for different types of traffic with varying demands [9]. The basic idea is to have one controller that could forward the incoming messages to the target servers for load balancing. To avoid single-point failure, a controller can take over the management task. Another advantage is that the controllers can provide many features that improve the workload and the network performance.

One limitation of this method is that it increases the number of switch-controller interactions, and one way to reduce this and subsequent ternary content accessible memory (TCAM) entries is to aggressively use wildcarded OpenFlow rules [10–11]. This mechanism removes most of the responsibilities from the controller to the switches. The central controller only identifies the significant flows from the efficient statistics collection mechanisms of the OpenFlow protocol.

*Network Operating System*

Network Operating System (NOX) is an operating system for networks [12] because it does not manage the network itself. Like other operating systems, it provides a programmatic interface that enables access to high-level network abstractions, such as users, topology, and services, for application programs to carry out complicated tasks safely and efficiently over a wide array of devices. It provides a flexible and scalable platform to write high-performance network applications that manage the network by installing flow entries in the switch forwarding tables. This is done in two ways, as follows.

First, Nox provides a global view of the actual state of the network. Applications use this view to make management decisions without worrying about its accuracy. It provides important APIs for the applications and maintains useful information, such as mapping between IP addresses and host names, etc. It also provides an abstraction of events to the controllers and reflects changes in the network conditions.

Second, high performance can be obtained from Nox by programming control switches in OpenFlow using a high-level language. When the switches encounter a new flow without a matching entry, it notifies the Nox servers. A new routing path is installed in the switches by the server. Thus, decisions need to be made only for the first packet of the flow, not the entire flow on a packet by packet basis. When a counter is exceeded, only actions such as rerouting a flow or sending it to an application are required. In this way, the server can perform the main computation in a high-level language within it, whereas OpenFlow can implement this effect inside a switch.

Despite numerous advantages, Nox does not possess the necessary functions that could guarantee QoS on various enterprise networks. Before we attempt to add QoS in Nox, let us first look at the main components of a Nox-based network. The main components of a Nox-based network primarily include a set of switches, one or more servers, and a database. Having the knowledge of the network conditions, the Nox software makes decision on management. The controller initially configures appropriate switches by establishing fine-grain flow. However, that incurs very high overhead to the forwarders. A good hierarchical flow aggregation and traffic engineering [13] is used to mitigate the overheads [14]. Such a multilayered networking can be defined by generalized multiprotocol label switching (GMPLS) layers: layer 3 packet switching/forwarding, layer 2.5 multiprotocol label switching (MPLS), layer 2 Ethernet switching, layer 1.5 synchronous optical networking (SONET)/synchronous digital hierarchy (SDH) cross-connect, layer 1 wavelength division multiplexing (WDM) lambda switching, and layer 0 fiber switching. MPLS is a part of OpenFlow for efficient and scalable traffic engineering, but not the GMPLS. The future Internet requires efficient and scalable traffic engineering as one of its major features. Hence, multilayer networking with optical transport network is very essential for the future Internet.

Other than GMPLS, improvements can be made in other Nox components, such as the service elements, the forwarding plane, the database, etc. Incorporating QoS features in Nox is at a very early stage. Tremendous opportunities lie in this area. The research community is actively engaged in the direction for providing scalability, QoS, and security to Nox and OpenFlow.

# References

1. Open Network Foundation. http://www.opennetworking.org/?p=358& option=com_wordpress&Itemid=72.
2. ONF White Paper. *Software-Defined Networking: The New Norm for Networks*. http://www.opennetworking.org/.
3. OpenFlow - Enabling Innovation in Your Network. http://www.open flow.org/.
4. Skoldstrom, P. and K. Yedavalli. 2012. Network virtualization and resource allocation in OpenFlow-based wide area networks. *SDN '12 Workshop on Software-Defined Networks; IEEE Conference on Communications (ICC)*.
5. Kim, W., P. Sharma, J. Lee, S. Banerjee, J. Tourrilhes, S. Lee, and P. Yalagandula. 2010. Automated and scalable QoS control for network convergence. *Proceedings of the INM/WREN '10*.
6. Ma, Q. and P. Steenkiste. 1999. Supporting dynamic interclass resource sharing: A multiclass QoS routing algorithm. *IEEE International Conference on Network Protocols*, pp. 649–660.
7. Nahrstedt, K. and S. Chen. 1998. Coexistence of QoS and best-effort flows—Routing and scheduling. *Proceedings of the 10th IEEE Tyrrhenian International Workshop on Digital Communications*.
8. Civanlar, S., M. Parlakisik, A. M. Tekalp, B. Gorkemli, B. Kaytaz, and E. Onem. 2010. A QoS-enabled OpenFlow environment for scalable video streaming. *2010 IEEE Globecom Workshop on the Network of the Future (FutureNet-III)*, pp. 351–356.
9. Koerner, M. and O. Kao. 2012. Multiple service load-balancing with OpenFlow. In *Proceedings of the 13th IEEE International Conference on High-Performance Switching and Routing (HPSR)*, pp. 210–214.
10. Curtis, A. R., J. C. Mogul, J. Tourrilhes, P. Yalagandula, P. Sharma, and S. Banerjee. 2011. DevoFlow: Scaling flow management for high-performance networks. In *Proceedings of the 2011 ACM SIGCOMM Conference* 41(4):254–65.
11. Wang, R., D. Butnariu, and J. Rexford. 2011. OpenFlow-based server load balancing gone wild. *Proceedings of the 11th USENIX Conference on Hot Topics in Management of Internet, Cloud, and Enterprise Networks and Services*.
12. Gude, N., T. Koponen, J. Pettit, B. Pfaff, M. Casado, N. Mckeown, and S. Shenker. 2008. Nox: Toward an operating system for networks. *Computer Communication Review (CCR)* 38(3):105–110.
13. Lehman, T., X. Yang, N. Ghani, F. Gu, C. Guok, I. Monga, and B. Tierney. 2011. Multilayer network: An architecture framework. *IEEE Communications Magazine* 49(5):122–130.
14. Jeong, K., J. Kim, and Y.-T. Kim. 2012. QoS-aware network operating system for software-defined networking with generalized OpenFlows. In *Proceedings of IEEE Network Operations and Management Symposium (NOMS)*, pp. 1167–1174.

# 12

# QoS-Oriented Design in OpenFlow

## XINGANG FU AND FEI HU

**Contents**

## Introduction

In the past three decades, the Internet Engineering Task Force (IETF) has explored several quality-of-service (QoS) architectures, but none has been truly successful and globally implemented. The reason is that QoS architectures, such as IntServ and Diffserv, are built on top of the current Internet's completely distributed hop-by-hop routing architecture, which is lacking a solid centralized control of overall network resources.

OpenFlow decouples the control and forwarding functionalities of routing, in which the forwarding function stays within the OpenFlow routers (forwarders), whereas the routing control function is handled by a separate controller layer, which is the brain of the network and could be centralized and/or possibly federated. Forwarding tables are dynamically uploaded to forwarders by the controller layer. Figure 12.1 shows that the controller layer controls many forwarders. OpenFlow can provide new QoS architectures over OpenFlow networks. Thus, it is possible to use different routing protocols (rather than the typical shortest path routing scheme) within the controller to generate flow tables that govern different isolated flows in the data plane. Much work is done to make OpenFlow better support QoS.

This chapter discusses the important improvements related to OpenFlow QoS. "QoS-Enabled OpenFlow Environment and Dynamic Rerouting Algorithm for Scalable Video Streaming" presents QoS-enabled OpenFlow environment and dynamic rerouting algorithm for scalable video streaming. "Automatic QoS Management on OpenFlow" talks about automatic QoS management in OpenFlow. "An OpenFlow Controller Design for Multimedia Delivery" deals with an OpenFlow controller design for multimedia delivery with QoS. "On QoS Support to Ofelia and OpenFlow" is about QoS support to Ofelia and OpenFlow. "QoS-Aware Network Operating System for GOF" presents the QoS-aware network operating system for generalized OpenFlow (GOFN). Finally, "Discussion" addresses

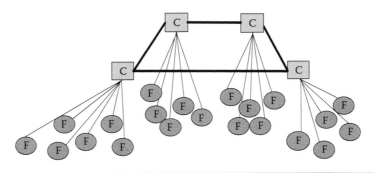

**Figure 12.1**   OpenFlow network with a federated controller layer and a forwarder layer.

QoS-related problems in multimedia service negotiation and path optimization.

## QoS-Enabled OpenFlow Environment and Dynamic Rerouting Algorithm for Scalable Video Streaming

Streaming media applications, such as videoconferencing, WebTV, etc., require steady network resources with little or no packet drop events and transmission delay. It could not always be achieved by the standard best-effort Internet. Scalable video coding (SVC) encodes the video in a base layer and one or more enhancement layers, and it is crucial that the base layer is streamed without any packet loss or delay variations.

However, the current best-effort–based Internet routing scheme applies shortest path algorithm that only considers the path length. When the shortest path route is congested, the video stream's performance would be poor because of data loss (especially for the SVC base layer) [1].

### Scalable Video Coding

According to SVC, the traffic stream in the network is decoupled into three distinct streams: (1) the SVC base layer video is defined as lossless QoS traffic (i.e., it cannot tolerate any packet losses); (2) SVC enhancement layers are defined as lossy QoS traffic (i.e., they can tolerate some packet losses); and (3) the rest of the traffic is defined as best-effort data.

If there are multiple enhancement layers for a video stream, they are treated as a bundle to be routed together. The base layer can be decoupled from enhancement layers and be separately routed (by simply using a different port number).

### Dynamic Rerouting for SVC

Dynamic rerouting of QoS flows using non–shortest paths for lossless and lossy QoS flows achieves significant improvement on the video's overall peak signal-to-noise ratio (PSNR) under network congestion. Egilmez et al. [1] proposed a dynamic rerouting solution, where the base layer of an SVC-encoded video is transmitted as a lossless-QoS flow, and the enhancement layers can be routed either as a lossy QoS flow or as a best-effort flow, respectively. Through experiments, they recommend optimizing routes for both the base and the enhancement layers in OpenFlow networks while keeping the best-effort traffic on its current shortest path route.

Generally, some basic assumptions for QoS routing are as follows:

1. QoS traffic should be carried with very low or even with zero packet loss.
2. QoS routing should optimize a particular cost function rather than a simple path length metric. The routes with larger capacity and even with longer distances may be more preferable to the shorter routes that may cause packet loss.
3. QoS traffic may have preemptive rights based on traffic priority and importance.
4. QoS routing may select optimal routes based on the traffic patterns and packet loss estimates of the best-effort traffic.

*QoS Optimization Model*

Assume that each OpenFlow node (forwarder) is controlled by one controller. Any pair of nodes are defined by $i$ and $k$, where $i = 1, 2,\ldots,$ $N, k = 1, 2,\ldots, N$, with $i \neq k$ and where $N$ is the total number of nodes in the network. Distinct routes between nodes $i$ and $k$ are denoted as $r_1, r_2 = 1, 2,\ldots, N_{ik}$, where $r_1$ is the route for the lossless QoS traffic (base layer), $r_2$ is the route for the lossy QoS traffic (bundle of enhancement layers), and $N_{ik}$ is the total number of distinct routes between nodes $i$ and $k$. We use $L_{ik}^r$ for the length and $C_{ik}^r$ for the capacity of route $r$ between nodes $i$ and $k$. The length of a route can be the number of hops or the propagation delay from the source to the destination. The capacity can be a measure of bandwidth in bits per second. Moreover, in the formulations, we have used three separate traffic variables—$Q_{ik}^r(t)$, $E_{ik}^r(t)$, and $B_{ik}^r(t)$—which are the amounts of lossless QoS traffic, lossy QoS traffic, and best-effort traffic on route $r$ at time $t$, respectively.

Suppose that we only reroute lossless QoS traffic under congestion conditions, whereas the rest of the traffic remains on their shortest path routes. The QoS optimization problem can be described as follows:

$$\min_{r_1} (1-\lambda)L_{ik}^{r_1} + \lambda PLT_{ik}^{r_1}(t)$$

subject to $L_{ik}^{r_1} < L_{\max}$,

$$PLT_{ik}^{r_1}(t) = \begin{cases} 0 & C_{ik}^{r}(t) \geq Q_{ik}^{r}(t) + B_{ik}^{r}(t) + E_{ik}^{r}(t) \\ \dfrac{Q_{ik}^{r}(t) + B_{ik}^{r}(t) + E_{ik}^{r}(t) - C_{ik}^{r}(t)}{B_{ik}^{r}(t) + E_{ik}^{r}(t)} & C_{ik}^{r}(t) < Q_{ik}^{r}(t) + B_{ik}^{r}(t) + E_{ik}^{r}(t) \end{cases}$$

$i = 1, 2,\ldots, N$, $k = 1, 2,\ldots, N$, $r_1, r_2 = 1, 2,\ldots, N_{ik}$, $0 \leq \lambda \leq 1$ and assuming that $r = r_1$ such that $C_{ik}^{r_1}(t) > Q_{ik}^{r_1}(t)$, which satisfies $PLT_{ik}^{r_1}(t) = 0$.

Now, let us suppose that lossless QoS traffic as well as enhancement layers are rerouted, whereas the best-effort traffic remains on its shortest path routes. The optimization problem can be modeled as follows:

$$\min_{r_1} (1-\lambda)\left\{L_{ik}^{r_1} + L_{ik}^{r_2}\right\} + \lambda\left\{PLT_{ik}^{r_1}(t) + PLT_{ik}^{r_2}(t)\right\}$$

subject to $L_{ik}^{r_1} < L_{\max}$, $L_{ik}^{r_2} < L_{\max}$

$i = 1, 2,\ldots, N$, $k = 1, 2,\ldots, N$, $r_1, r_2 = 1, 2,\ldots, N_{ik}$, $0 \leq \lambda \leq 1$ and assume that there is $r = r_1$ such that $C_{ik}^{r_1}(t) > Q_{ik}^{r_1}(t)$, which satisfies $PLT_{ik}^{r_1}(t) = 0$.

The simulation results in Ref. [1] show that the average quality of video streams is improved by 14% if only the base layer is rerouted. By rerouting the enhancement layer along with the base layer, the quality is further improved by another 6.5%.

*Controller Architecture: QoS Subsystems*

The controller proposed in Ref. [1] has the major functions of route calculation and route management. Figure 12.2 illustrates the controller architecture with various subfunctions.

To support QoS traffic, the controller has additional subfunctions, such as QoS contract management and QoS route management. The QoS routing algorithm determines the flow tables applicable only to QoS traffic. Underneath all the controller functions, a security layer provides secure communication between the controller and forwarders

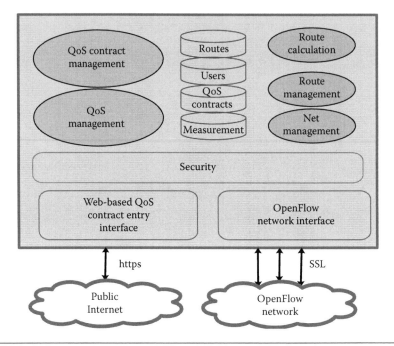

**Figure 12.2**   Controller subsystems to support QoS.

by using an secure socket layer (SSL) interface. Similarly, the users or system administrators can interact with the controller over the public Internet using a secure Web protocol such as HTTPS. Note that a policing function has to be implemented in the forwarder layer to make sure that the end points conform to the service-level agreements (SLAs) stated in their QoS contract.

The following additional functions are required in the network architecture:

1. *Resource monitoring.* Either forwarders keep track of how much of their local resources are available and report to the controller or, alternatively, the controller periodically polls the forwarders to determine the resource availability on routes.

2. *Resource signaling.* Forwarders signal their state of resource availability in real time so that necessary actions can be taken at the controller, for example, updating routing tables and/or adapting media encoding parameters by end terminals.

3. *Resource reservation.* The controller should instruct forwarders on the appropriate allocation of forwarder resources, such as the egress queues for QoS and best-effort traffic flows. Accordingly, it sends corresponding flow tables for different traffic flows.

## Automatic QoS Management on OpenFlow

In its current version, OpenFlow is not able to configure QoS parameters in a dynamic and on-demand manner (i.e., it does this manually). To deal with QoS problems in OpenFlow switches, a framework that enables QoS management in an OpenFlow environment is proposed by Ishimori et al. [2]. QoSFlow adds new QoS functions and allows the management of class and queues through rules or policies. These functionalities assure to manage QoS resources (e.g., bandwidth, queue size, or delay), without changing the software-defined network (SDN) architecture. In other words, all actions are invoked by an OpenFlow controller (control plane) and in a dynamic and on-demand manner (not manually).

### *QoSFlow Controller*

The QoSFlow controller is based on Nox, which is responsible for managing/monitoring actions and controlling signaling messages. The new controller, besides Nox application programming interface (API), contains the following new components: QoSFlow agent, QoSFlow manager, QoSFlow monitor, and database-QoSFlow client. These four modules have been designed to extend the Nox API with QoS features called QoSFlow API. The QoS agent is responsible for creating a communication module between an administrator management tool and the other two QoSFlow components—the manager and the monitor QoSFlow. By using the JavaScript object notation (JSON) interface, the agent is able to receive policies, manage, or monitor commands from a third-party administrator application. The QoSFlow monitor and manager components, respectively, monitor and manage the QoS of OpenFlow domains. Figure 12.3 shows the controller designs.

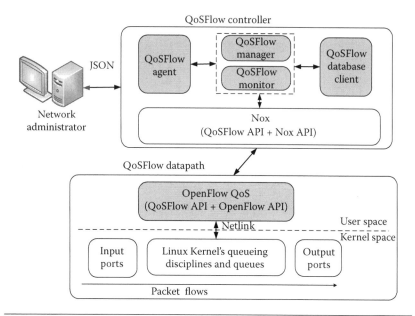

**Figure 12.3**   QoSFlow controller structure.

*QoSFlow Datapath*

The QoSFlow datapath component is called OpenFlow QoS, which is responsible for creating all low-level actions on the switch ports. This component allows OpenFlow to get all the required primitives to run management commands created either by the administrator's tool or through header packet information. In the QoS management tool, the actions are processed in the QoSFlow agent. When receiving those actions, it checks the type of the received requests to select the next procedure. This new message is automatically sent to OpenFlow QoS through Nox. The QoS actions can be applied automatically through the packet header information.

*QoSFlow Policies*

QoS policies allow administrators to manage an administrative domain, which includes tasks, such as mapping of configuration models in a low level, scaling the management of hundreds of entities, and controlling the behavior with end-to-end desirable features. Those policies are high-level rules abstraction. They define the behavior of a system, in which a set of instructions may be called when prearranged conditions are satisfied. QoSFlow has an administrative interface to set these policies, which can be converted in a low-level configuration. The policies follow the standards established by request for comments, such as 3703, 3060, 3460, and some drafts for policy schemes, besides other instances to store and create an SLA.

### An OpenFlow Controller Design for Multimedia Delivery

To support QoS, in Ref. [3], the incoming traffic is grouped as data flows and multimedia flows, where the multimedia flows are dynamically placed on QoS-guaranteed routes, and the data flows remain on their traditional shortest path routing approach. With these backgrounds, Egilmez et al. [3] proposed the OpenQoS design for multimedia delivery.

#### QoS Architectures and the OpenQoS Design

OpenQoS is an extension of the standard OpenFlow controller, which provides multimedia delivery with QoS. As depicted in Figure 12.4, OpenQoS offers various interfaces and functions to enable QoS.

The main interfaces of the controller design are as follows:

1. *Controller-forwarder interface.* The controller is attached to forwarders with a secure channel using the OpenFlow protocol to share necessary information. The controller is responsible for sending flow tables associated with data flows, requesting network state information from forwarders for discovering the network topology, and monitoring the network.

2. *Controller-controller interface.* The single-controller architecture does not scale well when the network is large. As the number of OpenFlow nodes increases, multiple controllers are required. This interface allows controllers to share the necessary information to cooperatively manage the whole network.

3. *Controller-service interface.* The controller provides an open and secure interface for service providers to set flow definitions for new data partitions and even to define new routing

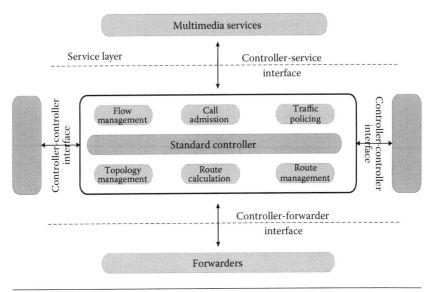

**Figure 12.4**  OpenQoS controller design.

rules associated with these partitions. It also provides a real-time interface to signal the controller when a new application starts a data flow.

The controller should manage the following several key functions:

- *Topology management*, which is responsible for discovering and maintaining network connectivity through the data received from forwarders
- *Route management*, which is responsible for determining the availability and the packet forwarding performance of routers to aid the route calculation. It requires the collection of the up-to-date network state from the forwarders on a synchronous or asynchronous basis.
- *Flow management*, which is responsible for collecting the flow definitions received from the service provider through the controller-service interface and for the efficient flow management by aggregation
- *Route calculation*, which is responsible for calculating and determining routes for different types of flows. Several routing algorithms can run in parallel to meet the performance requirements and the objectives of different flows. Network

**Table 12.1** Controller Comparison

| CONTROLLERS | FOR SVC | FOR MULTIMEDIA DELIVERY |
|---|---|---|
| Major functions | Two major functions: route calculation and route management. Several subfunctions: QoS contract management and QoS route management | Six key functions: topology management, route management, flow management, route calculation, call admission, and traffic policing |
| Algorithms | CSP problem | CSP problem |

topology and route management information are inputted to this function along with the service reservations.

- *Call admission*, which is a function that denies/blocks a request when the requested QoS parameters cannot be satisfied (i.e., there is no feasible route) and informs the controller to take necessary actions
- *Traffic policing*, which is responsible for determining whether data flows agree with their requested QoS parameters and for applying the policy rules when they do not agree with the parameters (e.g., by using preempting traffic or selective packet dropping)

Egilmez et al. proposed two controller designs—one for SVC and another for multimedia delivery. Basically, these controllers have some common parts, and the latter is, in some sense, more advanced (see Table 12.1).

*Per-Flow Routing in OpenQoS*

The current Internet does not allow routing on a per-flow basis. OpenFlow provides the flexibility of defining different types of flows to which a set of actions and rules can be associated. In OpenFlow, flows can contain same or different types of packets. For example, packets with transport control protocol (TCP) port number 80 (reserved for HTTP) can be a flow definition, or packets having a real-time transport protocol (RTP) header may indicate a flow that carries voice, video, or both. In essence, it is possible to set flows as a combination of header fields. However, the network operator should also consider the processing power limitations of the network devices (routers or switches). In OpenFlow, network devices store the flows and their associated rules in flow tables, which are processed as a pipeline. Its goal is to reduce the packet processing time.

*Optimization of Dynamic QoS Routing*

The dynamic QoS routing can be regarded as a constrained shortest path (CSP) problem. It is crucial to select a cost metric and constraints, where they both characterize the network conditions and support the QoS requirements. In multimedia applications, the typical QoS indicators are packet loss, delay, and delay variation (jitter). However, some QoS indicators may differ depending on the type of application, such as follows: (1) interactive multimedia applications that have strict end-to-end delay requirements (e.g., 150–200 ms for video conferencing). So, the CSP problem constraint should be based on the total delay; and (2) video streaming applications that require steady network conditions for continuous video playout. However, the initial start-up delay may vary from user to user. This implies that the delay variation is required to be bounded, so the CSP problem constraint should be based on the delay variation.

A network is represented as a directed simple graph $G(N,A)$, where $N$ is the set of nodes, and $A$ is the set of all arcs (also called links), so that arc $(i,j)$ is an ordered pair, which is outgoing from node $i$ and incoming to node $j$. Let $R_{st}$ be the set of all routes (subsets of $A$) from source node $s$ to destination node $t$. For any route $r \in R_{st}$, we define cost $f_C$ and delay $f_D$ measures as

$$f_C(r) = \sum_{(i,j) \in r} c_{ij} \quad f_D(r) = \sum_{(i,j) \in r} d_{ij}$$

where $c_{ij}$ and $d_{ij}$ are the cost and the delay coefficients for the arc $(i,j)$, respectively. The CSP problem can then be formally stated as finding

$$r^* = \arg\min_r \{ f_C(r) \mid r \in R_{st}, f_D(r) \leq D_{max} \}$$

to minimize the cost function $f_C(r)$ subject to the delay variation $f_D(r)$ to be less than or equal to a specified value $D_{max}$. We select the cost metric as follows:

$$c_{ij} = g_{ij} + d_{ij} \ \forall (i, j) \in A$$

where $g_{ij}$ denotes the congestion measure for the traffic on link $(i,j)$, and $d_{ij}$ is the delay measure. OpenQoS collects necessary parameters $g_{ij}$ and $d_{ij}$ using the route management function.

The aforementioned CSP problem is known to be NP complete. Thus, Egilmez et al. proposed to use the Lagrangian relaxation-based aggregated cost (LARAC) algorithm, which is a polynomial time algorithm that can efficiently find a good route without deviating from the optimal solution in $O([n + m \log m]^2)$ time [4]. The detailed solution can be found in Ref. [4]. When the route management function updates, the QoS indicating the parameters or the topology management function detects a topology change, and the route calculation function runs the LARAC algorithm to solve the CSP problem. Then, the controller updates the forwarders' flow tables accordingly. Hence, the QoS routes are dynamically set.

## On QoS Support to Ofelia and OpenFlow

During the evolution of the OpenFlow standard, some QoS capabilities have been added to the protocol. However, even the latest version has only a limited and not well-defined QoS framework. Hence, integrated QoS support is missing in current OpenFlow experimental testbeds including Ofelia. To enhance the Ofelia testbed with QoS support, Sonkoly et al. [5] proposed extensions to Ofelia.

Figure 12.5 shows the control framework architecture with the proposed QoS extensions. The Queue Manager plugin empowers uniform queue configuration capabilities. It hides equipment heterogeneity with consistent user interface to set up and manage the queues of switches in the testbed and to configure their properties.

Queue Manager translates users' request to vendor-specific configuration command sequences and executes them through different configuration/management interfaces, as shown in Figure 12.6. On one hand, available simple network management protocol (SNMP) interfaces can be used for hardware switches. On the other hand, for software switches, we have recently designed and developed an open-source Netconf-based interface, with the accompanying QoS extensions heavily relying on the traffic control (TC) Linux kernel module.

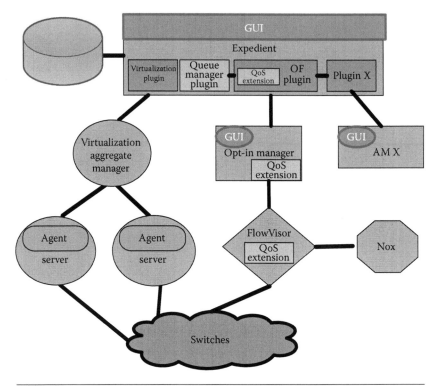

**Figure 12.5** General architecture of the OFELIA control framework extended with QoS support. GUI, graphical user interface; OF, OpenFlow.

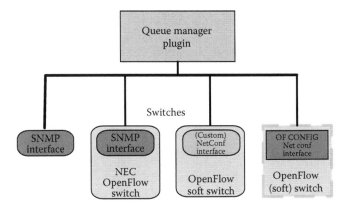

**Figure 12.6** Communication paths between the Queue Manager and the OpenFlow-capable switches. OF, OpenFlow; CONFIG, configuration.

Currently, the OF-Config protocol exists only as a protocol specification document, but OF-Config support can be easily added into Queue Manager when an implementation of soft switches or a new hardware firmware appears. In line with the findings in the QoS survey of OpenFlow devices deployed in Ofelia, Queue Manager can be connected with other devices as well. For example, the Netconf-based extensions might easily be ported to OpenWrt-based wireless routers. The necessary extensions and modifications to FlowVisor and Opt-In Manager have to be added as well so that the user-defined flow space could contain queues restricted from general access.

### QoS-Aware Network Operating System for GOF

OpenFlow switching and network operating system (Nox) have been proposed to support new conceptual networking trials for fine-grained control and visibility. However, Nox does not have the necessary functions for the QoS-guaranteed SDN service provisioning on carrier-grade provider Internet, such as QoS-aware virtual network embedding and end-to-end network QoS assessment.

Jeong et al. [6] proposed a QoS-aware network operating system (QNox) for an SDN with a GOF. GOF includes IP, multiprotocol label switching (MPLS)-transport profile (TP), and wavelength division multiplexing/automatically switched optical networks WDM/ASON. The MPLS-TP layer is mostly used for traffic engineering, whereas the WDM/ASON layer is used for long-distance transit networking.

Figure 12.7 depicts the proposed framework of QNox for SDN service provisioning with GOF. The major functional components of SDN/GOFN are service element (SE), control element (CE), management element (ME), and cognitive knowledge element (CKE). The forwarding element (FE) is a GOF-compatible switching node.

Clients request services using the QoS-aware open virtual network programming interface (QOVNPI). The SE receives service requests with attributes of the required computing and storage capacity, the location of the users' access points, the required performance and QoS parameters, the required fault restoration performance, and security level. The SLA and the service level specification (SLS) modules check and evaluate the availability of network resources and determine the QoS-guaranteed service provisioning.

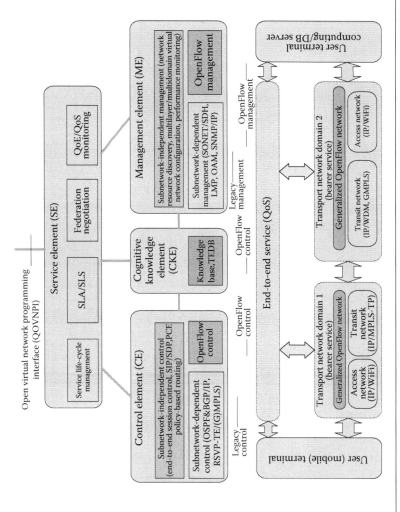

**Figure 12.7** QNox for SDN service provisioning with GOF.

The CE handles the end-to-end session control with path establishments on each transport network for connection-oriented services and responds to the flow table updates using CEFE interactions along the route of the GOF. The CE includes the session initiation protocol (SIP)/session description protocol (SDP) module for end-to-end QoS provisioning of real-time multimedia conversational services, such as voice over IP (VoIP) and multimedia conference call. To provide flexibility of interworking between GOFN and legacy networks (such as IP/SONET, IP/WiFi, and IP/generalized multi-protocol label switching (GMPLS)), the control plane is composed of transport network-independent control (including SIP/SDP, path computation element [PCE], and policy-based routing) and transport network-dependent control (including BGP, OSPF-ISIS BGP/IP, and RSVP-TE/MPLS).

The ME performs network resource discovery, multilayer/multidomain QoS-aware virtual overlay networking, and virtual network topology managements. The ME is also composed of the subnetwork-independent management function, the subnetwork-dependent management function (such as SNMP for IP network, LMP/OAM for ASON, and TMN for SONET/SDH transmission network), and the GOF management function for GOF-compatible GMPLS switch nodes. The ME also provides network QoS performance monitoring.

The CKE maintains a link state database (LSDB) and a traffic engineering database (TEDB) of transport network topology, and provides the decision making of mapping a virtual network topology for the SDN user's requested virtual topology onto the physical transport network topology. The CKE also supports traffic engineering for QoS-guaranteed service provisioning and network load balancing.

From the results of systematic experiments, the proposed QNox showed very good performance. The QNox provides network resource discovery in less than 1 s, route calculation for a network of 100 FEs with 4 to 10 links in less than 100 ms, and fault notification and fault restoration in less than 60 ms [6].

*Toward QoE-Driven Multimedia Service Negotiation, Path Optimization, and End-to-End Header Compression*

Kassler et al. [7] proposed a system for a quality-of-experience (QoE)–driven path assignment for multimedia services. The system aims to enable a negotiation of service and network communication parameters between end users and to assign the network paths. Actually, the solution is a centralized multiuser path assignments optimization, which maximizes QoE by considering service utility functions, network topology, link capacities, and delay. The proposed architecture includes the QoS matching and optimization function (QMOF) and the path assignment function (PAF). The QMOF works as a generic and reusable service capability, supporting optimized session delivery and controlled session adaptation. The role of PAF is to maintain a network topology database, which contains the network connectivity graph along with link capacities. They presented an example of path assignment for flows in a multimedia session to illustrate the proposed architecture [7].

Jivorasetkul and Shimamura [8] proposed a new header compression mechanism that can be deployed in end-to-end nodes using the SDN concept, instead of using hop-by-hop delivery. This mechanism can reduce both packet size and time delay to give better time performance, which is extremely useful in time-sensitive applications, such as VoIP, because high network latency can lead to voice quality problems. They proposed the new HC method in end-to-end nodes without requiring any compression/decompression cycles in intermediate nodes. Instead of transmitting the entire header, all subsequent packet headers are compressed and appended compressed ID by the compressor. Then, the compressed packets are forwarded by the forwarding entry in the OpenFlow switch and are decompressed based on the context information by the decompressor. In their experiments, end-to-end HC mechanism obtained the smallest round trip time compared with the hop-by-hop HC mechanism and the normal IP routing method.

## Discussion

QoS is an important issue in many applications, especially in streaming media, VoIP, video conferencing, and so on. Many experiments have been conducted to make OpenFlow support QoS. However, these designs are still under the testing phase and need to be examined further. Many designs are related to optimization problems, such as dynamic rerouting for SVC, dynamic QoS routing for multimedia flows, etc. Moreover, the solution needs heavy calculations in reality as the dimension increases. It also needs a comprehensive test before being applied to practical applications. Many experiments are actually under a small scale to test the proposed design, and no big-scale experiment is performed yet.

### Conclusion

In this chapter, all import improvements related to QoS in OpenFlow/SDN are summarized. Because of all the efforts of researchers, many useful designs have been proposed and tested. These designs show advantages and prospects, which could be added to the future OpenFlow/SDN. Many studies need to be done for the deep testing of proposed designs, such as the controller design, the dynamic routing algorithm, the QNox structure, and so on.

## References

1. Egilmez, H. E., B. Gorkemli, A. M. Tekalp, and S. Civanlar. 2011. Scalable video streaming over OpenFlow networks: An optimization framework for QoS routing. *2011 18th IEEE International Conference on Image Processing*, Brussels, Belgium, pp. 2241, 2244.
2. Ishimori, A., F. Farias, I. Furtado, E. Cerqueira, and A. Abelém. Automatic QoS management on OpenFlow software-defined networks. *IEEE Journal.* http://siti.ulusofona.pt/aigaion/index.php/publications/show/289.
3. Egilmez, H. E., S. T. Dane, K. T. Bagci, and A. M. Tekalp. OpenQoS: An OpenFlow controller design for multimedia delivery with end-to-end quality of service over software-defined networks. *2012 Asia-Pacific Signal & Information Processing Association Annual Summit and Conference (APSIPA ASC)*, Hollywood, CA, USA, pp. 1, 8.
4. Egilmez, H. E., S. Civanlar, and A. M. Tekalp. 2013. An optimization framework for QoS-enabled adaptive video streaming over OpenFlow networks. *IEEE Transactions on Multimedia*, vol. 15, no. 3, pp. 710, 715.
5. Sonkoly, B., A. Gulyás, F. Németh, J. Czentye, K. Kurucz, B. Novák, and G. Vaszkun. 2012. QoS support to Ofelia and OpenFlow. *2012 European Workshop on Software Defined Networking (EWSDN)*, Darmstadt, Germany, pp. 109, 113.
6. Jeong, K., J. Kim, and Y.-T. Kim. 2012. QoS-aware network operating system for software-defined networking with generalized OpenFlows. *2012 IEEE Network Operations and Management Symposium (NOMS)*, Maui, HI, USA, pp. 1167, 1174.

7. Kassler, A., L. Skorin-Kapov, O. Dobrijevic, M. Matijasevic, and P. Dely. Toward QoE-driven multimedia service negotiation and path optimization with software-defined networking. *2012 20th International Conference on Software, Telecommunications and Computer Networks (SoftCOM)*, Split, Croatia, pp. 1, 5.
8. Jivorasetkul, S. and M. Shimamura. 2012. End-to-end header compression over software-defined networks: A low-latency network architecture. *2012 4th International Conference on Intelligent Networking and Collaborative Systems (INCoS)*, Bucharest, Romania, pp. 493, 494.

# 13

# PROGRAMMABLE NETWORK TRAFFIC CLASSIFICATION WITH OPENFLOW EXTENSIONS

## SANPING LI, ERIC MURRAY, AND YAN LUO

### Contents

## Introduction

Network traffic classification refers to grouping network traffic into classes or categories on the basis of whether the traffic matches user-defined criteria. Identifying and categorizing traffic into application classes enables distinct handling for different types of traffic. It is the foundation for applying the appropriate quality-of-service (QoS) feature to that traffic [1,2], such as traffic shaping and traffic policing, enabling the allocation of network resources to deliver optimal performance for traffic. For example, high-priority traffic can be singled out for special handling and, thus, helps achieve peak application performance.

The research field of network traffic classification gains sustained attention while more and more applications emerge on the Internet; especially, the machine learning–based approach has been investigated intensively and extensively in recent years [3–5], primarily because of advantages such as that only packet header information is required and that sufficient classification accuracy is produced. It provides a promising alternative to identify traffic based on statistical flow features, such as total packet size and inter-arrival time, without raising privacy and security concerns, which is usually one of the major barriers for the practical deployment of traffic classification based on payload inspection. Flow features are essential parameters to traffic classification yet are difficult to obtain at line rate [6]. Moreover, the rapidly changing nature of diverse network applications leads to the ever-increasing volume and complexity of traffic on the Internet and, consequently, presents a high-performance challenge for traffic classification.

Because of the closed design of network devices and the ossification of network protocols, network architecture and management has not been flexible until OpenFlow brought the opportunity of software-defined networking (SDN) [7]. OpenFlow grows from a campus network experiment to a well-recognized specification accepted by a growing number of industrial partners. OpenFlow is based on an Ethernet switch, with an internal flow table and a standardized interface to add and remove flow entries. It is particularly useful in programmable flow analysis: an OpenFlow controller can update flow actions in an OpenFlow-capable device to extract a versatile set of flow features and apply policies on certain flows.

A major benefit of SDN is the ability to augment the functionality of network devices without compromising their integrity. Because the OpenFlow protocol is an open specification, researchers have the ability to extend the feature set of their OpenFlow devices to perform more complicated statistical analysis. Using the reference OpenFlow design as a starting point, the definition of the flow table can be expanded to include more information than just the header fields. In addition, the controller can be customized to collect flow features in addition to its main function of providing rules to its switches. Therefore, it is possible to demonstrate different traffic classification methods on top of the OpenFlow.

We propose to leverage the OpenFlow protocol to handle changing characteristics in traffic dynamics and enable programmable feature extraction. The OpenFlow controller can serve as an engine to fuse flow features and instruct the OpenFlow switches to collect more or less feature data because the machine learning model varies in response to concept drift, for instance, the changes in the characteristics of network traffic. On the OpenFlow switch, new feature extraction actions can then be dynamically enabled or introduced to provide the required input data to the traffic classification module.

We make the following several contributions in this work:

- First, we propose to obtain the changing characteristics of specific flows by dynamic feature extraction to handle concept drift in network traffic. The adequate amount of features can be requested to feed the machine learning classifiers.
- Second, an extension to OpenFlow protocol, named feature actions, is proposed to perform programmable flow feature extraction. To the best of our knowledge, this is the first protocol extension to OpenFlow for the purpose of online traffic classification.
- Third, we implement a prototype of OpenFlow with extensions to classify network traffic with network processor acceleration hardware. We perform comprehensive experiments to study and compare the performance of several design options.

This chapter is organized as follows: "Background and Motivation" describes the background and motivation. "Programmable Traffic Classification with Flow Features" introduces the discernibility of flow features and machine learning models. "OpenFlow Extensions for Flow Feature Extraction" elaborates the proposed extension to the OpenFlow protocol. "The Design" describes the design in detail. The performance results are presented in "Performance Evaluation." We discuss related works in "Related Work." This chapter ends with the "Conclusion."

## Background and Motivation

### Traffic Classification Methods

The research community has performed a plethora of work devoted to network traffic classification and proposed a variety of methods and tools. Next, we present a brief overview of existing traffic classification approaches, including port number based, payload based, and flow features based.

Historically, network traffic was classified based on source and destination port numbers, typically registered with the Internet Assigned Numbers Authority (IANA) to represent well-known applications. Although this approach is very simple and requires very little information that is acquired easily from packet headers, it is highly unreliable in the recent network environment. For example, most peer-to-peer (P2P) applications hide their identity by assigning ports dynamically and/or masquerading as well-known ports [8] to evade traffic filters and legal implications.

Payload-based classification is another well-researched approach [9,10], identifying traffic by searching packet payloads for signatures of known applications. It has potentially provided highly accurate results, given a complete set of payload signatures. Its drawbacks, however, are obvious. Besides the high effort of keeping the set of signatures updated, payload analysis is unable to cope with encrypted transmissions or increasingly high line rates. This approach relies on network traces, including packet payload that poses privacy and security concerns.

The flow feature-based approach has been substantially investigated for traffic analysis [11] and traffic classification with the help of machine learning algorithms [2,5]. It provides a promising alternative method based on application signature–independent statistical flow features. One of the major advantages is that only the packet header information is required to identify application categories for flows, without inspecting the packet payload. The packet header information is used to build flow feature statistics, which will be fed into machine learning classifiers to identify application types.

The increased capacity and availability provided by broadband connections has led to more complex behaviors arising from various emerging applications and services, inevitably imposing restrictions on its practical deployment. However, traditional classifiers constructed from an entire training data set with batch learning algorithms do not have the ability to adapt ever-changing application behavior presented in oncoming traffic stream. Just applying a static classifier to traffic with changing characteristics will only produce a growing number of identification errors. These changing data characteristics introduce the issue of concept drift in network traffic.

*Concept Drift*

Concept drift in machine learning means that the statistical property of a target variable changes over time in unforeseen ways, which would result in less accurate predictions produced as time passes. Most machine learning algorithms assume that a training data set is a random sample drawn from a stationary data distribution [12]. Constant feature collection is used to generate fixed decision rules, which will always be applied in classification tasks, but is not concerned about the emergence of new data characteristics different from the training data set. In this case, classifiers constructed from the entire training data set with batch learning algorithms do not have the capability to adapt the concept drifting presented in the oncoming data stream [12].

The issue of concept drift also appears in network traffic classification. Changing spatial-temporal characteristics is the inherent characteristic of large-scale network traffic. Because of ever-changing user behavioral patterns and continuously growing types of network applications, variant statistics certainly emerge in traffic data [2,13]. In the traditional machine learning–based traffic classification methods, statistics with a constant feature set is continuously fed into a fixed pretrained classifier. Using the statistics with all the features that can be obtained will necessarily increase the overhead of the classifier, whereas the reduction determined by feature selection techniques does not guarantee to capture the characteristics of concept drifting hidden in oncoming network traffic. To prevent deterioration in accuracy over time, the traffic classifier is required to be adaptable to concept drift; meanwhile, to minimize overhead of classification tasks, feature extraction is required to be tunable and programmable.

The operational deployment of the network traffic classification system has to search the solution to these problems to improve the portability and suitability of the classification model. However, there are very few studies currently available in the literature, which considers the programmability of the flow feature extraction required to represent the characteristics of concept drifting in network traffic. The OpenFlow protocol [7], as an example of SDN technology, provides a mechanism of enabling flow-based network programmability via separating the forwarding path of packets through switches from the external software controller. Thus, the programmable feature extraction can be enabled with the OpenFlow protocol.

*OpenFlow and SDN*

The SDN is an emerging network architecture, where network control is decoupled from network topology and is directly programmable. Network intelligence and state are logically centralized, and the underlying infrastructure is abstracted from applications and network services. As a result, campus and enterprises gain unprecedented programmability and network control, enabling them to build highly scalable, flexible networks that readily adapt to changing needs [14].

OpenFlow is an enabler of SDN. The essence of OpenFlow switching is the decoupling of packet switching and flow-table management. The centralized flow-table management at controllers and distributed flow tables embedded at switches make network management more flexible and less cumbersome for enterprise level networks. A 10-tuple defines a flow in OpenFlow devices, and a flow-table entry contains a 10-tuple definition and associated actions. There can be multiple switches and multiple controllers in an OpenFlow-enabled network. The controllers learn the network topology and serve as central network management points. Switches communicate with the controllers via the OpenFlow protocol. Each switch has a local flow table, which is used to match against incoming packets. A switch queries controllers for the actions on unknown flows.

## Programmable Traffic Classification with Flow Features

### Variant Discernibility in Features

The features used to describe the properties and the characteristics of an object are of fundamental importance for building classification models and performing classification tasks with machine learning algorithms. On one hand, we try to collect as many features as possible to fully describe the properties of an object. On the other hand, we have to select the most important features with respect to the specific classification model to accurately express the characteristics of an object, meanwhile reducing the resource consumption of classification tasks.

In network traffic classification, the objective of flow feature extraction is to collect adequate information with minimal resource consumption to ensure the accuracy and the efficiency of classification tasks. Flow features are extracted from the header information acquired from all or sampled $n$ packets captured belonging to a specific flow, excluding the payload part to avoid keeping individuals' communication under surveillance. Machine learning algorithms make use of flow features to generate characteristic models or decision rules in terms of network application category.

We are able to obtain a large number of flow features, for example, Ref. [15] listed 248 features for use in flow-based classification. However, a considerable part of the features is redundant, even noisy, for classification [16]. Generally, only a minor part of the features will be involved in the decision rules even without the use of the feature selection technique. Therefore, only these involved features need to be extracted to identify the application categories for oncoming traffic.

When concept drift is occurring, however, the decision rules generated with the original feature set are no longer able to accurately represent the changing characteristics of the specific application category. Thus, we need to extract new features and collect new statistics to adapt the changing concept to maintain the identification capability of the classification model. For example, when data set *entry03* [17] is used as the training data set to build the classifier with the decision tree method C4.5, the application category *P2P* can be identified with the decision tree branches shown in Figure 13.1. The set of features involved in the tree branches is {*attr_4*, *attr_8*, *attr_21*, *attr_120*, *attr_127*, *attr_172*, *attr_176*}, where the feature definitions [15] are listed in Table 13.1.

These classification rules, however, cannot almost completely recognize the *P2P* data in the data set *Day3*, which is captured at a different site from the *entry03* data set 3 years later [17]; the classification result is only 0.004 for precision and 0.001 for recall. In general, the measure of evaluating the classification performance and precision for a class is the number of true positives (i.e., the number of items correctly classified as belonging to the positive class) divided by the total number of elements classified as belonging to the positive class, whereas recall is defined as the number of true positives divided by the total number of elements actually belonging to the positive class. We observe that the characteristics of *P2P* in *Day3* cannot be identified with the previous feature set. Thus, the new feature set and new decision rules, as shown in Figure 13.2, are needed to be used to correctly

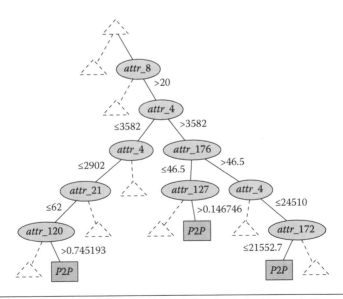

**Figure 13.1**   Branches of identifying *P2P* in the decision tree trained from *entry03*.

**Table 13.1**    Definition of Features

| FEATURE | DEFINITION |
|---------|------------|
| *attr_4* | Port number at server |
| *attr_8* | Port number at client |
| *attr_21* | Minimum of bytes in the (Ethernet) packet, using the size of the packet on the wire |
| *attr_27* | Variance of bytes in the (Ethernet) packet |
| *attr_71* | Count of all the packets seen with the PUSH bit set in the TCP header (client → server) |
| *attr_95* | Minimum segment size observed during the lifetime of the connection (client → server) |
| *attr_107* | Total number of bytes sent in the initial window, i.e., the number of bytes seen in the initial flight of data before receiving the first ACK packet from the other endpoint. Note that the ACK packet from the other endpoint is the first *ACK* acknowledging some data (the ACK's part of the three-way handshake do not count), and any retransmitted packet in this stage are excluded (client → server). |
| *attr_120* | Total data transmit time, calculated as the difference between the times of the capture of the first and the last packets carrying nonzero TCP data payload (server → client) |
| *attr_127* | Minimum RTT sample seen (client → server) |
| *attr_139* | Minimum full-size RTT sample (client → server) |
| *attr_172* | Variance of bytes in the (Ethernet) packet |
| *attr_176* | Mean of the total bytes in the IP packet |

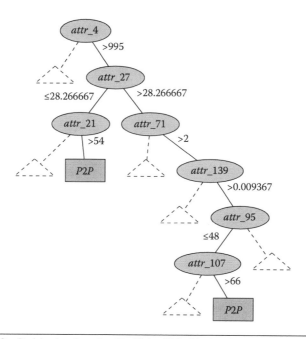

**Figure 13.2**    Decision tree branches identifying *P2P* with changing characteristics in the data set *Day3*.

classify the *P2P* in *Day3*. The set of features contained in the new tree branches is {*attr_4*, *attr_21*, *attr_27*, *attr_71*, *attr_95*, *attr_107*, *attr_139*}. The new rules adapting to *P2P* with concept drift in *Day3* are generated with an incremental learning algorithm.

*Traffic Classification with Features Feedback*

The approach of machine learning–based traffic classification provides a promising alternative based on flow feature statistics. Because of ever-changing user behavioral patterns and continuously growing types of network applications, variant statistics certainly emerge in traffic data [13]. Static classification models based on batch learning algorithm, however, do not have the ability to adapt the drifting concepts presented in oncoming network traffic. Thus, the classifier derived from the original training data set has to be updated over time to adapt to the changing characteristics in the specific applications to maintain the accuracy performance of traffic classification.

The decision tree–based incremental learning algorithm is a common and effective approach to address concept drifting by keeping the decision tree up-to-date. The concept-adapting very fast decision tree learner (CVFDT) system [12] is the most representative in this type of approach. It adapts to changing concepts with a sliding window of samples. When an old sample seems to be out-of-date, and the new one becomes more accurate, the incremental learning approach replaces the old one and grows an alternate subtree.

The process of network traffic classification based on machine learning algorithms is shown in Figure 13.3. In network traffic classification, the incremental learning algorithm generates the new classification rules to correctly identify the application types with changing characteristics. The new classification rules correspond to the new collection of features involved in the classification model. In this work, the new feature collection is fed back to the process of feature extraction from the classification model. Thus, the feature extraction is able to adapt to the changes in network applications and provide the input flow feature statistics for the classification task with minimal resource consumption, avoiding the waste of resources in extracting features that are useless for the identification of application types.

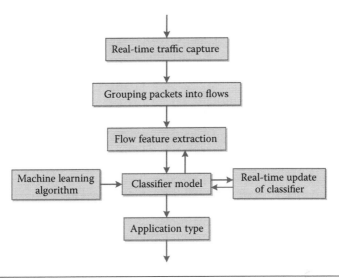

**Figure 13.3**    Network traffic classification based on machine learning methods.

### OpenFlow Extensions for Flow Feature Extraction

The proposed OpenFlow extension is to perform flow feature extraction to meet the specific requirement of application-aware traffic classification. The architecture of OpenFlow-enabled extensions for flow feature extraction is shown in Figure 13.4.

The machine learning model, running in a multicore environment, receives flow feature statistics from the OpenFlow controller to identify the application types for traffic. When the classification model is updated in real time, the changed feature set corresponding to certain application types, which resulted from incrementally updated classification rules, is fed back to the OpenFlow controller. The structure of these two OpenFlow extension messages is shown in Figure 13.5. The OpenFlow Feature eXtraction (OFX) feature statistics consists of one OpenFlow protocol header, one OpenFlow protocol flow match, one OpenFlow match field, and OFX action definitions of variable number; each OFX header defines the action type and length for the followed OFX action field. The message of the OFX feature feedback includes one OpenFlow protocol match and nine flag bits indicating what feature actions are required to be used for classification.

The controller has the primary function of assigning feature actions to flows, with a second function of queuing flow features for processing

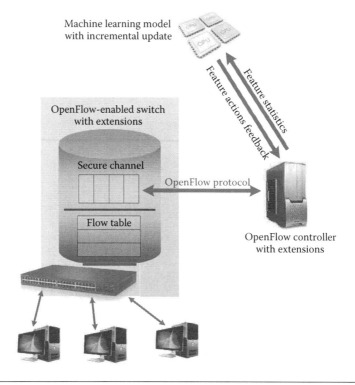

**Figure 13.4** OpenFlow extensions for flow feature extraction.

by the classification model. The job of the OpenFlow-enabled switch is to receive traffic arriving at specified interfaces and to execute feature actions on packets based on a flow table managed by the controller. Extensions have been made so that the switch collects the features defined with feature actions for each flow in the flow table and encapsulates the flow features to be sent to the controller.

To better use the system resource when extracting the flow features required for traffic classification, we define a set of actions, each of which can be performed upon a flow, to obtain one or more flow features. The definition of feature-related actions is determined by the group of network application categories and the features required by the corresponding decision rules. For example, we have a total of 55 features, as shown in Table 13.2, needed to be extracted for the classification procedure. We divide these flow features into nine actions, and each action contains some related flow features that are based on a single type of property and can be calculated by one scan through

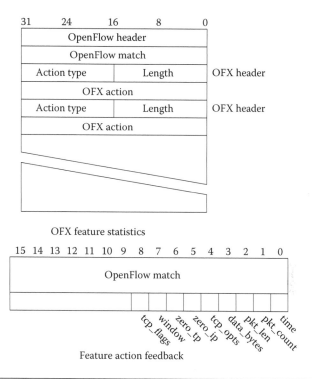

**Figure 13.5** Structure of OFX messages.

packets belonging to a certain flow entry. For example, the time action contains 12 flow features related with the inter-arrival time of packets, namely flow duration, minimum and maximum inter-arrival time, and the time interval between the $i$th packet and its previous packet ($2 \leq i \leq n$). These features can be calculated with the time stamp of packets.

The OpenFlow extension service relies on machine learning–based traffic classification model to define and update the features required for each flow entry in the flow table. With feature selection and incremental learning, variant feature sets corresponding to decision rules will be assigned for different flow entries. After being initialized or updated, the traffic classifier uses the latest classification rules and minimal feature collections to identify the incoming flows. The multicore processor executes the machine learning models and updates feature collections with actions necessary to monitor changing application types.

**Table 13.2** Features Contained in Feature Actions

| ACTIONS | NO. OF FEATURES | CONTAINED FEATURES |
|---|---|---|
| time | 12 | Flow duration |
| | | Minimum and maximum inter-arrival time |
| | | Time interval between the $i$th packet and the previous one ($2 \leq i \leq n$) |
| pkt_count | 1 | Number of packets |
| pkt_len | 13 | Total; the minimum and maximum packet size |
| | | Size of the $i$th packet ($1 \leq i \leq n$) |
| data_bytes | 13 | Total; the minimum and maximum data bytes excluding the transmission control protocol (TCP)/user datagram protocol (UDP) header |
| | | Data bytes of the $i$th packet ($1 \leq i \leq n$) |
| tcp_opts | 2 | The number of packets with nonzero option in the TCP header |
| | | The size of the option section in the TCP header |
| zero_ip | 1 | The number of packets without the IP data section |
| zero_tp | 2 | The number of packets without the transport layer data section |
| window | 3 | Total; the minimum and maximum size of the receive window for the TCP packet |
| tcp_flags | 8 | The number of TCP packets with specific flag bit, such as CWR, ECE, URG, ACK, PSH, RSY, SYN, FIN |

## The Design

We implemented the proposed OpenFlow extensions in a prototype supporting programmable traffic classification based on flow features rather than packet payloads. The design is described in this section.

### Top-Level Architecture

The high-level system architecture of OpenFlow-enabled network traffic classification, as shown in Figure 13.6, consists of two major components: the controller and the switch. The controller is composed of the machine learning module, with the capability of detecting concept drift, coupled with the OpenFlow controller with extensions. The switch is an OpenFlow switch with extensions, implemented with one of the two processor technologies: a software design on a ×86 multicore processor and a hardware-accelerated design on a multicore multithreaded network processor.

*Controller* The machine learning model, which uses extracted flow features to classify network traffic, is integrated with the OpenFlow

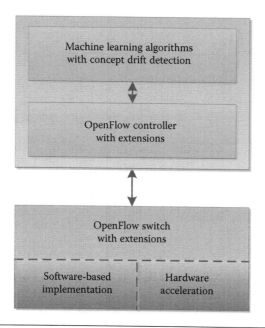

**Figure 13.6** Top-level design.

reference controller software as a separate thread. The OpenFlow controller has the primary function of assigning actions (such as egress port forwarding rules) to flows at an OpenFlow switch, with a secondary function of collecting network information from the switch. This process of collecting statistics has been expanded only from packets and octet counts to the advanced flow features (e.g., packet interarrival time) used by the machine learning models. In other words, it is the OpenFlow controller that instructs an OpenFlow switch to collect specific flow features on particular flows.

*Switch* While retaining the functionality of generic OpenFlow packet switching, the OpenFlow switch in our design is extended to perform feature extraction actions. Such a switch is built on top of the open-source OpenFlow reference design [18]. Our modified version of the OpenFlow reference switch receives network traffic arriving at specified interfaces and executes actions on those packets based on a flow table managed by the controller. Modifications have been made so that the switch collects the features of each flow in its flow table. After a flow has timed out (packets matching that flow have stopped arriving at the switch), those flow features will be encapsulated and sent to the controller. Compared with a generic OpenFlow switch, ours performs additional work to collect advanced flow feature statistics. As a result, our software version of the OpenFlow switch for traffic classification is subject to degraded forwarding performance if it does not have any special acceleration mechanism. To address this speed disadvantage, we are motivated to optimize the design by allowing the switch to receive packets and obtain flow features via hardware acceleration such as a network processor. The network processor implements its own flow table, creating a fast path for network traffic. Next, we make some comparisons of the software-based switch and the switch with hardware accelerations.

The behavior of the switch software without any hardware acceleration is demonstrated in Figure 13.7.* When a packet arrives at a switch interface, it will be forwarded up to the OpenFlow application, and its header information will be compared with the entries in the flow table. For the first packet of a flow, there will be no entry, and the packet will be forwarded out to an OpenFlow controller. The controller will then instruct the switch to create a new flow-table entry as well as specify the actions to execute when a match is found. As subsequent packets of that flow arrive, a match will be found in the flow table, and the specified action will be taken.

To run the switch software, a list of (physical or virtual) interfaces must be provided on the command line. The software will open up a raw socket on that interface to receive a copy of each packet that arrives. This process of copying every packet up to the user space and then performing different lookup operations adds significant network latency, which is why network processor acceleration was added to the design.

The performance of an OpenFlow switch can be substantially increased through the use of acceleration hardware, such as a NetFPGA or a network processor. The behavior of the switch with a network processor is given in Figure 13.8. The main flow table is stored in the network processor's memory space, allowing for faster access. When the first packet of a flow arrives, the procedure for forwarding the packet up to the host application and out to a controller is the same as the nonaccelerated case. However, once a flow entry

---

* We use the reference user space OpenFlow switch in this work.

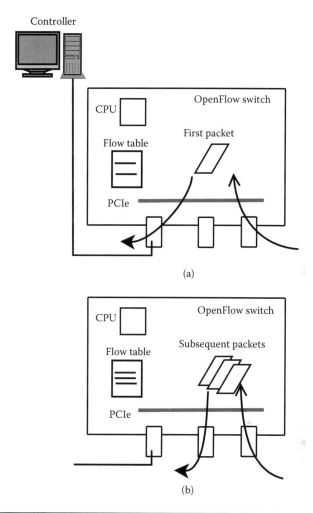

**Figure 13.7** OpenFlow switch software behavior. (a) The processing of the first packet of an unknown flow. (b) The processing packets of a known flow. Peripheral component interconnect express (PCIe).

has been added to the hardware's flow table, packets can be processed immediately without any expensive host memory copies.

*Software Design*

The OpenFlow reference switch (either the software version or the hardware acceleration) needs to be modified to perform the flow feature extraction for the traffic classifier and to interface with the network flow processor (NFP). This requires an entire new subsystem

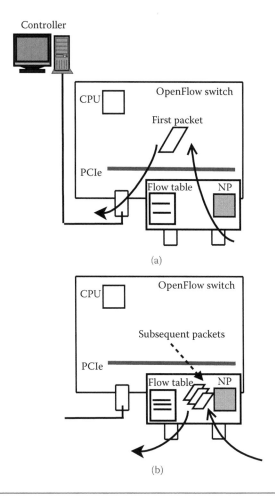

**Figure 13.8** OpenFlow switch behavior with an NP acceleration card. (a) The processing of the first packet of an unknown flow. (b) The processing of packets of a known flow. Network processors (NP).

to be added to the existing code while preserving the original functionality. This flow feature extraction system is given the name OFX for OpenFlow Feature eXtraction. We follow three steps. The first step is to extend the OpenFlow flow table so that each entry contained the flow features described in "OpenFlow Extensions for Flow Feature Extraction," in addition to its normal match and action structures. Step 2 is to extend the capabilities of the channel between the OpenFlow controller and the switch so that the controller can request the collection of specific flow features and the switch is able to send those statistics to the controller. The third major modification to the reference OpenFlow switch is an interface with the NFP network processor card—the NFP bridge. The NFP bridge is a software module that allows the switch to send and receive packets on the 10-GB interfaces of the Netronome NFP-3240 network processor card [19].

The traffic classification system is added into the OpenFlow controller's code as an independent thread. The work involves the establishment of a communication mechanism between the machine learning classification thread and the reference OpenFlow controller. This communication is bidirectional, as mentioned in "OpenFlow Extensions for Flow Feature Extraction." The traffic classifier needs to receive the flow information and the statistics that have been collected for that flow (if the flow already exists). After processing that information, the classifier may request that more or less information about a flow is collected.

*Acceleration with Multicore Multithreaded Network Processor*

A multicore network processor with hardware multithreading, Netronome NFP-3240, is used to accelerate the OpenFlow switching and flow feature extraction. NFP-3240 consists of 40 cores called microengines running at 1.2 GHz, each of which can be programmed with a stand-alone executable code. These cores are typically programmed to collaborate on packet receiving, processing, and transmitting. NFP-3240 is equipped with acceleration engines for crypto, hash, queue operations, and dual memory channels for off-chip static random-access memory (SRAM) and dynamic random-access memory (DRAM) units.

**Algorithm 1:** OpenFlow switching on NP

1  **begin**
2      extract 12-tuple from the received packet;
3      calculate the hash value (h/w based);
4      extract lower 20 bits *offset*;
5      *flow_entry_addr* =
6          *flow_table_base* + *offset* × *flow_ entry_size*;
7      read DRAM for flow-entry info;
8      compare if the flow entry matches the packet;
9      **if** *match* **then**

```
10        extract action;
11        perform action on the packet;
12   else
13        create a new flow-table entry with default action;
14        send packet to host via NFD message queue;
15   end
16   perform flow feature extraction;
17   update flow entry with new flow features;
18 end
```

Our NP acceleration code is built on top of a sample wire application that receives packets on one port and forwards them out via another port. This code was expanded to maintain an OpenFlow flow table and perform flow feature extraction. Each flow-table entry contains the flow features that are updated upon the arrival of each packet belonging to that flow.

In the packet processing microengines, we implement the OpenFlow switching by maintaining a flow table in the DRAM. More specifically, we organize the flow entries using a hash table. We calculate the hash value from the 12-tuple OpenFlow fields and use the hash value as an index to access the flow-table entry. Algorithm alg.of.on.np describes the algorithm executed on the NP side in more detail.

The algorithm on the host is similar to the original OpenFlow switching software, with the addition of flow feature extraction. Essentially, the host software will receive the packets (in fact, only the start-of-flow packets) handed over by the NP via the PCIe bus. The host software will ask the controller about the actions on this flow. After receiving the response, the host software will add/update the flow entry to its own flow table. The NP automatically creates a new flow-table entry for each packet that does not match an existing flow. However, the controller is capable of overriding this entry with new actions if desired.

The NFP OpenFlow and flow feature extraction operations are composed of several different microengine programs accessing common memory spaces. There are three major applications running simultaneously, which are called the OpenFlow application, the OFX action application, and the OFX timeout application. The use of the microengines of the NFP is given in Figure 13.9.

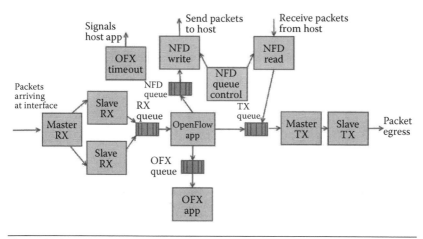

**Figure 13.9**   NFP OFX architecture.

Five microengines (Master RX, Slave RXes, Master TX, and Slave TX) are used to receive and send packets on the Ethernet interfaces. Netronome provides the software for these microengines. On the receive side, packets are assembled from the media switch fabric and stored in the DRAM. A handle to this packet is then placed on the RX queue. On the transmit side, packet handles are consumed from the TX queue so that each packet can be read from the DRAM and transmitted out.

Three microengines (OpenFlow App, OFX App, and OFX Timeout) are specifically programmed for OpenFlow flow-table processing and flow feature extraction. The OpenFlow App receives a handle of each incoming packet and then reads in the packet header and performs a flow-table lookup. If a match is found, the handle is forwarded to the TX queue. If no match is found, then the software needs to be aware of its arrival, and the handle is pushed onto the NFD queue. Next, the microengine creates a new flow-table entry automatically. In both cases, a copy of the packet header information will be pushed onto the OFX queue so that the OFX App microengine can perform the necessary statistics collection. The statistics collected by the OFX App are written back into the corresponding flow entry of the packet being processed.

The OFX Timeout microengine inspects the timestamps of every entry in the flow table to determine which flows have timed out. A flow has timed out when no matching packets have been observed for more than the predefined timeout threshold. The Timeout microengine sends the list of timed-out flows up to the host. Next, the host will read in the flow features for all timed-out flows, send that information to the controller, and delete the flow-table entry.

The NFD-related microengine code is Netronome's proprietary system for copying packets to and from the host memory over the PCIe bus. The host application uses the NFD application programming interfaces (APIs) to access the packets sent from the NFD Write microengine and to send packets out to the NFD Read microengine. The NFD Queue Controller microengine is responsible for signaling the other two engines and for preserving transaction ordering.

### Performance Evaluation

*Experiment Setup*

The experiments involve the use of three machines: one machine generating/sending traffic, one processing the traffic using the traffic classification system, and one sinking the packets and measuring the performance metrics. For the software-only tests, a 1-Gbps network interface card (NIC) is used. On the other hand, the NFP uses 10-Gbps interfaces. However, this would only be a factor if the software solution was able to perform beyond the 1-Gbps line rate (and, in fact, it is not).

In the experiments, two main performance metrics are latency and throughput. It is important not only to determine what the absolute capabilities of the NFP-accelerated traffic classification system were, but also to examine the improvement over the software-based solution. Therefore, two separate sets of tests are performed, both of which use the same machines in the same layout, the only difference being the presence of an NFP-3240 instead of a generic 1-Gbps NIC.

We start with the performance tests of the OpenFlow switch (both software and the NP version of it). These tests present us a baseline performance to investigate the overheads of flow feature extraction and traffic classification, which are implemented on top of the OpenFlow controller and switch code base. We report and analyze their throughput and latency results. We also present the execution time and accuracy of traffic classification operations.

*OpenFlow Experiment Results*

*Latency*  The latency performance is measured by using the Unix ping command. The first packet processed is expected to have a much larger round-trip time (RTT) than the subsequent packets of that flow. This is because extra work is performed to add this new flow entry into the flow table. Once the flow has been successfully added, the RTT is expected to be consistent until the flow has timed out and the flow entry has been removed from the table.

For the system using the regular NICs, different average RTTs are observed for different packet transmission rates, as shown in Figure 13.10. When sending one packet per second, the first packet takes approximately 10.3 ms to arrive, whereas subsequent packets show an average of 0.53 ms. However, when the transmission rate is increased to four packets per second, the average RTT is decreased to 0.354 ms. The improvement is likely caused by the scheduling in the user space OpenFlow application. When the packet arrives at the host, it is queued until the user space application resumes executions and then is processed. It is possible that the speed-up is a result of the application dequeuing and the processing of multiple packets within a single time slice.

With the presence of NFP on the NIC, a significant latency improvement is observed. The RTT of the first packet is approximately

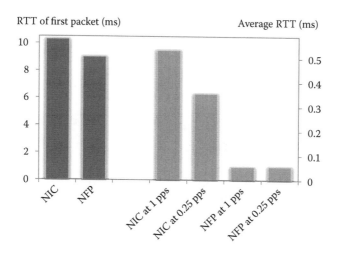

**Figure 13.10**  Latency results.

9 ms, just 1 ms faster than the software-only case. The rest of the packets have an average RTT of 0.055 ms. The latency is decreased by a factor of 10. Furthermore, the average RTT does not change as the transmission rate is adjusted. The reason for such a significant improvement is caused by the fact that the NFP-based solution does not have to perform multiple copy operations because copying a packet from the kernel space to the user space is itself an expensive operation. A packet arriving on the NFP is copied into the card's DRAM only once and then sent back out on the wire. Furthermore, a network processor's resources are dedicated to packet forwarding, whereas Linux applications on the host compete for resources with other processes.

*Throughput*   To measure the throughput, packets of a specific size is sent across the system. All packets being sent through belong to the same flow. Originally, only one application, Iperf, is used to generate and measure the throughput. Iperf is a performance measurement tool that runs as a client-side and server-side user space application [20]. However, when trying to measure the bandwidth of the system with NFP acceleration, Iperf could not generate enough packets to exhaust the system's resources. As an alternative to Iperf, two different tools are used. The Linux kernel's *pktgen* module [21] is used to generate packets. *Pktgen* is able to generate packets at a much higher rate. As a kernel module, *pktgen* receives a higher scheduling priority than user space applications and does not have to go through the time-consuming process of copying data between the kernel space and the user space. On the packet sink machine, *tcpstat* [22] is used to measure the throughput.

For each packet size, the NFP-based system exhibits a significantly higher throughput as expected. The throughput with an NFP-based system acquired an improvement of at least 10 times in comparison with the throughput with the software-based system, as illustrated in Figure 13.11.

These experimental results demonstrate that network processors can substantially improve the performance of complex network applications. By moving most of the packet processing from general-purpose cores to dedicated network processor cores, time-consuming procedures such as flow feature extraction can possibly be performed close to line rates.

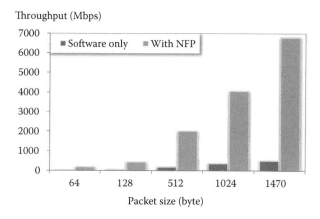

**Figure 13.11**   Throughput comparison.

*Traffic Classification Performance*

One experiment is performed to see how the system could handle many different flows in real network traffic. The packet traces used in the experiments are obtained from the 2009 Inter-Service Academy Cyber Defense Exercise data sets [23] captured on April 21, 2009, containing many diverse flows and traffic types [24]. We choose two trace files, as listed in Table 13.3, to be used in the experiments. The packet data contained in the trace file are sent out to the traffic classification system through the *tcpreplay* application [25]. *Tcpreplay* takes a trace file and exports its packets through a given host interface. It also allows configuration of the speed at which the packets are sent out. To test the classification system, the maximum transmission rate under 1-Gb and 10-Gb Ethernet links is selected. In addition, the software-based flow feature extraction with multithreaded pipelined implementation, developed in our previous work [6], is tested for performance comparison.

**Table 13.3**   Traffic Trace

| TRACE | dmp.113 | dmp.114 |
|---|---|---|
| Capturing time, April 21, 2009 | | |
| Size (MB) | 953 | 480 |
| Number of packets | 1,276,181 | 1,195,894 |

*Throughput*   The throughput comparison of various software implementations under the 1-Gb Ethernet link is shown in Figure 13.12. Because too much control traffic created by the OpenFlow, flow setup, and flow statistics pulling competes for the CPU cycles, the throughput of software-based original OpenFlow and OpenFlow extension is far worse than the multithreaded pipelined implementation, which does not perform any OpenFlow operations. Similar experimental observations of excessive overheads imposed by the software implementation of OpenFlow are also presented in Ref. [26].

Figure 13.13 shows the system performance under the 10-Gbps Ethernet link. Because *tcpreplay* is unable to exhaust the resources of the NFP-enabled system (up to 4 Gbps when sending packets using the trace file dmp.113), the only results obtained from this test are that the system is able to successfully process the packets in the trace file at the application's maximum transmission rate. Even so, we are able to observe that the throughput performance of the NFP-enabled system is significantly better than the software-based implementation.

*Classification Time*   The experimental results on the execution time of classification tasks are shown in Figure 13.14. For classification tasks, the use of actions with feedback reduced the execution time to a certain extent than the whole action sets. We divided the execution

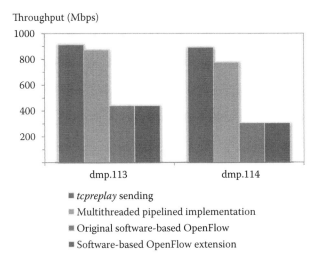

**Figure 13.12**   Throughput at 1-Gbps link.

Throughput (Mbps)

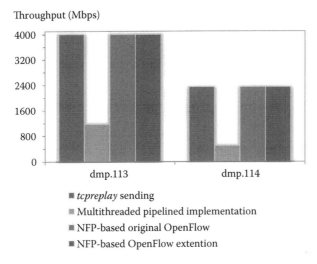

- ■ *tcpreplay* sending
- ■ Multithreaded pipelined implementation
- ■ NFP-based original OpenFlow
- ■ NFP-based OpenFlow extention

**Figure 13.13**   Throughput at 10-Gbps link.

time of classification tasks into two parts: the time of obtaining features and the identification time. Getting features in the classification thread consists of the processes of receiving feature action sets from the OpenFlow controller, and translating and passing them through the classifier. The identification part pertains to the process of identifying the application type for flows using the machine learning classifier. All action sets means that the switch always collects the data of all feature actions for each flow, whereas actions feedback means that the classifier sends an actions feedback for a flow to the controller

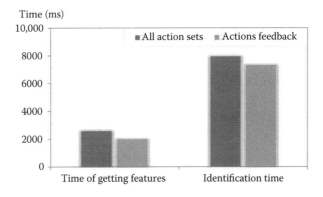

**Figure 13.14**   Execution time of classification tasks.

after performing classification tasks so that the switch only needs to collect features that are required by the classification task. The use of the actions feedback technology reduces the time of getting features by 22% and the identification time by approximately 8%, in comparison with using the whole feature actions of flows.

*Classification Accuracy* Applying the technology of the OpenFlow extensions would not lower the classification accuracy. The machine learning model with the capability of concept drift detection incrementally updates the classifier in real time during the classification process while it sends the modified feature actions definition for the specific flow back to the OpenFlow controller. The actions feedback ensures that the switch will not collect the features that are useless for identifying the application type. The classification model still gets all the feature values required to perform the classification operation; therefore, the accuracy performance would not be reduced. In certain circumstances, the application of the dynamic feature set could improve the classification accuracy slightly because of the reduction of the noise data contained in the entire feature collection, similar with the feature selection technique [16]. Figure 13.15 shows the accuracy results with the whole features set and the reduction set.

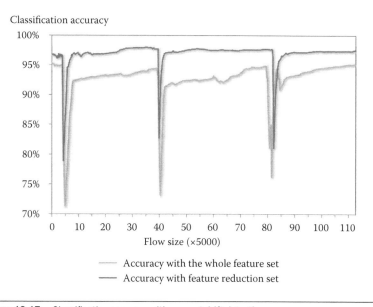

Figure 13.15   Classification accuracy with concept drift detection.

Note that we propose a new incremental learning algorithm in another work (under review) to handle the concept drift hidden in traffic dynamics. In this work, just one existing method, the CVFDT learner [12], is used in the classification experiments.

## Related Work

Network traffic classification gains continuous attention in recent years. The research community has studied and proposed several approaches to implement tasks of traffic classification [3,4] or traffic identification (of a specific application type, such as *P2P* [27,28]). Machine learning–based methods, in particular, have been extensively investigated because only packet header information is required and competitive recognition accuracy is yielded. A series of surveys [1,2,13] reviewed the progress of traffic classification [29–31] and presented performance comparison and evaluation of different methods. While improving in accuracy and efficiency, ever-emerging network applications and ever-changing application behaviors are among the reasons why traffic classification remains one of the open issues in research.

The flow feature–based methods for network traffic classification, traffic analysis, and intrusion detection have similarities in implementations, such as the process of feature extraction. Therefore, the OpenFlow protocol was also applied to these issues. An article [32] presented a flow feature–based distributed denial-of-service (DDoS) attack detection method, which was implemented over the OpenFlow-based network. However, in this work, just a fixed number of predetermined flow features were extracted with the OpenFlow switch. Moreover, we could not get the throughput performance of the proposed detection method, although the article mentioned that the experimental setup involved receiving data online.

Both traffic classification and intrusion detection systems need to analyze packet headers and even content at higher level protocols. It is also required that the function implemented by the system be updated with new procedures because of the evolving characteristics of network traffic and attacks. Because it requires high-performance processing capabilities and flexibility, this application is a good candidate to be implemented in a network processor. A work [33] proposed an architecture of the DDoS protection system on the IXP2400 network processor, allowing flow monitoring at a 2.4-Gbps rate.

In our previous work [6], we proposed multiple software-based designs of flow feature extraction, including serial, parallel, pipelined, and hybrid implementation, and the hybrid implementation, i.e. the multithreaded pipelined implementation, presented the best throughput performance. So, the hybrid feature extraction was selected to perform the throughput comparison in this work. Another previous work [34] presented a complementary design to OpenFlow's reference designs, applying network processor–based acceleration to perform OpenFlow switch.

## Conclusion

This article presents the design and performance studies of a programmable traffic classification system that accurately classifies traffic without affecting the latency or bandwidth of the network. We propose to leverage the OpenFlow protocol to implement an extension, named as feature actions, to enable programmable flow feature extraction and adapt concept drift in traffic dynamics. The OpenFlow controller serves as an engine to fuse flow features and instruct the OpenFlow switches to collect more or less feature data because the machine learning model varies in response to concept drift. On the OpenFlow switch, feature extraction actions can then be dynamically enabled or introduced to provide the required input data to the traffic classification model. Because of the excessive overheads imposed by the software implementation of OpenFlow, network processor acceleration is used to increase the system performance, with the implementation of OpenFlow and feature extraction applications in microengine programs.

The experimental results demonstrate that network processors can substantially improve the performance of complex network applications. By moving most of the packet processing from general-purpose cores to dedicated network processor cores, time-consuming procedures such as flow feature extraction can be performed at increasing line rates. Furthermore, the supported line rates for enterprise networks will continue to increase, and the dedicated packet processing hardware will become necessary for existing applications to scale to such rates.

# References

1. Callado, A., C. Kamienski, G. Szabo, B. Gero, J. Kelner, S. Fernandes, and D. Sadok. 2009. A survey on internet traffic identification. *Communications Surveys Tutorials (IEEE)* 11(3):37–52.

2. Nguyen, T. and G. Armitage. 2008. A survey of techniques for Internet traffic classification using machine learning. *Communications Surveys and Tutorials (IEEE)* 10(4):56–76.

3. Jin, Y., N. Duffield, J. Erman, P. Haffner, S. Sen, and Z.-L. Zhang. 2012. A modular machine learning system for flow-level traffic classification in large networks. *ACM Trans. Knowl. Discov. Data* 6(1):4:1–4:34, http://www.doi.acm.org/10.1145/2133360.2133364.

4. Jing, N., M. Yang, S. Cheng, Q. Dong, and H. Xiong. 2011. An efficient SVM-based method for multiclass network traffic classification. In *2011 IEEE 30th International Performance Computing and Communications Conference (IPCCC)*, pp. 1–8.

5. Zhang, J., Y. Xiang, Y. Wang, W. Zhou, Y. Xiang, and Y. Guan. 2012. Network traffic classification using correlation information. *IEEE Transactions on Parallel and Distributed Systems* 24(1):104–117.

6. Li, S. and Y. Luo. 2010. High-performance flow feature extraction with multicore processors. In *NAS*, pp. 193–201.

7. McKeown, N., T. Anderson, H. Balakrishnan, G. Parulkar, L. Peterson, J. Rexford, S. Shenker et al. 2008. Openflow: Enabling innovation in campus networks. *SIGCOMM Comput. Commun. Rev.* 38(2):69–74, http://www.doi.acm.org/10.1145/1355734.1355746.

8. Karagiannis, T., A. Broido, M. Faloutsos, and K. Claffy. 2004. Transport layer identification of *P2P* traffic. In *Proceedings of the 4th ACM SIGCOMM Conference on Internet Measurement (IMC '04)*, pp. 121–134.

9. Karagiannis, T., K. Papagiannaki, and M. Faloutsos. 2005. Blinc: Multi-level traffic classification in the dark. In *Proceedings of the 2005 Conference on Applications, Technologies, Architectures, and Protocols for Computer Communications (SIGCOMM '05)*, pp. 229–240.

10. Park, B.-C., Y. Won, M.-S. Kim, and J. Hong. 2008. Toward automated application signature generation for traffic identification. In *2008 IEEE Network Operations and Management Symposium (NOMS '08)*, pp. 160–167.

11. Li, W., M. Canini, A. W. Moore, and R. Bolla. 2009. Efficient application identification and the temporal and spatial stability of classification schema. *Comput. Netw.* 53:790–809, http://www.portal.acm.org/citation.cfm?id=1517860.1518209.

12. Hulten, G., L. Spencer, and P. Domingos. 2001. Mining time-changing data streams. In *Proceedings of the 7th ACM SIGKDD International Conference on Knowledge Discovery and Data Mining (KDD '01)*, pp. 97–106.

13. Dainotti, A., A. Pescape, and K. Claffy. 2012. Issues and future directions in traffic classification. *Network (IEEE)* 26(1):35–40.

14. Foundation, O. N. 2012. Software-defined networking: The new norm for networks. White Paper, http://www.opennetworking.org/images/stories/downloads/openflow/wp-sdn-newnorm.pdf.

15. Moore, A., M. Crogan, A. W. Moore, Q. Mary, D. Zuev, and M. L. Crogan. 2005. Discriminators for use in flow-based classification. Department of Computer Science, Queen Mary University of London, Technical Report RR-05-13.

16. Li, S. and Y. Luo. 2009. Discernibility analysis and accuracy improvement of machine learning algorithms for network intrusion detection. In *ICC (IEEE)*, pp. 1–5.

17. Moore, A. W. and D. Zuev. 2005. Internet traffic classification using Bayesian analysis techniques. In *Proceedings of the 2005 ACM SIGMETRICS International Conference on Measurement and Modeling of Computer Systems (SIGMETRICS '05)*, pp. 50–60, http://www.doi.acm.org/10.1145/1064212.1064220.

18. OF Consortium. http://www.openflowswitch.org/.

19. Netronome, Product Brief—NFP-32XX Network Flow Processor. 2008. http://www.netronome.com/.

20. Iperf. http://www.dast.nlanr.net/Projects/Iperf/.

21. pktgen. http://www.linuxfoundation.org/collaborate/workgroups/networking/pktgen.

22. tcpstat. http://www.frenchfries.net/paul/tcpstat/.

23. ITOC CDX 2009 Data Sets. http://www.itoc.usma.edu/research/dataset/.

24. Sangster, B., T. J. O'Connor, T. Cook, R. Fanelli, E. Dean, W. J. Adams, C. Morrell et al. 2009. Toward instrumenting network warfare competitions to generate labeled data sets. In *Proceedings of the 2nd Conference on Cyber Security Experimentation and Test (CSET '09)*, pp. 9–9, http://dl.acm.org/citation.cfm?id=1855481.1855490.

25. tcpreplay. http://tcpreplay.synfin.net/. [Online]. Available: http://tcpreplay.synfin.net/.

26. Curtis, A. R., J. C. Mogul, J. Tourrilhes, P. Yalagandula, P. Sharma, and S. Banerjee. 2011. Devoflow: Scaling flow management for high-performance networks. *SIGCOMM Comput. Commun. Rev.* 41(4):254–265, http://www.doi.acm.org/10.1145/2043164.2018466.

27. Kettig, O. and H. Kolbe. 2011. Monitoring the impact of *P2P* users on a broadband operator's network over time. *IEEE Transactions on Network and Service Management* 8(2):116–127.

28. Valenti, S. and D. Rossi. 2011. Identifying key features for *P2P* traffic classification. In *2011 IEEE International Conference on Communications (ICC)*, pp. 1–6.

29. Kim, H.-C., K. Claffy, M. Fomenkov, D. Barman, M. Faloutsos, and K. Lee. 2008. Internet traffic classification demystified: Myths, caveats, and the best practices. In 2008 *ACM CoNEXT*.

30. Soysal, M. and E. G. Schmidt. 2010. Machine learning algorithms for accurate flow-based network traffic classification: Evaluation and comparison. *Performance Evaluation* 67(6):451–467.

31. Williams, N., S. Zander, and G. Armitage. 2006. A preliminary performance comparison of five machine learning algorithms for practical IP traffic flow classification. *SIGCOMM Comput. Commun. Rev.* 36(5):5–16.

32. Braga, R., E. Mota, and A. Passito. 2010. Lightweight DDoS flooding attack detection using Nox/OpenFlow. In *2010 IEEE 35th Conference on Local Computer Networks (LCN)*, pp. 408–415.

33. Siradjev, D., Q. Ke, J. Park, and Y.-T. Kim. 2007. High-speed and flexible source-end DDoS protection system using IXP2400 network processor. In *Proceedings of the 7th IEEE International Conference on IP Operations and Management (IPOM '07)*, pp. 180–183, http://www.portal.acm.org/citation.cfm?id=1775321.1775343.

34. Luo, Y., P. Cascon, E. Murray, and J. Ortega. 2009. Accelerating OpenFlow switching with network processors. In *Proceedings of the 5th ACM/IEEE Symposium on Architectures for Networking and Communications Systems (ANCS '09)*, pp. 70–71, http://www.doi.acm.org/10.1145/1882486.1882504.

35. Haas, R., L. Kencl, A. Kind, B. Metzler, R. Pletka, M. Waldvogel, L. Frelechoux, P. Droz, and C. Jeffries. 2003. Creating advanced functions on network processors: Experience and perspectives. *Network (IEEE)* 17(4):46–54.

36. Zhao, L., Y. Luo, L. N. Bhuyan, and R. Iyer. 2006. A network processor-based content-aware switch. *IEEE Micro* 26(3):72–84.

37. Xinidis, K., K. Anagnostakis, and E. Markatos. 2005. *Design and Implementation of a High-Performance Network Intrusion Prevention System* 181:359–374.

38. Bos, H., L. Xu, K. Reeuwijk, M. Cristea, and K. Huang. 2005. Network intrusion prevention on the network card. In *IXA Education Summit*.

39. Luo, J., J. Pettit, M. Casado, J. Lockwood, and N. McKeown. 2007. Prototyping fast, simple, secure switches for ethane. In *IEEE Symposium on High Performance Interconnects*, pp. 73–82.

40. Naous, J., D. Erickson, G. A. Covington, G. Appenzeller, and N. McKeown. 2008. Implementing an OpenFlow switch on the NetFPGA platform. In *ACM Symposium on Architecture for Networking and Communications Systems*, pp. 1–9.

# PART IV
## ADVANCED TOPICS

# 14

# OpenFlow/SDN for Metro/Backbone Optical Networks

## LEI LIU, HONGXIANG GUO, AND TAKEHIRO TSURITANI

**Contents**

## Motivation

Optical networks are typically composed of nodes such as reconfigurable optical add-drop multiplexers (ROADM), wavelength cross-connects (WXC), and photonic cross-connects (PXC). In the current commercial metro and backbone networks, these optical nodes are controlled and managed through the element management system (EMS) and the network management system (NMS) in a manual and semistatic style for lightpath provisioning, as shown in Figure 14.1. Although this approach is very reliable, the network carriers require a control plane technique for dynamic and intelligent control of wavelength paths in metro/backbone optical networks to save operational expense, reduce the processing latency, and handle the rapid increase of dynamic network traffic.

A well-known choice is the generalized multiprotocol label switching (GMPLS) [1], which is a stable protocol suite to automatically provision end-to-end connections in a fully distributed manner. However,

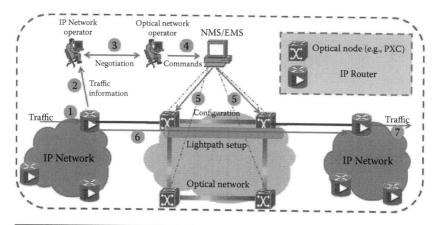

**Figure 14.1**   NMS/EMS-based lightpath provisioning in current metro/backbone optical networks.

the GMPLS-based control plane has not been widely deployed, and more importantly, most network carriers seem to lack the confidence to promote the commercial deployment of GMPLS in the optical backbone. The reasons for this situation are complicated [2,3]. One of the most important reasons is that the GMPLS-based distributed control plane is too complex, especially for dynamic control of both IP and optical layers as a unified control plane (UCP).

On the other hand, the software-defined networking (SDN) [4] architecture, in particular, the OpenFlow protocol [5], has been proposed in recent years. Although the original motivation of OpenFlow is to develop virtualized and programmable networks, it has also been widely regarded as a promising candidate for a UCP technique in heterogeneous networks [6,7]. OpenFlow can provide satisfactory flexibility for operators to control a network (i.e., software defined) and can arguably match carriers' preference given its simplicity and manageability. Because of the advantages of OpenFlow, as well as its potential capability to perform the unified control functionality in carrier-grade networking, it has received extensive attention from both academia and industry worldwide [8–26]. However, despite massive progress of OpenFlow in terms of research, product prototype, and standardization, it is still at the starting stage for optical networking.

In light of this, in this chapter, we will review and present several key techniques for OpenFlow-based control and management in metro and backbone optical networks. The network scenarios that we introduced include wavelength switched optical networks (WSON), optical burst switching (OBS) networks, multilayer multigranularity optical networks, multidomain optical networks, and elastic optical networks (EON). In addition, we briefly present the comparison between OpenFlow and other control plane techniques for optical networks. We also introduce the interoperability/interworking between OpenFlow and the path computation element (PCE) [27], and between OpenFlow and GMPLS for intelligent optical network control.

The rest of this chapter is organized as follows. "OpenFlow-Based Wavelength Path Control in Transparent Optical Networks" proposes solutions for OpenFlow-based wavelength path control in transparent WSON. "OpenFlow and Other Control Plane Techniques for Optical Networks" presents the comparison and interoperability between OpenFlow and GMPLS/PCE. "OpenFlow for Multidomain Optical Networks" introduces the OpenFlow-based control plane for multidomain optical networks. "OpenFlow for Other Optical Switching Paradigms" presents OpenFlow for other optical switching paradigms. Finally, "Conclusion" is given at the end of this chapter.

## OpenFlow-Based Wavelength Path Control in Transparent Optical Networks

In this section, we experimentally present OpenFlow-based wavelength path control in transparent optical networks. We propose two different approaches (sequential and delayed approaches) for lightpath setup and two different approaches (active and passive approaches) for lightpath release in optical networks by using the OpenFlow protocol. The overall feasibility of these approaches was experimentally validated, and the network performances were quantitatively evaluated. More importantly, all the proposed methodologies were demonstrated and evaluated on a real transparent optical network testbed with both an OpenFlow-based control plane and a data plane, which is beneficial for verifying their feasibility and effectiveness and for obtaining valuable insights for proposed solutions when deploying into real OpenFlow-controlled optical networks.

*A Brief Introduction to OpenFlow*

We briefly outline the main features of OpenFlow. A more detailed and exhaustive description is available in Ref. [5]. An OpenFlow-based network consists of at least one OpenFlow controller (e.g., Nox [28]), several OpenFlow switches, a secure channel that interconnects the Nox with the OpenFlow switch, and the OpenFlow protocol for signaling between the Nox and the switches. In an OpenFlow-controlled network, packet forwarding is executed in the OpenFlow switch according to a flow table, and the Nox is responsible for the routing decision. When the OpenFlow switch receives a packet that does not match any entry in the flow table, it sends the packet to the Nox. The Nox may drop the packet or may add a new flow entry in the flow table to force the OpenFlow switch to forward packets belonging to the same flow on a given path. Each entry in the flow table of an OpenFlow switch contains a packet header that defines the flow, the action that defines how to process the packets, and flow statistics, as shown in Figure 14.2.

*OpenFlow-Based Optical Nodes*

We adopt the PXC as an example to illustrate how to use the OpenFlow protocol to control an optical node. Figure 14.3a shows the structure of a PXC. To control this node through the OpenFlow protocol, we introduce virtual Ethernet interfaces (veths) to an OpenFlow switch. These veths are virtualized from the physical interfaces of the PXC, and each veth exactly corresponds to a physical interface of the

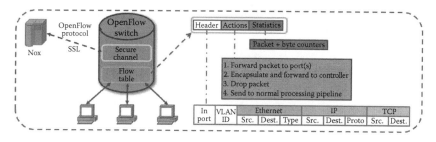

**Figure 14.2**  Architecture of an OpenFlow-enabled network. SSL, secure sockets layer; VLAN, virtual local area network; Proto, protocol; Src, source; Dest, destination. (From The OpenFlow Switch Consortium. http://www.openflow.org/.)

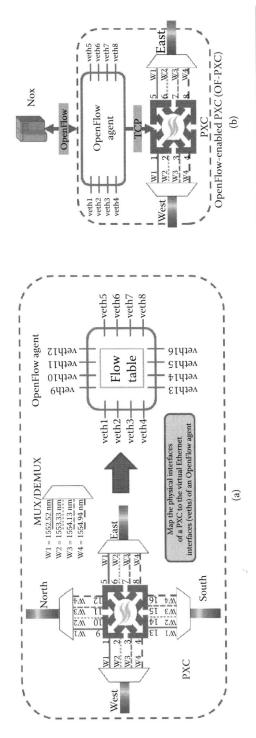

**Figure 14.3** OpenFlow-based optical nodes: (a) Virtualization of the physical interfaces of a PXC to the virtual Ethernet interfaces (veths) of an OFA; (b) The architecture of an OF-PXC. MUX, multiplexer; DMUX, demultiplexer.

PXC, as shown in Figure 14.3a. By using this approach, the virtual OpenFlow switch obtains a virtualized view of the physical structure of the PXC, which is greatly beneficial for the Nox to control the cross-connections within the PXC by using the OpenFlow protocol. For simplicity, this virtual OpenFlow switch is referred to as the OpenFlow agent (OFA) in the rest of this chapter.

The combination of an OFA and the corresponding PXC is referred to as an OpenFlow-enabled PXC (OF-PXC), which is controlled by a Nox controller through the OpenFlow protocol, as shown in Figure 14.3b. Once the first IP packet of a new flow is received by the Nox, the Nox obtains the source and destination IP addresses of this flow and then performs routing and wavelength assignment (RWA) based on its knowledge of the whole network. After that, according to the results, the Nox inserts a new flow entry in the flow table of the OFA. In turn, based on the flow entry, the OFA automatically sends standard transaction language 1 (TL1) commands to cross-connect the corresponding ports of PXC through the transmission control protocol (TCP) interface to establish the lightpath, thanks to the same virtualized view of the PXC structure in the OFA. For example, as shown in Figure 14.3b, if the RWA result calculated by the Nox is a path from the west side to the east side through the PXC and the assigned wavelength is W3, the Nox first inserts a flow entry in the OFA with the input port veth3 and the output port veth7. After that, based on this flow entry, the OpenFlow switch sends a TL1 command to the PXC to cross-connect the physical ports between 3 and 7. Note that the optical cross-connection information (input and output ports, wavelength, etc.) can be encapsulated in the OpenFlow protocol, as proposed in Ref. [29].

*Signaling Procedures for Lightpath Control in an OpenFlow-Based Optical Network*

To dynamically control a wavelength path, we present two approaches for lightpath setup, referred to as the sequential approach and the delayed approach. Moreover, we also present two approaches for lightpath release, which are referred to as the active approach and the passive approach.

*Lightpath Setup*

*The Sequential Approach for Lightpath Setup*   The sequential approach refers to the solution by which the OpenFlow protocol controls the ingress IP router and optical nodes in sequential order. The signaling procedure is depicted in Figure 14.4 and summarized as follows:

Step 1: IP traffic (an IP flow) arrives at the ingress OpenFlow-enabled IP router (OF-R)/switch (OF-R1).

Step 2: If the IP flow does not match any flow entries in the flow table of OF-R1, OF-R1 forwards the first packet of this flow to the Nox.

Step 3: The Nox calculates the route (e.g., the path computation result is OF-R1→OF-PXC1→OF-PXC2), assigns the wavelength in the optical network, and then adds a new flow entry in OF-R1, OF-PXC1, and OF-PXC2.

Step 4: The OFAs of OF-PXC1 and OF-PXC2 send TL1 commands to set up the cross-connects of the PXC.

Step 5: The lightpath is successfully established, and the flow arrives at the OF-R2 through the lightpath.

Step 6: If the flow does not match any flow entries in the flow table of OF-R2, OF-R2 forwards the first packet of this flow to the Nox.

Step 7: Nox inserts a new flow entry in the flow table of OF-R2.

Step 8: The IP flow is forwarded to the destination.

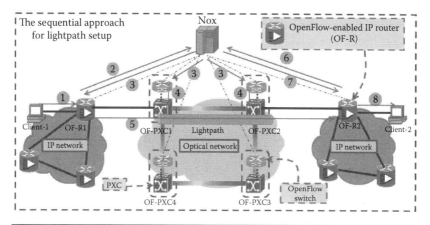

**Figure 14.4**   Signaling procedures of lightpath setup by using the sequential approach.

The sequential approach is the most straightforward solution for lightpath provisioning. However, considering the configuration latency of the PXC for cross-connection and the signaling latency within each OF-PXC (i.e., between each OFA and the corresponding PXC), there is no guarantee that the lightpath in the optical domain is completely provisioned before the flow arrives at the ingress OF-PXC (e.g., OF-PXC1 in Figure 14.4). Assume that the signaling propagation latency between Nox and the ingress OF-R is $t_1$, the time latency for the flow entry insertion in the ingress OF-R is $t_2$, the latency for the IP data flow transmission from the ingress OF-R to the ingress OF-PXC (e.g., OF-R1 to OF-PXC1 in Figure 14.4) is $t_3$, and the optical connection setup latency is $T_{opt}$, which comprises the signaling propagation latency among the Nox, the OFAs, and the PXC; the internal processing latency in the OFAs; and the latency for cross-connection in the PXC. In this case, packet losses will occur when $T_{opt}$ is longer than the sum of $t_1$, $t_2$, and $t_3$. In other words, the packet loss possibility of the sequential approach is equal to the possibility of the case that $T_{opt} > t_1 + t_2 + t_3$. When packet loss occurs, if we further assume that the total amount of data for the IP flow is TD and the transmission rate is $V$, the amount of lost packets can be calculated as $V(T_{opt} - t_1 - t_2 - t_3)$, and in turn, the packet loss probability can be measured as $V(T_{opt} - t_1 - t_2 - t_3)/TD$. Clearly, when packet losses occur, the service performance will degrade even if some higher layer mechanisms (e.g., TCP) are used to retransmit the lost packets. In light of this, we present the delayed approach for lightpath provisioning to address this issue.

*The Delayed Approach for Lightpath Setup*    The signaling procedure of lightpath setup by using the delayed approach within an OpenFlow-based optical network is shown in Figure 14.5. Different from the sequential approach, by using the delayed approach, the Nox first controls the OF-PXC to establish a lightpath in the optical domain. After that, the Nox waits for a deliberate time delay and then inserts a new flow entry in the ingress IP router. With this approach, the control of the ingress IP router is delayed. Therefore, this approach is referred to as the delayed approach. The operations of this approach in an OpenFlow-enabled optical network are summarized as follows:

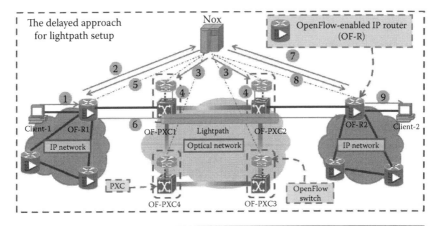

**Figure 14.5**   Signaling procedures of lightpath setup by using the delayed approach.

Step 1: IP traffic (an IP flow) arrives at the ingress OF-R (OF-R1).

Step 2: If the IP flow does not match any flow entries in the flow table of OF-R1, OF-R1 forwards the first packet of this flow to the Nox.

Step 3: The Nox calculates the route (e.g., the path computation result is OF-R1→OF-PXC1→OF-PXC2), assigns the wavelength in the optical network, and then adds a new flow entry in OF-PXC1 and OF-PXC2.

Step 4: The OFAs of OF-PXC1 and OF-PXC2 send TL1 commands to set up the cross-connections of the PXC.

Step 5: The Nox waits for an appropriate time delay for the successful lightpath provisioning in the optical domain and then inserts a new flow entry in the ingress OF-R (OF-R1).

Step 6: The flow arrives at the OF-R2 through the lightpath.

Step 7: If the flow does not match any flow entries in the flow table of OF-R2, OF-R2 forwards the first packet of this flow to the Nox.

Step 8: Nox inserts a new flow entry in the flow table of OF-R2.

Step 9: The IP flow is forwarded to the destination.

The advantage of the delayed approach is that this approach can guarantee a successful end-to-end IP flow transmission because the optical connection is well established before the IP flow arrives at the ingress optical node, at the expense of longer signaling latency caused by the waiting time in the Nox controller.

*Lightpath Release*

*The Active Approach for Lightpath Release* The active approach is a special solution for lightpath release. It is only applicable when the amount of data for the arriving traffic is known in advance. In this case, the connection holding time in the optical domain can be predicted. Therefore, the Nox controller can insert a hard timeout field in the corresponding flow entry in the flow table. This field can be used to indicate the connection holding time. After the hard timeout expires, the OFA automatically deletes the corresponding flow entry and, in turn, releases the cross-connection of the PXC. Because this approach releases a lightpath directly from each OFA without any external requests or indications, it is referred to as the active approach.

*The Passive Approach for Lightpath Release* The passive approach releases a lightpath by the Nox controller after a lightpath release indication is received. The signaling procedure of the passive approach for lightpath release is shown in Figure 14.6. By using this approach, after a flow is successfully transmitted, the source client sends a packet to notify the completeness of the service. After the Nox controller receives this packet, the Nox controller can determine the corresponding flow entry and then delete this existing entry. The detailed

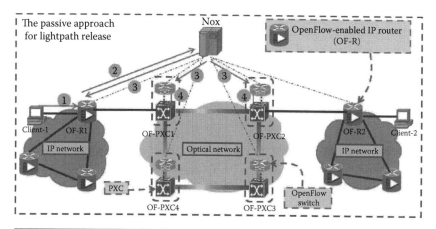

**Figure 14.6**  Signaling procedures of lightpath release by using the passive approach.

operations of the passive approach for lightpath release are summarized as follows:

Step 1: Client-1 sends a specified notification packet after the completeness of flow transmission.

Step 2: This notification packet does not match any flow entries in the flow table of OF-R1. Then, OF-R1 forwards this packet to the Nox.

Step 3: Nox receives this notification packet, and the Nox deletes the corresponding flow entries in OF-R1, OF-PXC1, and OF-PXC2.

Step 4: The OFAs of OF-PXC1 and OF-PXC2 send TL1 commands to delete the corresponding cross-connections in the PXC.

Clearly, different from the NMS/EMS-based solution presented in Figure 14.1, the approaches and procedures proposed in the subsection are fully dynamic. The OpenFlow-based control plane can dynamically set up the lightpath according to the IP traffic demands, achieving the unified control functionality of both IP and optical layers. In an optical network with an OpenFlow-based control plane, the NMS system is also expected to be introduced. In this case, it can be used to send a request to Nox to establish or delete a connection from the network operator.

*Experimental Demonstration and Evaluation Results*

To verify the aforementioned methodologies and evaluate the solutions, we constructed a transparent optical network testbed with four OF-PXC in a mesh topology, as shown in Figure 14.7. Each optical link was configured with four wavelengths (1552.50, 1553.33, 1554.13, and 1554.94), and an optical spectrum analyzer was installed in the link OF-PXC1–OF-PXC2. Four OF-Rs were assumed to be connected to the four OF-PXC. In addition, a PC was attached to each OF-R as a client, through an Ethernet cable. However, because of the lack of suitable transponders (TPNDs), four distributed feedback laser diode (DFB-LD) modules as quasi-TPNDs were connected to OF-PXC1, which worked at the aforementioned wavelengths as light sources, as shown in Figure 14.7.

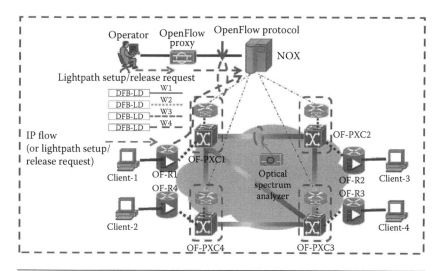

**Figure 14.7** Experimental setup for OpenFlow-based dynamic wavelength path control.

If an IP flow from the client is received by the OF-R, the OF-R sends the first packet of this IP flow to the Nox controller to insert a flow entry and establish a lightpath according to the sequential approach or delayed approach. Alternatively, to evaluate the lightpath setup and release, only the first packet of the forthcoming IP flows needs to be sent to the OF-R because only the first packet of a flow is required by the Nox for path computation, flow entry modification, and lightpath setup. The rest of the packets of this flow will transmit through the established lightpath, without any communication with the Nox controller. In this sense, the first IP packet of a new flow can be regarded as a lightpath setup request in a dynamic provisioning scenario. Furthermore, the operator can also send this lightpath setup request to the Nox controller in a static provisioning scenario. However, note that a proxy of the OpenFlow switch must be inserted between the operator and the Nox to encapsulate this request with the OpenFlow protocol.

In our current implementation, the centralized Nox obtained a copy of the network topology and wavelength availabilities for each optical link. Moreover, in the optical domain, all the feasible routes between a pair of OF-Rs were precomputed through the conventional k-shortest-path algorithm in the Nox. For a new IP flow (or a lightpath setup request), the Nox always selects the route with least hops among all the feasible routes, unless the wavelength continuity constraint (WCC) could not be satisfied on this route. After the route was determined, the first fit mechanism was adopted to assign a wavelength for this route.

In the experiment, we first assumed that all the wavelengths in links OF-PXC1–OF-PXC3 and OF-PXC1–OF-PXC4 were occupied, and we updated the information of wavelength availabilities in the Nox. Then, four IP flows (or lightpath setup requests) were sent from Client-1, as summarized in Table 14.1. Flows 1 and 2 were routed on the shortest path on wavelengths W1 and W2, respectively. For flow 3, because the WCC could not be satisfied on the shortest path, Nox selected the second shortest path to route this request and assigned wavelength W3 for this lightpath. Similarly, flow 4 was routed on the third shortest path by using the wavelength W4. Figure 14.8a to d shows the optical spectrum on link OF-PXC1–OF-PXC2 when the four lightpaths were established one by one.

Figure 14.9 shows printed information from the Nox controller during the lightpath setup for flow 1 by using the sequential approach. After a new packet is received by the Nox, the Nox performs the RWA algorithm and then installs a new flow entry in OF-R1, OF-PXC1, and OF-PXC2, in sequential order. In Figure 14.9, the dpID is short for datapath identifier (ID), which is the OpenFlow terminology used to represent an OpenFlow switch. In our experimental setup, dpID 1, dpID 2, and dpID 3 refer to OF-R1, OF-PXC1, and OF-PXC2, respectively. By using Wireshark [30], a packet-analyzing software that monitors network traffic and displays operating system-based time-stamped packets, we obtained the protocol capture files, as shown in Figure 14.10. From Figure 14.10, we can clearly observe the OpenFlow-based signaling procedures when the sequential approach

**Table 14.1**  Summary of Flows and Lightpaths

| NO. | SOURCE | DESTINATION | CALCULATED LIGHTPATH | WAVELENGTH |
|-----|--------|-------------|----------------------|------------|
| 1 | Client-1 | Client-3 | OF-PXC1→OF-PXC2 | W1 (1552.52 nm) |
| 2 | Client-1 | Client-3 | OF-PXC1→OF-PXC2 | W2 (1553.33 nm) |
| 3 | Client-1 | Client-4 | OF-PXC1→OF-PXC2→OF-PXC3 | W3 (1554.13 nm) |
| 4 | Client-1 | Client-2 | OF-PXC1→OF-PXC2→ OF-PXC3→OF-PXC4 | W4 (1554.94 nm) |

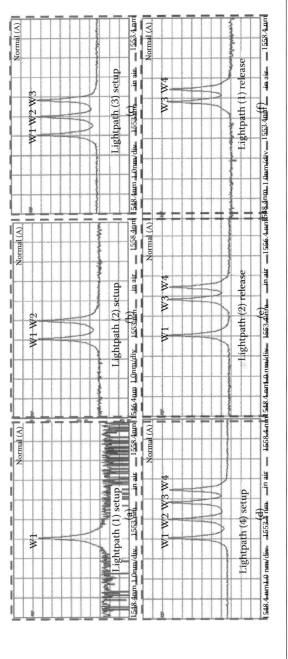

**Figure 14.8**   Optical spectrum observed on link 0F-PXC1–0F-PXC2.

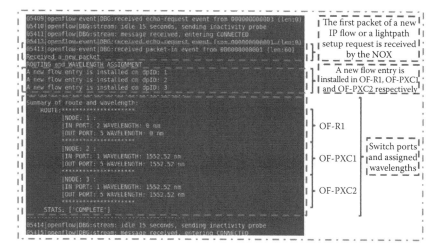

**Figure 14.9** Printed information from the Nox during lightpath setup with the sequential approach.

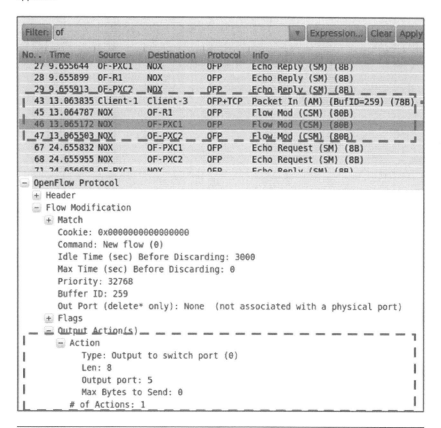

**Figure 14.10** Wireshark capture of the OpenFlow protocol during lightpath setup with the sequential approach.

is adopted for lightpath setup. The signaling latency among Nox and each OFA is very short—approximately 1.67 ms in total—as shown in Figure 14.10.

On the other hand, Figures 14.11 and 14.12 show the printed information from the Nox controller and the Wireshark capture, respectively, during lightpath setup for flow 1 by using the delayed approach. As shown in both figures, once a new packet is received by the Nox controller, it first inserts a new flow entry in OF-PXC1 and OF-PXC2 to set up a lightpath in the optical domain. After that, the Nox waits for 300 ms and then inserts a flow entry in the OF-R1 to start the flow transmission. Note that the waiting time depends on the signaling latency between the OFA and the optical node, as well as the configuration latency of the optical node. It may vary with different hardware (e.g., optical nodes) or different lightpaths (e.g., different hops). If this

**Figure 14.11** Printed information from the Nox during lightpath setup with the delayed approach.

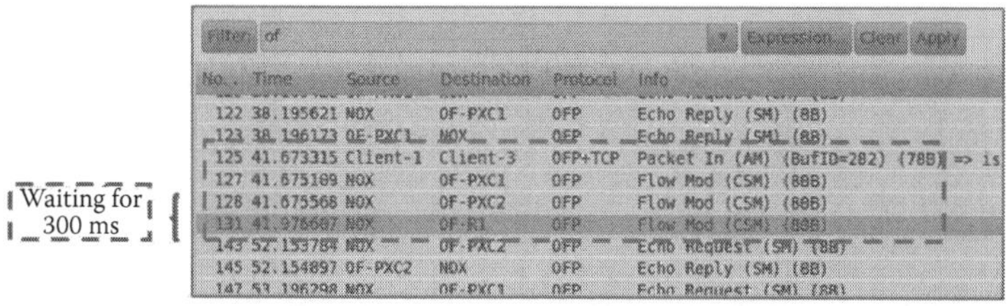

**Figure 14.12** Wireshark capture of the OpenFlow protocol during lightpath setup with the delayed approach.

```
┌─────────────────────────────────────────────────────────────────────────────┐
│                              Flow table of OF-PXC1                            │
│  [root@Openflow-1 ~]# dpctl dump-flows unix:/var/run/dp0                      │
│  stats reply (xid=0xe05599hd): flags=none type=1(flow)                        │
│     cookie=0, duration_sec=36s, duration_nsec=259000000s, table_id=1, priority=327868, n_packets=0 │
│  n_bytes=0, idle_timeout=0, hard_timeout=2000, in_port=4, actions=output : 8  │
│                                                            Flow entry for lightpath (4) │
│     cookie=0, duration_sec=43s, duration_nsec=826000000s, table_id=1, priority=327868, n_packets=0, │
│  n_bytes=0, idle_timeout=0, hard_timeout=5000, in_port=3, actions=output : 7  │
│                                                            Flow entry for lightpath (3) │
│     cookie=0, duration_sec=50s, duration_nsec=240000000s, table_id=1, priority=327868, n_packets=0, │
│  n_bytes=0, idle_timeout=0, hard_timeout=500, in_port=2, actions=output : 6   │
│                                                            Flow entry for lightpath (2) │
│     cookie=0, duration_sec=64s, duration_nsec=613000000s, table_id=1, priority=327868, n_packets=0, │
│  n_bytes=0, idle_timeout=0, hard_timeout=1000, in_port=1, actions=output : 5  │
│                                                            Flow entry for lightpath (1) │
│  [root@Openflow-1 ~]#                                                         │
└─────────────────────────────────────────────────────────────────────────────┘
```

**Figure 14.13**  Flow table of OF-PXC1 after four lightpaths are successfully established.

waiting time is set too short, it cannot guarantee successful flow transmission through the lightpath. On the contrary, if this time is too long, it will introduce long end-to-end signaling latency. In an actual operational scenario, a precise waiting time can be obtained by performance monitoring in the TPND at the receiver side.

Figure 14.13 shows the flow entries in the flow table of OF-PXC1 for these four lightpaths when the active approach is used for lightpath release. In this experiment, we assume that the connection holding times for flows 1 to 4 are 1000, 500, 5000, and 2000 s, respectively. In this case, a hard timeout field is required to insert for the corresponding flow entry, as previously mentioned. Take the flow entry for flow 1 as an example. The hard_timeout = 1000 and duration_sec = 64s indicate that the total connection holding time is 1000 s and that this connection has already been established for 64 s. The in_port = 1, actions = output:5 shows that the input port and output port of the OFA in OF-PXC1 are veth1 and veth5, respectively, corresponding to the cross-connection from port 1 to port 5 in the optical node. After the hard timeout expires, the OFA automatically deletes the corresponding flow entry and releases the cross-connection in the PXC. Because the optical connection for flow 2 is with the shortest holding time in this experiment, it is released first, and then the optical connection for flow 1 is released, as shown in Figure 14.8e to f.

The experimental results of lightpath release for flow 1 with the passive approach are shown in Figures 14.14 and 14.15, which are the printed information from the Nox controller and the Wireshark capture, respectively. Similarly, a request is sent from Client-1 after the completeness of flow traffic transmission. The results in Figures 14.14 and 14.15 show that this approach is feasible and efficient for releasing a lightpath.

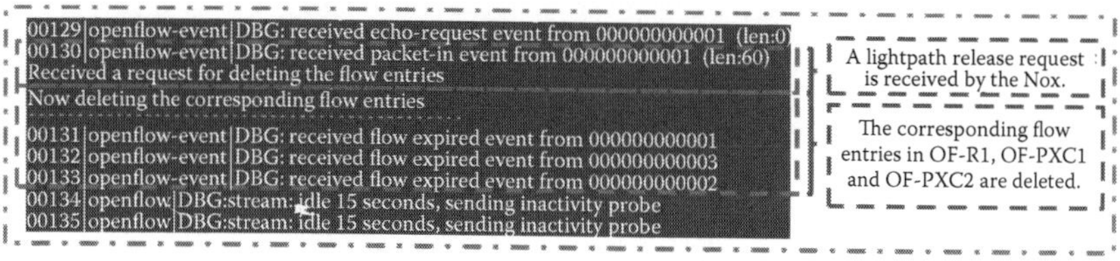
00129 openflow-event DBG: received echo-request event from 000000000001 (len:0)
00130 openflow-event DBG: received packet-in event from 000000000001 (len:60)
Received a request for deleting the flow entries
Now deleting the corresponding flow entries
00131 openflow-event DBG: received flow expired event from 000000000001
00132 openflow-event DBG: received flow expired event from 000000000003
00133 openflow-event DBG: received flow expired event from 000000000002
00134 openflow DBG:stream: idle 15 seconds, sending inactivity probe
00135 openflow DBG:stream: idle 15 seconds, sending inactivity probe
A lightpath release request is received by the Nox.
The corresponding flow entries in OF-R1, OF-PXC1 and OF-PXC2 are deleted.

**Figure 14.14** Printed information from the Nox during lightpath release with the passive approach.

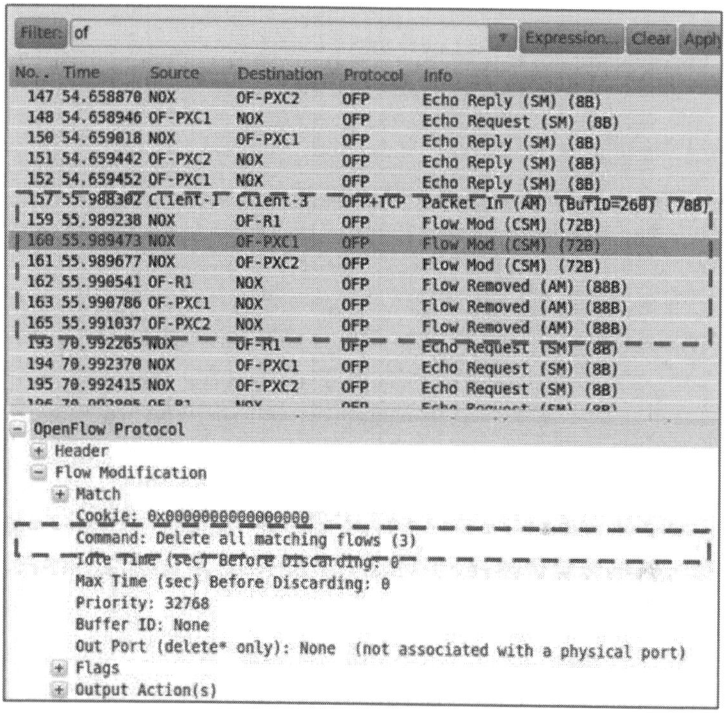
Filter: of
Expression... Clear Apply
No. . Time Source Destination Protocol Info
147 54.658870 NOX OF-PXC2 OFP Echo Reply (SM) (8B)
148 54.658946 OF-PXC1 NOX OFP Echo Request (SM) (8B)
150 54.659018 NOX OF-PXC1 OFP Echo Reply (SM) (8B)
151 54.659442 OF-PXC2 NOX OFP Echo Reply (SM) (8B)
152 54.659452 OF-PXC1 NOX OFP Echo Reply (SM) (8B)
157 55.988302 Client-1 Client-3 OFP+TCP Packet In (AM) (BufID=268) (78B)
159 55.989238 NOX OF-R1 OFP Flow Mod (CSM) (72B)
160 55.989473 NOX OF-PXC1 OFP Flow Mod (CSM) (72B)
161 55.989677 NOX OF-PXC2 OFP Flow Mod (CSM) (72B)
162 55.990541 OF-R1 NOX OFP Flow Removed (AM) (88B)
163 55.990786 OF-PXC1 NOX OFP Flow Removed (AM) (88B)
165 55.991037 OF-PXC2 NOX OFP Flow Removed (AM) (88B)
193 70.992265 NOX OF-R1 OFP Echo Request (SM) (8B)
194 70.992370 NOX OF-PXC1 OFP Echo Request (SM) (8B)
195 70.992415 NOX OF-PXC2 OFP Echo Request (SM) (8B)
OpenFlow Protocol
Header
Flow Modification
Match
Cookie: 0x0000000000000000
Command: Delete all matching flows (3)
Idle Time (sec) Before Discarding: 0
Max Time (sec) Before Discarding: 0
Priority: 32768
Buffer ID: None
Out Port (delete* only): None (not associated with a physical port)
Flags
Output Action(s)

**Figure 14.15** Wireshark capture of the OpenFlow protocol during lightpath release with the passive approach.

Moreover, we quantitatively measured the path setup and release latency in our demonstrated scenario by using the proposed solutions. Note that the path setup latency evaluated in this chapter not only includes the lightpath provisioning latency in the optical domain, but also consists of the control of the ingress OF-R. In this regard, the contributors to the overall path setup latency include the packet (request) propagation latency from the client to the Nox; the packet processing latency in the Nox and OFAs; the signaling latency among the Nox, the OFAs, and the PXC; and the internal cross-connection time of the PXC. Similarly, the path release latency includes the deletion of the corresponding flow entries in the ingress OF-R as well as the lightpath release latency in the optical domain. In the case of the active approach, the path release latency consists of the processing latency in OFAs, the signaling latency between the OFAs and the PXC, and the operational latency within the PXC for deleting the cross-connection. In the case of the passive approach, the packet (request) propagation latency from the client to the Nox, the packet processing latency in the Nox, as well as the OpenFlow signaling latency among the Nox and each OFA must be added.

The results are summarized in Table 14.2. For the internal cross-connection time of the PXC, we simply used the reference value of 10 ms. The results in Table 14.2 show that the time for provisioning paths for flows 1 to 4 is approximately 243 to 254 ms by using the sequential approach. If the delayed approach is adopted for lightpath provisioning, this time increases to 314 to 316 ms, which is longer than that of the sequential approach. This is because the Nox controller has to wait for a certain period (in this experiment, for 300 ms) to cover the lightpath setup latency in the optical domain. For the sequential approach, the path setup times for flows 3 and 4 are slightly longer than those for flows 1 and 2, mainly because of the longer processing latency in the Nox (e.g., RWA) and the signaling latency. Note that, although the sequential approach outperforms the delayed approach in terms of the path setup latency, it may result in packet losses if the optical connection is not completely established before the flow arrives, as previously mentioned.

In addition, because the lightpath release requests are directly sent from each OFA after the timeout, the path release times for the four lightpaths are almost the same—approximately 123 ms—when the active approach is deployed. On the other hand, if the passive

**Table 14.2**  Summary of Path Setup and Release Latency

| FLOW NO. | SEQUENTIAL APPROACH FOR LIGHTPATH SETUP (ms) | DELAYED APPROACH FOR LIGHTPATH SETUP (ms) | ACTIVE APPROACH FOR LIGHTPATH RELEASE (ms) | PASSIVE APPROACH FOR LIGHTPATH RELEASE (ms) |
|---|---|---|---|---|
| 1 | ~243 | ~314 | ~123 | ~126 |
| 2 | ~243 | ~314 | ~123 | ~126 |
| 3 | ~249 | ~315 | ~124 | ~127 |
| 4 | ~254 | ~316 | ~124 | ~128 |

approach is adopted, the lightpath release latency is approximately 126 to 128 ms for flows 1 to 4, which is slightly longer than that of the active approach. The reason for this is that, compared with the active approach, the packet (request) propagation latency from the client to the Nox, the packet processing latency in the Nox, as well as the OpenFlow signaling latency among the Nox and each OFA are added for overall path release latency.

When the network becomes more complex, the path setup latency by using both the sequential and delayed approaches will increase accordingly because of the longer processing time in the Nox controller for RWA computation and the longer latency of the OpenFlow-based signaling for the insertion of flow entries in all the related OF-PXC. Moreover, the path setup latency of the delayed approach is still expected to be longer than that of the sequential approach because of the Nox waiting time. For an optical connection with more hops, a relatively longer waiting time must be introduced in the Nox to cover the provisioning time of the lightpath. On the other hand, the active approach will not be significantly affected in terms of path release latency even if there are many nodes involved in a lightpath. This is because, by using the active approach, the cross-connections of the PXC are directly and simultaneously released from each OFA after the timeout. Therefore, each OFA is fully independent from the Nox controller and other OFAs during the lightpath release, which implies that the overall path release latency will not be remarkably affected with the increasing numbers of OF-PXC along a lightpath. However, the path release latency of the passive approach will increase when the network becomes more complex, and thus, it will be longer than that of the active approach. This is because, in this case, the centralized Nox controller has to inform more OFAs to release the cross-connections through the OpenFlow protocol, which will increase the overall signaling latency.

Finally, based on the proposed solutions and the obtained results, we compared the conventional NMS/EMS-based and the OpenFlow-based path control schemes, as summarized in Table 14.3. Compared with the conventional NMS/EMS-based solution, the OpenFlow-based scheme is able to dynamically set up and release an end-to-end path for IP flow transmission within the optical domain, with many advantages, such as the reduction of operational cost and path provisioning latency. These advantages, as well as the centralized feature of the OpenFlow

**Table 14.3**    Comparison of the Conventional NMS/EMS-Based and the OpenFlow-Based Schemes

| ITEMS | NMS/EMS-BASED SOLUTIONS | OPENFLOW-BASED SOLUTIONS |
|---|---|---|
| Features | Mature technique with the deployment in real operational scenarios | In the early stage, with many issues to be investigated |
| | Manual and semistatic styles for path setup and release | Dynamic setup and release of end-to-end paths |
| | High operational cost | Low operational cost |
| | Separated control of IP and optical domains | Unified control of IP and optical domains |
| Path setup latency | Long (several hours to several days) | Short (hundreds of milliseconds) |
| Path release latency | Long (several hours to several days) | Short (hundreds of milliseconds) |

protocol, are attractive and would be greatly beneficial for the carriers, if it were deployed as an intelligent control plane for optical networks.

**OpenFlow and Other Control Plane Techniques for Optical Networks**

*Comparison of OpenFlow, GMPLS, and GMPLS/PCE*

The control plane plays a key role in enabling an intelligent and dynamic WSON, with a great reduction of operational expense and processing latency. In recent years, there has been much advance in this area, ranging from the traditional GMPLS [1] to a PCE/GMPLS-based architecture [31]. Compared with the fully distributed GMPLS-based control plane, the PCE/GMPLS decouples the path computation functionality from GMPLS controllers (GCs) to a dedicated PCE, which can bring several advantages for optical network control, as detailed in Ref. [27].

Table 14.4 shows the comparison of the GMPLS-based, PCE/ GMPLS-based, and OpenFlow-based control planes for WSON. It can be seen that the technical evolution from GMPLS to PCE/ GMPLS to OpenFlow is a process wherein the control plane evolves from a fully distributed architecture to a fully centralized architecture. The PCE/GMPLS can be regarded as an intermediate case because, in this control plane, the control functionality is also distributed, but the path computation is centralized. This evolution trend may match the carrier's preference because a centralized control scheme is easier to migrate and update from the current NMS and EMS architectures that

**Table 14.4**  Comparison of GMPLS-Based, PCE/GMPLS-Based, and OpenFlow-Based Control Planes for WSON

| CONTROL PLANE | FEATURE | COMPLEXITY | FLEXIBILITY/ MANAGEABILITY | TECHNICAL MATURITY | STANDARDIZATION |
|---|---|---|---|---|---|
| GMPLS | Fully distributed | High | Low | High | The architecture/ framework was standardized in 2004 [1]. Remaining issues are being discussed with the IETF CCAMP working group. |
| PCE/ GMPLS | Centralized path computation, distributed control | High | Medium | High | The architecture/ framework was standardized in 2011 [31]. Remaining issues are being discussed with the IETF CCAMP and PCE working groups. |
| OpenFlow | Fully centralized | Low | High | Low | At a starting stage |

*Note:*  IETF, Internet Engineering Task Force; CCAMP, Common Control and Measurement Plan.

have been established, with huge investments already made. Moreover, the GMPLS-based control plane is complicated especially when it is deployed as a UCP for IP/dense wavelength division multiplexing (DWDM) multilayer networks because of its distributed nature, the number of protocols, and the interactions among different layers. The flexibility and manageability of the GMPLS-based control plane is low because, for example, if the carrier wants to update the operational policy or path computation algorithm, all the GCs have to be modified and updated. The PCE/GMPLS-based control plane is also complex because of the introduction of the PCE and PCE Communication Protocol (PCEP) [32]. However, it outperforms the GMPLS-based control plane in terms of flexibility and manageability because of the flexible placement of the path computation function and the easy update of path selection policies. On the other hand, the OpenFlow-based control plane provides maximum flexibility and manageability for carriers because all the functionalities are integrated into a single

OpenFlow controller. More importantly, the OpenFlow-based control plane is a natural choice for a UCP in IP/DWDM multilayer networks because of its inherent feature, as investigated in "OpenFlow-Based Wavelength Path Control in Transparent Optical Networks."

However, note that, although the OpenFlow-based control plane brings benefits, such as it is simple, flexible, and manageable, especially for IP/DWDM multilayer optical networks, it is still immature for controlling an optical network compared with the GMPLS-based and PCE/GMPLS-based control planes. With a decade of research, development, and standardization efforts, the technical maturity of both GMPLS and PCE is high. Despite some remaining issues that are still being discussed within the IETF CCAMP and PCE working groups, the architecture/frame of GMPLS and PCE for controlling a WSON has been standardized [1,31]. On the other hand, the OpenFlow-based control plane is still at a starting stage for optical networks. A more detailed comparison for these control plane techniques can be found in Ref. [33–36].

*Interworking between OpenFlow and PCE*

The scalability is a major issue of an OpenFlow-based UCP because, in a centralized architecture, the Nox has to perform both advanced path computations (e.g., impairment-aware RWA [IA-RWA]) and OpenFlow-based signaling for path provisioning. Several previous studies have verified that IA-RWA computations in optical networks are CPU intensive [37]. Therefore, considering the potential deployment of an OpenFlow-based UCP, it is very important to offload the Nox and thus enhance the network scalability. Note that it is possible to have more CPUs for the Nox to scale up its computation capability. However, this solution is not cost efficient because of the low per-dollar performance as the CPU's price goes up quickly with more sockets, and an upgrade of the motherboard and additional memory installation are also required to have more CPUs in a single machine, as investigated in Ref. [38].

Another straightforward solution to address this issue is to decouple the IA-RWA computation to a dedicated PCE. Besides the computation offload, this solution will also bring a lot of advantages, such as flexible path computation in a multidomain network, the ease of integration with business support system (BSS) and operations support system

(OSS), etc., because of the mature, well-defined, and feature-complete PCEP [32]. However, because the Nox/OpenFlow is responsible for lightpath setup/release and, consequently, the PCE is not necessarily aware of the up-to-date network resource information (e.g., available wavelengths), it is unable to compute an optimal path, especially if, given the optical technology constraints, it needs to consider the WCC [39].

In light of this, a more efficient solution is proposed in Ref. [16] for the seamless and dynamic interworking between a Nox/OpenFlow controller and a PCE by introducing a topology server. The PCE can then request the up-to-date topology from the server, and, in this work, the PCE obtains the topology at each PCEP request. The overall feasibility and efficiency of the proposed solution are experimentally verified on an IP/DWDM multilayer network testbed with both control and date planes, whereas its performance metrics are quantitatively evaluated.

*OpenFlow/PCE Integrated Control Plane*   In this section, the network architecture is introduced first. After that, the proposed extensions for the OpenFlow protocol, the Nox, and the PCE are presented. Finally, the procedure for dynamic end-to-end path provisioning by using the proposed OpenFlow/PCE integrated control plane is investigated in detail.

*Network Architecture*   Figure 14.16a shows the proposed network architecture, in which we consider an IP/DWDM multilayer network.

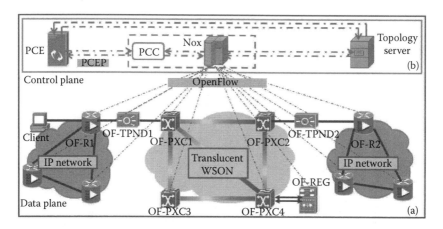

**Figure 14.16**   (a) Network architecture; (b) modules of the OpenFlow/PCE integrated control plane. OF-REG, OpenFlow-enabled regenerator.

In the IP layer, the IP routers are enhanced to support the OpenFlow protocol and are referred to as OF-Rs. A translucent WSON, with sparsely but strategically equipped 3R regenerators, is deployed in the optical layer. Such a translucent network is a promising paradigm for industrial deployment because extensive studies indicate that it can provide an adequate trade-off between network cost and service provisioning performance [40]. An OpenFlow/PCE integrated control plane, as detailed next, is deployed to control this multilayer network through the OpenFlow protocol, with its functional modules shown in Figure 14.16b.

*OpenFlow Extensions*   To control a translucent WSON through the extended OpenFlow protocol, OpenFlow-enabled optical switching nodes (e.g., PXC), TPNDs, and 3R regenerators are required, which are referred to as OF-PXC, OF-TPND, and OF-REG, respectively. The OpenFlow-based PXC is introduced in "OpenFlow-Based Wavelength Path Control in Transparent Optical Networks." The design of the OF-TPND (or OF-REG) is shown in Figure 14.17. TPND or regenerator groups are connected to an extended OpenFlow module that is able to trap the failure alarms from TPNDs and then convert the alarm into an extended OpenFlow Packet In message to notify the Nox for restoration path provisioning. On the other hand, if this OpenFlow module receives extended (or specified) Flow Mod

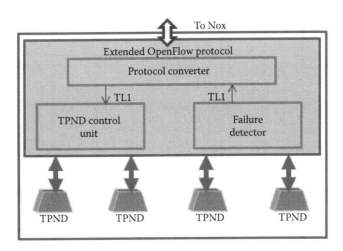

**Figure 14.17**   Functional architecture of the OF-TPND (or OF-REG).

messages (as detailed next) from the Nox, it translates them into TL1 commands to control the related TPNDs or REGs.

For the extended Packet In message, we insert an alarm indication into the standard OpenFlow Packet In messages. If a Packet In message with this alarm indication is received by the Nox, the Nox controller can realize that this Packet In message is not the first packet of a new flow but an alarm notification. In this case, the Nox will start the restoration path creation to recover the failed services. Note that a new message type can be introduced into the standard OpenFlow protocol to indicate the lightpath failure. Moreover, the Nox has to use Flow Mod messages to deliver the TPND control information (e.g., required wavelength) to the OF-TPND or OF-REG. In general, there are two approaches to achieve this functionality. The first approach is to introduce an extended Flow Mod message, adding some new fields to the standard Flow Mod message, to carry the TPND or regenerator control information. The second approach is to introduce a new flow entry–TPND control information translation table into the Nox and each OF-TPND. The content in this table is predetermined by the carrier as an operational policy. In this case, the Nox first translates the TPND control information into a specified flow entry based on the flow entry–TPND control information translation table, and then, by using standard Flow Mod messages, this specified flow entry is inserted into the OF-TPND. In turn, the OF-TPND translates this specified flow entry into the original TPND control information to control the TPND. Clearly, the first approach is more standard and flexible, whereas the second approach is simpler and more cost efficient.

*Nox Extensions*  In the proposed control plane, a new entity is introduced, namely a topology server. It is responsible for managing (i.e., gathering, storing, and serving) the traffic engineering (TE) information of the network (including topology, wavelength, and 3R regenerator availability) as well as information related to physical impairments (e.g., optical signal-to-noise ratio [OSNR]), as shown in Figure 14.16b. Although the proposed architecture is very flexible, in this work, we proceed as follows: Upon successful path setup or release operations, the Nox updates the TE information (e.g., available wavelengths for each link) and then automatically generates an eXtensible Markup Language (XML) file to encode the traffic engineering database (TED), as detailed next. After that, the Nox automatically sends the generated XML file to the topology server to update the TE information of the whole network by using a TED Update message. In addition, the Nox is extended with path computation client (PCC) [27] capabilities for communications with the PCE by means of the PCEP. After an initial session handshake, the PCC may send a path computation request (PCReq) message to the PCE requesting an IA-RWA computation. The PCE replies with a path computation reply (PCRep) containing either a path composed of an explicit route object (ERO) in the case of a successful IA-RWA computation, along with the path attributes, or with a NOPATH object if a satisfied path could not be found. If either the Nox (PCC) or the PCE does not desire to keep the connection open, it ends the PCEP session by using a Close message. After the IA-RWA computation is completed, the results are passed to the Nox to control the corresponding OF-PXC, OF-TPNDs, and OF-REGs to set up the lightpath by using the aforementioned OpenFlow extensions.

*PCE Extensions*    Once a PCReq message is received, the PCE is extended to first send a TE database request (TED Req) message to the topology server. In turn, the topology server replies with a TED reply (TED Rep) message, encoded using the XML format, which describes the up-to-date network information. Figure 14.18 shows the selected parts of an XML file for TED encoding. For each node, the TED encoding includes the node ID, node OSNR, and 3R regenerator information. For each link, the TED encoding file contains

**Figure 14.18**    Selected parts of an XML file for TED encoding.

the link ID, link OSNR, and wavelength availabilities information. Such a TED encoding file can be automatically generated by the Nox, as aforementioned. Because the PCE requests the latest topological information on a per-request basis, the server is referred to as a per-request–based dynamic topology server. The PCE performs the IA-RWA computation by using a translucent-oriented Dijkstra algorithm, with the objective function [41] code 32768. This algorithm has been detailed in our previous work in Ref. [42], which is able to compute the path, the regeneration points, and the wavelengths at each transparent segment, minimizing the path cost and fulfilling both WCC and OSNR requirements.

*Procedure for End-to-End Path Provisioning*  The procedure for end-to-end path provisioning is shown in Figure 14.19. If a new flow arrives at the ingress OF-R (OF-R1), and if this flow does not match

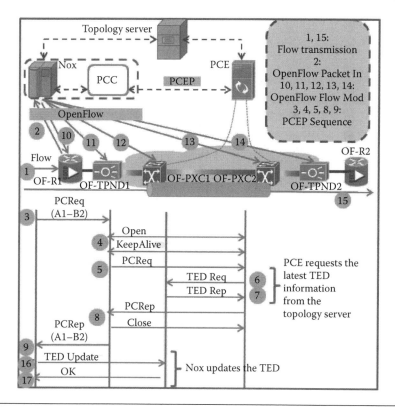

**Figure 14.19**  Procedure for end-to-end path provisioning by using the OpenFlow/PCE integrated control plane.

any existing flow entry in the flow table of OF-R1, OF-R1 forwards the first packet of this flow to the Nox, which requests a path computation to the PCE. After the initial session handshake, the Nox sends a PCReq message to the PCE through the PCC. In turn, the PCE sends a TED Req message to the topology server, and obtains a latest copy of the TED from the topology server for IA-RWA computation. The IA-RWA result (the network path, associated wavelengths, and regeneration points, if needed) are returned to the Nox, and subsequently, the Nox proceeds to set up an end-to-end path by controlling all the OF-Rs, OF-TPNDs, OF-PXC, and OF-REGs along the computed path by using the OpenFlow protocol, in a sequential order, starting from the ingress OF-R1. After that, the Nox sends a TED Update message to the topology server to update the TED.

Detailed information encapsulated in the PCRep and PCReq messages, as well as the OpenFlow Packet In and Flow Mod messages, is depicted in Figure 14.20. In the Packet In message, the source and destination addresses of the incoming flow is encapsulated, and the Flow Mod messages contain the input/output ports and the wavelength information that is used to control OF-PXC/OF-TPNDs/OF-REGs for lightpath provisioning. In the PCReq message, the source and destination addresses as well as the objective function code are encapsulated for the PCE to compute route by using the translucent-oriented Dijkstra algorithm [42]. The path computation results, including ERO and path attributes, are included in the PCRep message.

*Experimental Setup, Results, and Discussions*   We set up an IP over a translucent WSON multilayer network testbed composed of both control

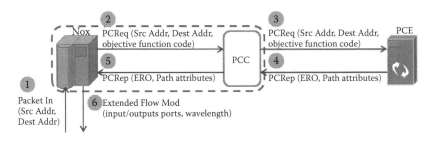

**Figure 14.20**   Detailed information encapsulated in the PCRep and PCReq messages, as well as the OpenFlow Packet In and Flow Mod messages. Src Addr, source address; Dest Addr, destination address.

and data planes, and the data plane of the optical layer was equipped with real hardware including two TPNDs and four PXC, as shown in Figure 14.21. In the data plane, four OF-PXC nodes integrated with DWDM multiplexer (MUX)/demultiplexer (DEMUX) optical filters were used. Two optical transport channel unit 2 (OTU2) (10.7 Gbps)-based OF-TPNDs were attached at OF-PXC1/PXC2, and a shared OF-REG was deployed at OF-PXC4. Two OF-Rs were connected to two TPNDs. Each link was deployed with 10 wavelengths. The link/node OSNR values (as defined in Ref. [42]) were statically configured as shown in Figure 14.21. Both the Nox and PCE were deployed in PCs with 3.2-GHz CPU and 1-GB memory. In Figure 14.21, the number close to each network element (NE) is their ID from the view point of the Nox (i.e., datapath ID in OpenFlow terminology).

In general, the topology server can be located either inside or outside the Nox. Figure 14.22 shows the performance comparison of both cases, in terms of CPU use of the Nox when the flow request intervals are 2 and 5 s. It can be seen that colocating the topology server at the Nox has a very limited negative effect on the load of the Nox. Therefore, it is better to deploy the topology server inside the Nox. When the topology server is integrated into the Nox, the Nox can update the TED locally, which is able to reduce the overall processing latency. Table 14.5 shows the reduction of processing time for TED update when the topology server is integrated with the Nox compared with the case where the topology server is separated. As expected, we verified that the more complex the network topology,

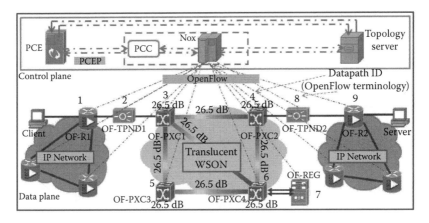

**Figure 14.21**  Experimental setup for OpenFlow/PCE controlled translucent WSON.

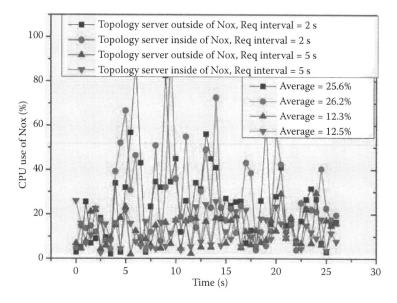

**Figure 14.22**   Comparison of the CPU use of the Nox when the topology server is deployed inside or outside of the Nox. Req, request.

the greater the benefits obtained by integrating the topology server with the Nox.

Therefore, a colocated topology server was the retained deployment model. To verify the feasibility of the proposed solution, we first sent out a flow from the client to the server (as shown in Figure 14.21). Then, according to the procedure depicted in Figure 14.19, path 1 was calculated by the PCE and provisioned by the Nox via the OpenFlow protocol, following the route shown in Table 14.6. Figure 14.23 shows the Wireshark capture of the message sequence. It can be seen that, because the XML file for describing the TED is large, it is divided into several TCP packets for transmission from the topology server to the PCE. The message latency between a TED Req and a TED Rep was

**Table 14.5**   Reduction of Processing Latency when the Topology Server Is Integrated with the Nox

| NETWORK TOPOLOGY | REDUCTION OF PROCESSING LATENCY (ms) |
| --- | --- |
| 3 nodes + 2 links | 5.94 |
| 4 nodes + 5 links | 10.46 |
| 7 nodes + 10 links | 19.98 |
| 10 nodes + 12 links | 29.82 |
| 14 nodes + 22 links | 41.26 |

**Table 14.6**  Provisioning Latencies for Different Paths

| PATHS | ROUTE | ACPL (ms) | OPPL (ms) |
|-------|-------|-----------|-----------|
| 1 | 1-2-3-4-8 | 146.1 | 390.3 |
| 2 | 1-2-3-6-4-8 | 149.9 | 393.6 |
| 3 | 1-2-3-5-6-7-4-8 | 153.7 | 398.1 |

approximately 11.7 ms, and the overall path computation latency was approximately 16.4 ms, as shown in Figure 14.23. Although absolute values may vary depending on hardware and optimization settings, we sent 100 successive requests for path 1 setup to get the average latency, which is more meaningful than the result with a single request. Figure 14.24a depicts the distribution of message latency between a TED Req and a TED Rep, and Figure 14.24b shows the distribution of the overall path computation latency for these repeated path 1 setup requests. We measured that the average message latency between a TED Req and a TED Rep was 11.52 ms, and the average path computation latency for path 1 was 15.71 ms. Figure 14.25a and b show the Wireshark capture of the PCReq and PCRep messages. As previously mentioned in Figure 14.20, it can be seen that the source and destination addresses as well as the objective function code are encapsulated in the PCReq message, and the path computation results, including ERO and path attributes, are included in the PCRep message.

After the provisioning of path 1, we manually set all the wavelengths on link 3–4 to the occupied state, and we started a new flow from the

| No. | Time | Source | Destination | Protocol | Info |
|-----|------|--------|-------------|----------|------|
| 1 | 0.000000 | Client | Server | OFP+TCP | Packet In (AM) (BufID=269) (78B) |
| 6 | 0.058114 | PCE | Nox | PCEP | Open message |
| 8 | 0.064024 | Nox | PCE | PCEP | Open message |
| 10 | 0.064044 | Nox | PCE | PCEP | KeepAlive message |
| 12 | 0.064253 | PCE | Nox | PCEP | KeepAlive message |
| 13 | 0.065715 | Nox | PCE | PCEP | Path computation request message |
| 16 | 0.069820 | PCE | Topology server | TCP | 59786 > 4895 (SYN) Seq=0 Win=584 |
| 17 | 0.070070 | Topology server | PCE | TCP | 4895 > 59786 (SYN, ACK) Seq=0 Ac |
| 18 | 0.070112 | PCE | Topology server | TCP | 59786 > 4895 (ACK) Seq=1 Ack=1 W |
| 19 | 0.070665 | Topology server | PCE | TCP | 4895 > 59786 (PSH, ACK) Seq=1 Ac |
| ··· | 11.7 ms | ··· | | 16.4 | ··· |
| 105 | 0.081458 | Topology server | PCE | TCP ms | 4895 > 59786 (ACK) Seq=100713 Ac |
| 106 | 0.081469 | PCE | Topology server | TCP | 59786 > 4895 (RST) Seq=1 Win=0 L |
| 107 | 0.081474 | Topology server | PCE | TCP | 4895 > 59786 (ACK) Seq=102161 Ac |
| 108 | 0.081484 | PCE | Topology server | TCP | 59786 > 4895 (RST) Seq=1 Win=0 L |
| 109 | 0.082083 | PCE | Nox | PCEP | Path computation reply message |
| 124 | 0.084082 | Nox | PCE | PCEP | Close message |
| 128 | 0.144069 | Nox | OF-R1 | OFP | Flow mod (CSM) (80B) |
| 129 | 0.144074 | Nox | OF-TPND1 | OFP | Flow mod (CSM) (80B) |
| 130 | 0.144946 | Nox | OF-PXC1 | OFP | Flow mod (CSM) (80B) |
| 132 | 0.144983 | Nox | OF-PXC2 | OFP | Flow mod (CSM) (80B) |
| 133 | 0.144986 | Nox | OF-TPND2 | OFP | Flow mod (CSM) (80B) |

TED Req and TED Rep

Path provisioning

**Figure 14.23**  Wireshark capture of the message sequence for end-to-end path provisioning. TED Req, traffic engineering database request; TED Rep, traffic engineering database reply.

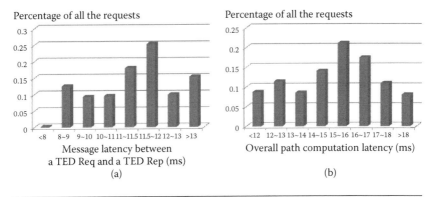

Message latency between
a TED Req and a TED Rep (ms)
(a)

Overall path computation latency (ms)
(b)

**Figure 14.24** Distribution of message latency between a traffic engineering database request (TED Req) and a traffic engineering database reply (TED Rep) message (a) and overall path computation latency for 100 repeated path 1 setup requests (b).

client to the server. Because of the WCC, path 2 was selected by the PCE for transmitting this flow, as shown in Table 14.6. Similarly, the wavelengths on link 3–6 were set to the occupied state, and then a new path 3 was calculated and provisioned for the next incoming flow. Note that, because of the OSNR constraints, an OF-REG was allocated to compensate physical impairments, as shown in Table 14.6.

We repeated the setup/release of paths 1, 2, and 3 more than 100 times to measure the path provisioning latency. Table 14.6 shows the

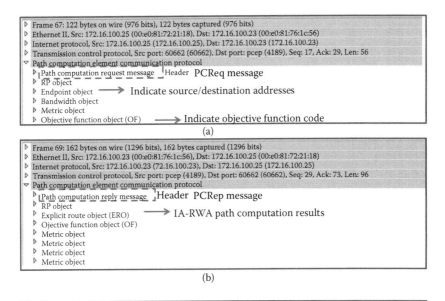

**Figure 14.25** Wireshark capture of the PCReq message (a) and the PCRep message (b).

average control plane latency (ACPL) and the overall path provisioning latency (OPPL) for creating these paths. The ACPL is the average control plane message (i.e., OpenFlow, PCEP, TED Req, and TED Rep) latency to complete the procedure in Figure 14.19. The OPPL consists of both the ACPL and the configuration latency of the data plane hardware. From Table 14.6, it can be seen that, by using the proposed OpenFlow/PCE integrated control plane, the end-to-end path provisioning can be completed within 400 ms in our tested scenario.

In this section, we present an OpenFlow/PCE integrated control plane for IP over translucent WSON, with the assistance of a per-request–based dynamic topology server. This work refines an OpenFlow–based control architecture that relies on SDN principles by adding two entities—a PCE and a topology server—and where the entities communicate using mature, open, and standard interfaces. This architecture allows more flexibility in deployment models, integrating the advances done in the IETF PCE working group and leveraging on a mature protocol.

The overall feasibility and efficiency of the proposed solution was verified on a real network testbed with both control and data planes. The experimental results indicate that a better performance can be obtained if the topology server is integrated within the Nox. In this case, the path provisioning latencies are approximately 400 ms, including both the processing latency in the control plane and the configuration latency of NEs in the data plane. In addition, we believe that a more efficient encoding of the TED can be implemented to further reduce the overall latency. In this line, the architecture would benefit from standard information and data models of the controlled network, as well as a standard protocol for topology management and exchange. Finally, if we compared with GMPLS/PCE-based control planes for translucent WSON, as demonstrated in Refs. [43–46], we can further verify that the OpenFlow is more simple and flexible for control plane design and optical network control.

*Interworking between OpenFlow and GMPLS*

As previously mentioned, in today's commercial IP/DWDM multilayer optical networks, the IP and optical layers are separately operated without dynamic interaction, which leads to a high operational cost, a low network efficiency, and a long processing latency for end-to-end path provisioning. Therefore, a UCP for both IP and optical layers, as one of the key challenges for network carriers, is very important to address the aforementioned issues. GMPLS, with a decade of development and standardization work, is a mature control plane technique for optical transport networks. However, GMPLS-based UCP for IP/DWDM multilayer networks is overly complex for deployment in a real operational scenario. In this context, some studies concluded that GMPLS has devolved from being a UCP to one that is meant only as a control plane for optical transport networks [47]. On the other hand, OpenFlow has been verified as a promising solution for a UCP, but it is not mature enough to control optical switching nodes so far. Therefore, as an intermediate step toward a fully OpenFlow-based UCP, a logical choice at present is to introduce an OpenFlow/GMPLS interworking control plane that is able to use GMPLS to control the optical layer and to dynamically coordinate between IP and optical layers. Another advantage is that GMPLS can offload the OpenFlow controller, which is beneficial for improving the network scalability. In light of this, in this section, we experimentally present three interworking solutions (parallel, overlay, and integrated) between OpenFlow and GMPLS for intelligent wavelength path control in IP/DWDM multilayer optical networks. The overall feasibility of these solutions is assessed, and their performance is quantitatively evaluated and compared on an actual network testbed.

*Solutions for OpenFlow and GMPLS interworking*

*Parallel Solution* Figure 14.26a shows the parallel solution for OpenFlow/GMPLS interworking. It is referred to as parallel solution because the OpenFlow and GMPLS control planes are fully separated to control IP and optical layers (e.g., optical switching nodes and TPNDs), respectively. The interface between the OpenFlow controller (e.g., Nox) and the GMPLS control plane is based on the vendor's proprietary implementations. The procedure for end-to-end path provisioning by using the parallel solution is shown in Figure 14.27. If a new flow arrives at the ingress OF-R, and if this flow does not match any existing flow entry, the first packet of this flow is forwarded to the Nox for processing. According to the source and destination addresses, the Nox informs the GMPLS control plane to set up a label

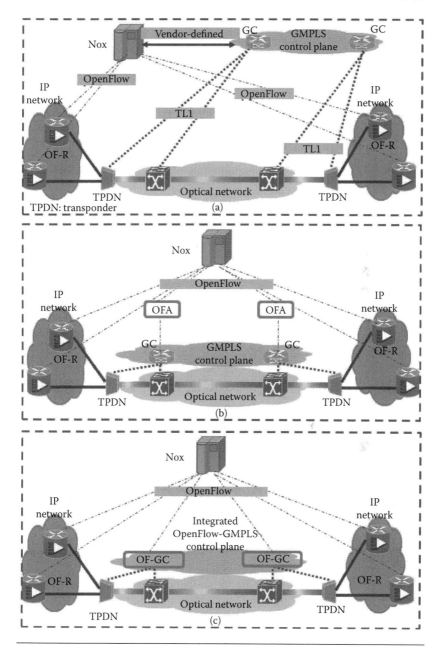

**Figure 14.26**    (a) Parallel solution; (b) overlay solution; (c) integrated solution.

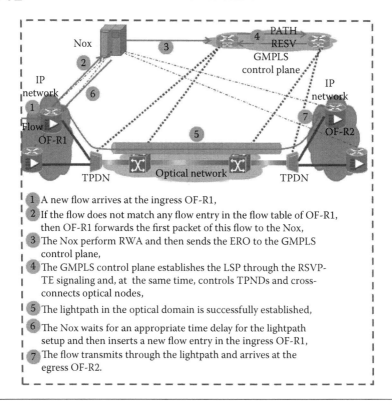

**Figure 14.27** Procedures for path provisioning by using the parallel solution.

switched path (LSP) in the optical domain through the resource reservation protocol (RSVP) with TE (RSVP-TE). Note that the RWA
can be conducted by the Nox given the complete network information
that is available in the Nox. Alternatively, the GMPLS control plane
is able to perform the RWA computation that has been widely studied
in previous works [39]. In both cases, to guarantee an LSP that is fully
established before the flow arrives at the ingress optical node, the Nox
has to wait for a deliberate time latency by considering the RSVP-TE
signaling latency and the configuration latency of the optical nodes
and TPNDs, and then insert a new flow entry in the ingress OF-R to
start the flow transmission. The parallel solution is the most straightforward way for the OpenFlow/GMPLS interworking. However, in
this case, the interface between the Nox and the GMPLS control
plane is not based on a standard protocol. Such nonstandard interface
may introduce serious multivendor interoperability issues because vendors may make different interfaces to satisfy their own requirements.

Moreover, new messages and protocols have to be introduced into the Nox, which increases the design complexity. To address these issues, the overlay solution is proposed, as detailed in the next subsection.

*Overlay Solution*  Figure 14.26b shows the overlay solution for OpenFlow/GMPLS interworking. With this solution, an OFA is introduced between the Nox and each GC, as shown in Figure 14.26b. By introducing the OFA, the Nox is able to control both IP and optical layers through the OpenFlow protocol, which is beneficial for addressing the aforementioned multivendor interoperability issues and simplifying the Nox design. It is referred to as the overlay solution because the OpenFlow-based control plane exists as a higher layer of the GMPLS control plane. The procedure for end-to-end path provisioning by using the overlay solution is shown in Figure 14.28. If the first packet of a new flow is received by the Nox, the Nox performs the RWA computation and then inserts a new flow entry in each OFA

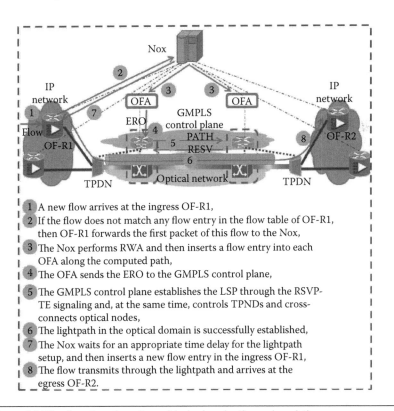

1 A new flow arrives at the ingress OF-R1,
2 If the flow does not match any flow entry in the flow table of OF-R1, then OF-R1 forwards the first packet of this flow to the Nox,
3 The Nox performs RWA and then inserts a flow entry into each OFA along the computed path,
4 The OFA sends the ERO to the GMPLS control plane,
5 The GMPLS control plane establishes the LSP through the RSVP-TE signaling and, at the same time, controls TPNDs and cross-connects optical nodes,
6 The lightpath in the optical domain is successfully established,
7 The Nox waits for an appropriate time delay for the lightpath setup, and then inserts a new flow entry in the ingress OF-R1,
8 The flow transmits through the lightpath and arrives at the egress OF-R2.

**Figure 14.28**  Procedures for path provisioning by using the overlay solution.

along the calculated path in the optical domain by using Flow Mod messages, starting from the ingress OFA. In this case, protocol extensions are required for OpenFlow to send the ERO to the GMPLS control plane through Flow Mod messages. Alternatively, if the RWA is conducted by the GMPLS control plane, the Nox only needs to insert a new flow entry in the ingress OFA, and then the ingress OFA informs the GMPLS control plane to complete RWA computation and lightpath provisioning according to the RWA computation results.

*Integrated Solution*   Figure 14.26c shows the integrated solution for OpenFlow/GMPLS interworking. Different from the overlay solution, a GC and an OFS are integrated in the same controller, which is referred to as an OpenFlow-enabled GMPLS controller (OF-GC). The OF-GC is able to communicate with the Nox by using the standard OpenFlow protocol. The procedure for end-to-end path provisioning by using the integrated solution is shown in Figure 14.29. This procedure is similar to that of the overlay solution. The only difference

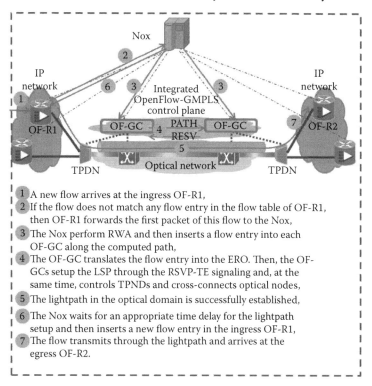

**Figure 14.29** Procedures for path provisioning by using the integrated solution.

is that all the operations between an OFA and the GMPLS control plane (e.g., ERO delivery) are all carried out inside the same controller (i.e., OF-GC), which is helpful for reducing the overall processing latency, at the expense of a higher processing load within an OF-GC because both the GMPLS protocol suite and the OpenFlow module are implemented in a single controller.

*Experimental Setup, Results, and Discussion*   To verify the feasibility and efficiency of the proposed solutions, we constructed a transparent optical network testbed with four PXC in a mesh topology, as shown in Figure 14.30a. Two OTU2-based (10.7-Gbps) TPNDs were attached to PXC1 and PXC2, and two OF-Rs were assumed to be connected to these TPNDs. In addition, two PCs were attached to OF-Rs as clients through Ethernet cables. The Nox statically obtained and dynamically updated the network and resource information, including the topology, wavelength availabilities, and the node address information. In this experiment, we only considered the case that the RWA computation is performed by the Nox by using the RWA algorithm presented in Ref. [48]. The RWA computation by the GMPLS control plane is not experimentally investigated in this work. It is because, in this case, the Nox is not aware of the route of the lightpath computed by the GMPLS control plane, which may raise additional issues, such as information synchronization between the Nox and the GMPLS control plane.

In the experiment, Client-1 first sent out a flow to the destination Client-2. Because this flow did not match any flow entry in OF-R1, OF-R1 forwarded the first packet of this flow to the Nox as a standard OpenFlow Packet In message, as shown in Figure 14.30b, c, and d. Then, the Nox performed the RWA and informed the GMPLS control plane to set up a lightpath according to the RWA results (PXC1→PXC2) by using the parallel, overlay, and integrated solutions. Finally, the Nox waited for a suitable time delay considering the ERO delivery time, the RSVP-TE signaling latency, and the configuration time for PXC and TPNDs, and then inserted a new flow entry into OF-R1 to start the flow transmission by using the Flow Mod message. Figure 14.30b, c, and d shows the Wireshark capture of the OpenFlow and RSVP-TE sequences for this lightpath setup by using the parallel, overlay, and integrated solutions. It can be seen that the OpenFlow message latency is very short. The major contributor to the

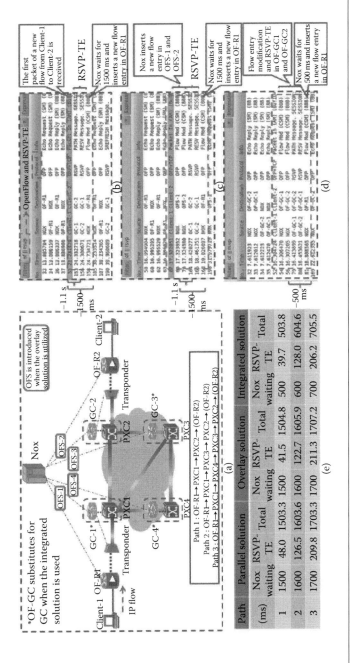

**Figure 14.30** (a) Experimental setup; (b) Wireshark capture of OpenFlow and RSVP-TE procedures for path provisioning when the parallel solution is used; (c) Wireshark capture when the overlay solution is used; (d) Wireshark capture when the integrated solution is used; (e) Summary of path provisioning latencies.

overall latency is the ERO delivery time from the Nox (or OFA) to the GMPLS control plane. Specifically, the parallel and overlay solutions had a similar performance in terms of the OPPL, which was much longer than that of the integrated approach. It is because, in the parallel and overlay solutions, the Nox (or OFA) and GC are separated. Therefore, the time for the Nox (or OFA) to send/configure the ERO to the remote GC was very long (~1.1 s, as shown in Figure 14.30b and c). We further compared experimentally the performance of these solutions by creating paths 2 and 3 with different hops, as shown in Figure 14.30a. The results are summarized in Figure 14.30e. It can be seen that, similarly, the integrated solution outperforms the parallel and overlay solutions in terms of path provisioning latency.

In this section, we present experimentally three interworking solutions (parallel, overlay, and integrated) between OpenFlow and GMPLS for intelligent wavelength path control in IP/DWDM multilayer optical networks. The overall feasibility of these solutions is experimentally assessed, and their performance is quantitatively evaluated and compared on an actual transparent network testbed with four PXC nodes in a mesh topology. The experimental results show that the integrated solution outperforms the parallel and overlay solutions in terms of the OPPL, at the expense of higher design complexity and processing load within a single controller (i.e., OF-GC).

### OpenFlow for Multidomain Optical Networks

In this section, we present an architecture of an OpenFlow-based control plane for a multidomain WSON (shown in Figure 14.31a), focusing on the scenario where such network is operated by a single carrier. The deployment of a single Nox controller in a multicarrier context raises security, confidentiality, and manageability issues and is not addressed in this work. In this architecture, a centralized OpenFlow controller (e.g., Nox) is introduced, which is capable of controlling the OpenFlow-enabled optical nodes through the OpenFlow protocol. Although a centralized Nox can be used to compute an end-to-end path by using an IA-RWA algorithm, provided that the complete network information (e.g., topology, wavelength availability, physical impairments, etc.) is available at the Nox, such an approach becomes infeasible in an actual operational scenario because of two main

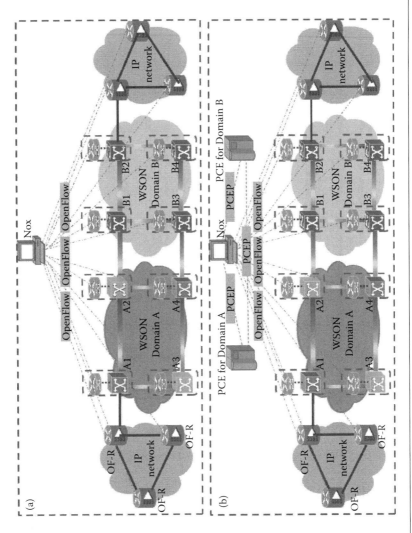

**Figure 14.31** (a) OpenFlow-based control plane for multidomain WSON; (b) Interworking between OpenFlow and PCE for multidomain WSON.

reasons: first, the scalability issues regarding the Nox to know and update all the detailed topology information of the multidomain network and, second, the fact that the IA-RWA computation is a CPU-intensive task [37], potentially harder in a multidomain scenario and should not be carried out in a single controller.

On the other hand, the PCE [27] and the PCEP [32] have been recently proposed and standardized by IETF, targeting, in particular, interdomain path computations. Therefore, a promising solution involves the deployment of collaborating PCEs in the OpenFlow-based control plane for intelligent path computation in multidomain WSON, as shown in the architecture below (Figure 14.31b). In light of this, in this section, we experimentally investigate this issue in detail on a multidomain WSON testbed.

*Interworking between PCE and OpenFlow for Multidomain Optical Networks*

The interworking between the PCE and the Nox is shown in Figure 14.32. The Nox is extended with PCC capabilities, so it can send PCReq messages to the PCE, getting the PCRep after the path computation is completed, by means of the PCEP.

Considering that the IA-RWA computation is CPU intensive, to balance the load in the OpenFlow/PCE integrated control plane, we separate the IA-RWA computation in our proposed solution: the PCE is only responsible for impairment-aware routing (IA-R), whereas the

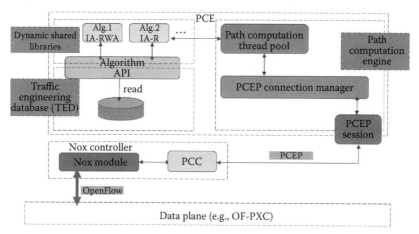

**Figure 14.32** Interworking between the PCE and the Nox controller. API, application programming interface; Alg., algorithm; IA-R, impairment-aware routing.

Nox performs the wavelength assignment (WA). A new IA-R algorithm, with the objective function code [41] 32770, is implemented in the PCE for intra-/interdomain path computation. This algorithm, which is developed based on Ref. [42], is able to compute a path with minimum cost and satisfied OSNR inside each domain. The multidomain path computation is based on the backward recursive path computation (BRPC) [49]. The destination domain PCE computes a virtual shortest path tree (VSPT) from the domain ingress nodes to the destination node and then sends the computed VSPT to the upstream PCE to compute its own VSPT. The upstream PCE recursively applies this procedure up to the source domain, obtaining an optimal path from the source node to the destination node.

*OpenFlow and PCEP Procedures for Dynamic Path Computation and Lightpath Provisioning*

*Solution 1: PCE-Based BRPC-Enabled IA-R Combined with Nox-Based WA* The procedure of solution 1 is shown in Figure 14.33a. If a new flow arrives at the ingress OF-R (OF-R1), and if this flow does not match any flow entry in the flow table of OF-R1, OF-R1 forwards the first packet of this flow to the Nox, which requests a path computation to the PCE in the first domain (PCE A). After the initial session handshake, the Nox/PCC sends a PCReq message. Upon completion of the aforementioned BRPC procedure, in which PCE A and PCE B collaborate to compute an end-to-end path considering the OSNR requirements, the result is sent to the Nox. After that, the Nox performs the WA for the computed path and then proceeds to set up the end-to-end lightpath by using the approach presented in "OpenFlow-Based Wavelength Path Control in Transparent Optical Networks." Note that, to guarantee that the lightpath is successfully established before the flow arrives at the ingress optical node, the Nox has to wait for a suitable time latency for cross-connections within each OF-PXC and then insert a new flow entry in OF-R1 to start the flow transmission.

*Solution 2: PCE-Based Parallel IA-R Combined with Nox-Based WA* Considering that, in most operational cases, the domain boundary nodes are deployed with 3R regenerators, and to reduce the end-to-end path computation latency, we propose the alternative solution 2, as shown in

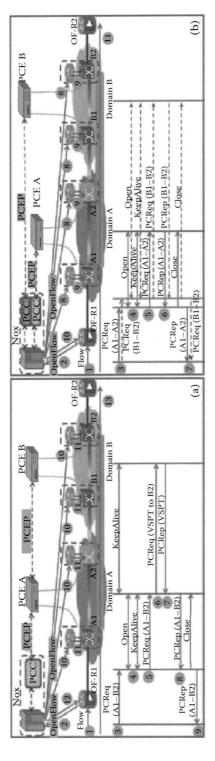

**Figure 14.33**  (a) OpenFlow and PCEP procedures of solution 1; (b) OpenFlow and PCEP procedures of solution 2.

Figure 14.33b. In this solution, the Nox (considering interdomain links and domain connectivity) decomposes an interdomain path request into multiple intradomain requests and sends these requests to different PCEs simultaneously (parallelization). Once these intradomain paths are computed by the different PCEs, the Nox combines the results as an end-to-end path and assigns the wavelengths for each segment. This approach is conceptually similar with the hierarchical PCE (H-PCE) [50] architecture, in which the Nox acts as a parent PCE.

### Experimental Setup, Results, and Discussions

To evaluate the efficiency of the proposed solutions, we set up a WSON testbed with two domains, as shown in Figure 14.34. Domain A models the Japan core network topology, and domain B models a Pan European mesh network topology. Each domain was deployed with a dedicated PCE, and permanent PCEP adjacencies existed between them. The topology and the OSNR values of each domain were configured in the TED of the corresponding PCE, and a simplified network topology with wavelength availability information was statically configured in the Nox. Because of the lack of hardware, only four nodes were equipped with real data plane, including OF-PXC-A1/A2 in domain A and OF-PXC-B1/B2 in domain B, as shown in Figure 14.34. The rest of the nodes were emulated within the PCE and Nox.

In this experiment, Client-1 sent out a flow to the destination Client-2, as shown in Figure 14.34. This flow did not match any flow entry in OF-R1, so OF-R1 forwarded the first packet of this flow to the Nox as a standard OpenFlow Packet In event (as shown

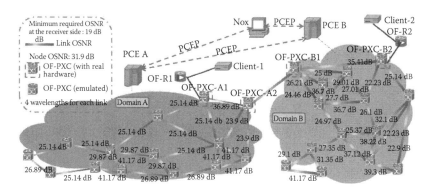

**Figure 14.34**   Experimental setup.

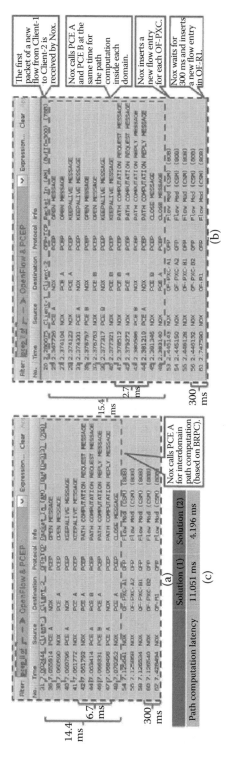

**Figure 14.35**  (a) Wireshark capture of solution 1 for dynamic path computation and lightpath provisioning; (b) Wireshark capture of solution 2 for dynamic path computation and lightpath provisioning; (c) Comparison of the path computation latency of solutions 1 and 2.

in Figure 14.35a, b). Then, according to the source/destination IP addresses encapsulated in the IP packet, as well as the topology information in the Nox, the Nox sent a PCReq to the PCE through the PCC. When solution 1 was used, the PCReq was sent to PCE A, for computing an end-to-end path from the source OF-PXC-A1 to the destination OF-PXC-B2 by using the BRPC algorithm. On the other hand, when solution 2 was used, two PCReq messages were sent to PCE A and PCE B simultaneously to calculate a path from OF-PXC-A1 to OF-PXC-A2 and from OF-PXC-B1 to OF-PXC-B2, respectively. Upon reception of the path computation results, the Nox assigned the wavelength and then inserted new flow entries into the OF-PXC along the calculated path to set up the lightpath. Figure 14.35a and b shows the Wireshark captures of solutions 1 and 2, respectively, during the lightpath provisioning, which are the same as the procedures that we depicted in Figure 14.33. The overall lightpath provisioning latency was similar with both solutions, approximately 450 ms, including the Nox waiting time of 300 ms, as shown in Figure 14.35a and b. However, note that, as we expected, the path computation latency of solution 2 was approximately 2.7 ms, which was shorter than that of solution 1 (~6.7 ms). However, in solution 2, the Nox had to use more time for the session handshake with two PCEs; therefore, the overall processing latency between PCE and Nox by using solutions 1 and 2 was similar (14.4 and 15.4 ms, respectively). Because the only difference between the two solutions is the path computation approach, to get more insightful values, we sent 1000 random interdomain PCReqs on our testbed and measured the average path computation latency. The results are summarized in Figure 14.35c. It can be observed that the path computation latency of solution 2 is approximately 7 ms shorter than that of solution 1. Combined with the relatively longer session handshake latency of solution 2, we can conclude that both solutions are with a similar performance in terms of the lightpath path provisioning latency in the OpenFlow/PCE integrated control plane. Finally, let us note that the longer the domain sequence or the longer the intradomain path computation delay, the greater the benefits of solution 2 regarding end-to-end path computation latency because of parallelization. However, in this case, the Nox has to insure optimality when selecting interdomain links.

## OpenFlow for Other Optical Switching Paradigms

*OpenFlow for OBS Networks*

After decades of development, optical networking has witnessed great improvements, especially the gradual maturity of subwavelength optical switching technologies. Among those, OBS [51] network is regarded as one of the most promising paradigms for future optical Internet deployment because it is capable of providing on-demand transparent optical bandwidth to bursty data traffic by using commercially available fast switches. Until now, several network testbeds have been successfully demonstrated, and currently, some commercial products are also available.

Conventionally, the multiprotocol label switching (MPLS) scheme is usually used to dynamically maintain label switching paths (LSP) for the optical bursts, forming a so-called labeled OBS (LOBS) [52] network. However, such an MPLS-based control plane manipulates the OBS layer solely, whereas a service-oriented cross-layer optimization is rarely considered because of the complexity of interoperations between the optical layer and upper layers; thus, an efficient cross-layer UCP is still necessary to leverage the best of optical switching by realizing self-adapting optical transmission provisioning in an intelligent way.

With great simplicity and layer-free controllability, OpenFlow is right for the cross-layer UCP solution for LOBS networks. In the following subsections, the principles of the OpenFlow-controlled LOBS networks are described briefly from the aspects of network architecture, node structure, signaling procedure, and protocol extensions. Finally, an experimental demonstration is presented as well.

*Network Architecture*  Figure 14.36 depicts an overview of the OpenFlow-enabled LOBS network. Traditionally, the optical

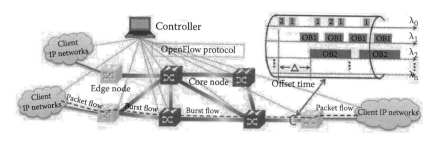

**Figure 14.36**  Architecture of the OpenFlow-based LOBS Network.

switching networks lay in the bottom layer and are transparent to the IP layer, but because OpenFlow innately supports cross-layer forwarding control, the client IP networks and the LOBS layer may operate under the instruction of the same network controller, forming a flat and unified switching network. Compared with the distributed control plane, where the shortest path first (OSPF)–based routing and resource RSVP–based resource reservation can usually be used and where each node take part in routing computation and RSVP processing for end-to-end LSP provisioning, the OpenFlow-based network architecture would be much easier to operate with a centralized controller (e.g., Nox), taking all the responsibilities for maintaining network status and making configuration decisions. The optical switching nodes are only required to support the OpenFlow protocol and to conduct switching operations based on the specified flow entries.

In IP networks, OpenFlow switches perform packet-forwarding actions by looking up flow tables written by the controller, and the first IP packet of an unrecognized packet flow would be forwarded to the controller for routing decisions. In LOBS networks, however, the switching units are optical bursts (OBs), which are the aggregation of bundles of IP packets belonging to a certain forwarding equivalent class (such as packets with the same destination address) and are transparently transmitted on a wavelength channel without any O/E/O conversion. Therefore, for the controller, the burst header packets (BHP) forwarded on the control channel, each identifying a specific optical burst, would compose burst flows in the LOBS domain just like packet flows in the IP domain. In other words, at the boarder of the IP and optical domain, namely at the LOBS edge nodes, packet flows are aggregated into burst flows.

Considering that the flow conversion procedure is essentially the burst assembly procedure and that packets of different service types may be applied with different assembly strategies to reflect quality of service (QoS) differentiation on latency and bit error rate (BER) requirements, when the first packet of a newly arrived flow reaches the LOBS edge node, it might be sent to the controller for assembly/scheduling parameter consultancy (such as burst generation threshold, offset time, output wavelength, modulation format, etc.). Upon receiving this packet, the controller may automatically execute policy-based flow configurations, including setting appropriate burst assembly parameters and calculating a switching path, and then issue flow entries to the nodes.

Depending on their functionalities, the OpenFlow-enabled LOBS nodes can be divided into edge nodes and core nodes. Edge nodes are responsible for grooming packet flows into optical burst flows and scheduling them onto wavelength channels according to the controller's instructions, whereas core nodes simply perform label switching of BHPs and all-optical switching of the corresponding OBs. Both edge nodes and core nodes need an interface to communicate with the controller via the OpenFlow protocol and interpret flow entries into local configurations for processing.

More specifically, as shown in Figure 14.37 (left), the edge node consists of an OpenFlow protocol interface, a burst assembler, a burst scheduler, and a burst header generator. The assembler, in turn, consists of a packet classifier and a burst generator (queuing buffer). For each flow, the flow table contains all the configuration information set by the controller depending on its service type and QoS level. This information is mapped into burst assembly parameters (such as time/length threshold for burst aggregation) and burst scheduling parameters (such as offset time, wavelength, bit rate, and modulation format) in the assembler and scheduler, respectively. The packet classifier sorts client traffic into different queuing buffers, where bursts are generated with those parameters. Meanwhile, the burst header generator creates a BHP if a certain burst is ready to be sent out.

As for the core node (Figure 14.37, right), the burst header processor conducts label switching according to the burst forwarding table

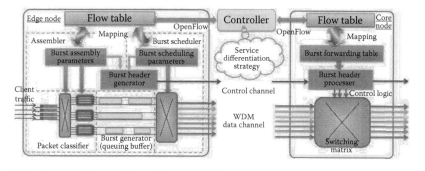

**Figure 14.37** Architecture of the OpenFlow-Enabled LOBS Edge Node and Core Node. WDM, wavelength division multiplexing.

that is mapped from the flow table written by the controller as well as OB switching by the switching matrix to guide the OBs onto specific output ports after an exact offset time.

Based on the aforementioned network architecture and OpenFlow-enhanced LOBS nodes, the typical signaling procedure for a flow setup and transmission would be as depicted in Figure 14.38 and can be summarized as follows:

Step 1: The source OBS edge node buffers the incoming packet flow and checks against existing flow configurations. If it is an unrecognized flow, then continue to step 2; if it matches a certain existing flow, then skip to step 5.

Step 2: The source edge node sends the first packet of the unrecognized flow as a link request message (LRQ) to the controller and buffers the packets when waiting for response.

Step 3: The controller examines the packet and decides whether to reject the transmission request or calculate an end-to-end switching path and related burst configurations, and inserts appropriate flow entries into all the nodes along the calculated path.

Step 4: The OBS edge/core nodes that receive flow entries would convert them into local configurations (e.g., burst assembly/scheduling/forwarding parameters). Optionally, the nodes can reply the controller with a confirmation message to indicate operation success/failure.

Step 5: The source OBS edge node examines the queuing buffer against flow configuration parameters and sends out BHPs on the control channel when burst generation thresholds are reached to reserve resource for the corresponding OBs.

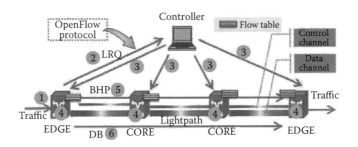

**Figure 14.38**  Signaling procedure of the OpenFlow-based LOBS network. DB, data burst.

Step 6: After an offset time, the OBs are sent out from the source OBS edge node and transported to the destination edge node transparently.

*Protocol Extensions*    Considering that the original OpenFlow protocol is not preferably supportive for optical networks, and especially, in the aforementioned network scenario, where the flow conversion from packets to optical bursts needs a series of parameters not included in current flow-entry specifications, some minor extensions for the OpenFlow protocol may be needed.

In the OpenFlow protocol, the flow control functionalities are basically realized by the nodes that execute flow entries, which are issued by the controller with Flow Mod messages, and for the flow-entry structure, there is a flexible layer-free matching field (Struct OFP_MATCH) to distinguish a specific traffic flow and a corresponding action field to indicate the operations that are going to apply for that flow. For the protocol versions released before version 1.1, the Flow Mod message does not provide MPLS label fields in the OFP_MATCH struct so that an alternative solution is to use other matching fields to be mapped into the input and output label for BHP forwarding (as illustrated by the example in the upper-left corner of Figure 14.39, where the TCP source and destination ports are used to indicate the labels in the LOBS nodes). Moreover, for both versions 1.0 and 1.1, the action field is only capable of giving an output_port for the BHP but cannot be extended to carry all the burst assembly and scheduling parameters that the edge nodes need; thus, an additional vendor-defined message could be exploited as a compensation to the flow entry and could be sent to the LOBS edge nodes to set the burst assembly/scheduling configurations (as depicted in Figure 14.39).

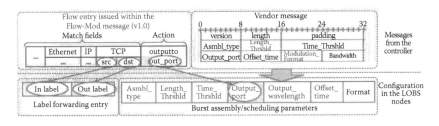

**Figure 14.39**    Flow-entry mapping and protocol extension for OpenFlow version 1.0 to support LOBS networks.

Because OpenFlow version 1.2 and above versions have claimed to provide expandability in the matching field as well as the action field by the type-length-value (TLV) structure, the burst assembly and scheduling parameters can be implemented reasonably by extending certain signaling frames in the action field.

*Experimental Demonstration*  The OpenFlow-based LOBS testbed consists of two self-developed LOBS edge nodes and one core node [18,53], constructing a linear topology for a proof-of-concept demonstration. The detailed setup is shown in Figure 14.40.

The control channel of the LOBS networks used a 1310-nm wavelength. Two other wavelengths (1557.8 and 1553.2 nm) were used for data channels. Both the control and data channels worked at a 1.25-Gbps data rate. Linux-based PCs were deployed on top of each LOBS node to serve as protocol interfaces between the nodes and the controller. The controller was running the Nox.

In our experiment, IP packet flows were injected into the LOBS edge node 1 that, in turn, sent the first packet as an LRQ to the Nox controller. The controller was preconfigured to hold the full information about the whole network, including topology and label/wavelength availabilities. Therefore, the Nox could make burst assembly/scheduling strategy and routing decision immediately when a link-request message was received. Then, we tested the signaling procedure for LSP creation and measured their latency from the controller's view.

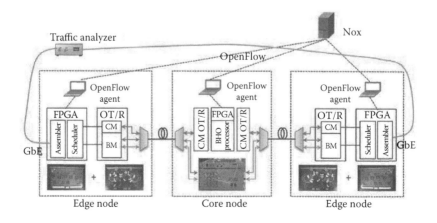

**Figure 14.40**  The demonstrated Openflow-enabled LOBS networks. FPGA, field programmable gate array; OT/R, optical transmitter/receiver; CM, control packet module; BM, burst module.

| Time (s) | From | To | Protocol | Message type |
|---|---|---|---|---|
| 138.681095 | OBS-EDGE2 | Nox | OFP | Echo Reply (SM) |
| 138.681181 | Nox | OBS-EDGE2 | OFP | Echo Reply (SM) |
| ① 141.316115 | OBS-EDGE1 | Nox | OFP | Packet In (AM) |
| 141.316902 | Nox | OBS-EDGE1 | OFP | Flow Mod (CSM) |
| ② 141.317469 | Nox | OBS-CORE | OPF | Flow Mod (CSM) |
| 141.317638 | Nox | OBS-EDGE2 | OFP | Flow Mod (CSM) |
| | | | | Link request message |

**Figure 14.41**  The signaling messages captured on the Nox controller.

Figure 14.41 shows the signaling messages captured on the Nox controller with the Wireshark software. It can be observed that, after the Nox controller received the LRQ message from the OBS edge node 1, it inserted flow entries into all the nodes along the computed path (i.e., OBS-EDGE1→OBS-CORE→OBS-EDGE2) in a sequential order (as shown in Figure 14.41). The latency of the whole OpenFlow signaling procedure was approximately 1.5 ms.

The experimental results verified the feasibility of OpenFlow-based UCP for LOBS networks, but if a large-scale network is considered, the single centralized Nox controller may face a serious scalability issue, and the link setup latency would probably grow to an unacceptable degree, which is obviously a drawback of the centralized control plane.

*OpenFlow for Multilayer Multigranularity Optical Switching Networks*

Despite a decade of development and standardization efforts, with three different interconnection models (i.e., peer, overlay, and border peer) [54,55] proposed for a GMPLS-based UCP in IP and optical multilayer networks, there are no commercial deployments of these models to date, and the debate for their practicability in a real operational scenario grows in intensity [2]. The first development, feasibility validation, and performance evaluation of a GMPLS-based UCP handling both MPLS transport profile (MPLS-TP) and WSON were experimentally conducted in Ref. [56]. Nevertheless, because the GMPLS standardization significantly lags behind the product development, vendors make private protocol extensions to satisfy their own requirements. As a result, the interworking among GMPLS products from different vendors is very difficult [57–59], which seriously reduces flexibility in network construction.

On the other hand, SDN and, in particular, the OpenFlow architecture, which allows operators to control the network using software running on a network operating system (e.g., Nox) within an external controller, provides maximum flexibility for the operator to control a network and matches the carrier's preferences given its simplicity and manageability. In light of this, in this section, we present an OpenFlow-based UCP for multilayer multigranularity optical switching networks. We verify the overall feasibility and efficiency of the proposed OpenFlow-based UCP and quantitatively evaluate the latencies for end-to-end path creation and restoration through a field trial among Japan, China, and Spain.

*Enabling Techniques of OpenFlow-Based UCP for Multilayer Multigranularity Optical Switching Networks* Figure 14.42 shows the network architecture. In this chapter, we consider a multilayer optical switching network consisting of three different layers: IP, OBS, and optical circuit switching (OCS), where the switching granularities are packet, burst, and wavelength, respectively. This network architecture has been recently proposed by the Open Grid Forum (OGF) [60] and experimentally investigated in Refs. [61,62], where small-scale

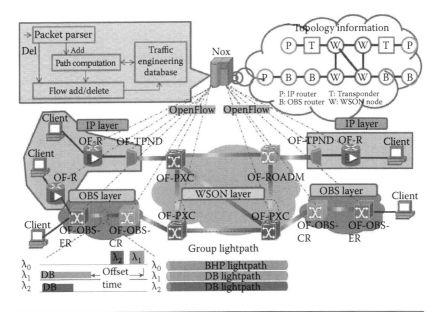

**Figure 14.42** Proposed network architecture for an OpenFlow-based multilayer multigranularity optical switching network under the control of a single Nox controller.

OBS networks are strategically deployed at some edges of a WSON to access and assemble client traffic. The motivations for an OBS layer are two-fold. First, it can improve the resource use and network flexibility with OBS statistical multiplexing [62]. Second, it illustrates that the OpenFlow-based UCP is applicable to various switching paradigms. All the nodes in the different layers and TPNDs are enhanced with OpenFlow capabilities, as detailed in previous sections, and an external Nox controller is introduced to control the network through the OpenFlow protocol, as shown in Figure 14.42.

Once a new IP flow arrives at the ingress OF-R (or OF-OBS-ER), the ingress router forwards the first packet of this flow (or a copy of the BHP) to the Nox. The Nox calculates the route, assigns the wavelength in the optical domain, and then adds a flow entry in all the nodes along the computed route in a sequential order, starting from the ingress router, as shown in Figure 14.43. In turn, according to the inserted flow entry, each OFA within the OBS/WSON nodes automatically controls the underlying hardware to cross-connect the corresponding ports. Note that, if BHP/data burst (DB) pairs are required to transmit through the WSON, the Nox has to insert a pair of flow entries to establish a group lightpath consisting of at least one BHP lightpath and one DB lightpath for end-to-end BHP/DB delivery, as shown in Figure 14.42. Moreover, upon link failure, the OF-TPND detects the loss-of-signal (LOS) and then forwards an alarm to the Nox to trigger the restoration path establishment, as the procedure shown in Figure 14.44.

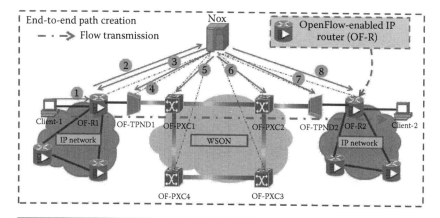

**Figure 14.43** Procedure for end-to-end path creation.

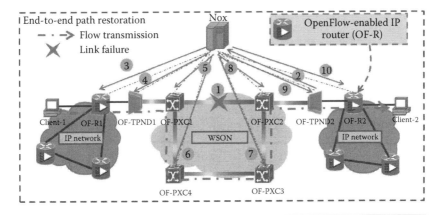

**Figure 14.44** Procedure for end-to-end path restoration.

*Field Trial Setup, Results, and Discussion* The field trial setup is shown in Figure 14.45. An IP/OBS/WSON network with the aforementioned extensions was constructed in Beijing (China), Tokyo (Japan), and Barcelona (Spain). For the OpenFlow-based UCP, a Nox was deployed in Japan, communicating with all the nodes by using the OpenFlow protocol through public Internet. The OpenFlow protocol used in this field trial was based on version 1.0, and the version of the Nox controller was Nox-Classic [28]. In the data plane, the hardware of OF-TPND1 and OF-ROADM1 was emulated by using a

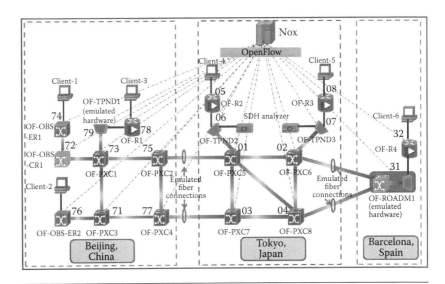

**Figure 14.45** Field trial setup.

commodity PC, which can emulate the failure alarm similar to a real TPND. The remaining nodes were equipped with real hardware. The whole network is assumed to be operated by a single carrier in Japan, representing a scenario where a carrier rents network resources from other carriers in different countries to provide international services. In Figure 14.45, the number close to each NE is their ID from the viewpoint of the Nox (i.e., datapath ID in OpenFlow terminology). The network topology was statically configured in the Nox controller. Each link was deployed with four wavelengths, including W1 (1552.52 nm), W2 (1553.33 nm), W3 (1554.13 nm), and W4 (1554.94 nm). The Nox was equipped with a 3.2-GHz CPU (Core 2 Duo) and 1-GB of RAM memory.

We first verified the dynamic cross-layer path creation by setting up five paths, one by one, as summarized in Table 14.7. Specifically, for path 1 provisioning between Client-1 and Client-2, a path across the OBS and WSON layers was provisioned. In this case, the Nox inserted two flow entries in OF-PXC1 and OF-PXC3, corresponding to one BHP lightpath and one DB lightpath. The signaling procedure was captured by Wireshark, as shown in Figure 14.46. Considering that lightpaths are usually bidirectional in most operational scenarios, different wavelengths were selected for paths 3, 4, and 5 because of the WCC. We also used the passive approach presented in "OpenFlow-Based Wavelength Path Control in Transparent Optical Networks" to release these paths. In Table 14.7, the path setup/release latency in the control plane is shown, which consists of the OpenFlow message propagation time and the Nox processing latency. The overall latency includes the aforementioned control plane processing latency, the interaction time between an OFA and its controlled hardware, and the internal processing time of optical nodes/TPNDs. Because this internal processing time is difficult to measure and varies depending on the hardware and the vendor, for simplicity, we used a reference value of 10 ms for all nodes/TPNDs. For the other latency contributors, we measured their average values by repeating the experiment more than 100 times. We observed that the processing time in the Nox was relatively short and that the key contributor to the path setup latency in the control plane was the OpenFlow message propagation time (i.e., Packet In and Flow Mod).

**Table 14.7**  Summary of Paths

| NO. | SRC. | DEST. | CALCULATED PATH | WAVELENGTH | SL (CP) (ms) | SL (OVERALL) | RL (CP) (ms) | RL (OVERALL) (ms) |
|---|---|---|---|---|---|---|---|---|
| 1 | Client-1 | Client-2 | 74→72→73→71→76 | W1, W2 | ~149 | ~432 | ~141 | ~298 |
| 2 | Client-3 | Client-4 | 78→79→73→75→01→06→05 | W1 | ~138 | ~388 | ~132 | ~257 |
| 3 | Client-4 | Client-5 | 05→06→01→02→07→08 | W2 | ~13 | ~257 | ~11 | ~131 |
| 4 | Client-4 | Client-6 | 05→06→01→02→31→32 | W3 | ~181 | ~424 | ~170 | ~290 |
| 5 | Client-6 | Client-3 | 32→31→02→01→75→73→79→78 | W4 | ~332 | ~578 | ~321 | ~445 |

*Note:* SL/RL (CP), path setup/release latency in the control plane; SL/RL (overall), overall path setup/release latency.

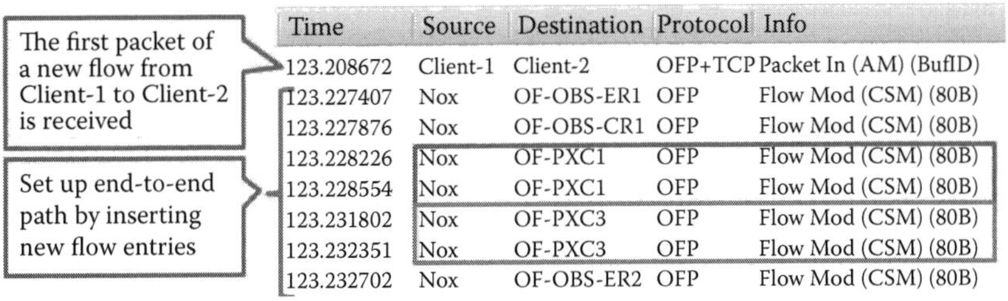

| | Time | Source | Destination | Protocol | Info |
|---|---|---|---|---|---|
| The first packet of a new flow from Client-1 to Client-2 is received | 123.208672 | Client-1 | Client-2 | OFP+TCP | Packet In (AM) (BufID) |
| | 123.227407 | Nox | OF-OBS-ER1 | OFP | Flow Mod (CSM) (80B) |
| | 123.227876 | Nox | OF-OBS-CR1 | OFP | Flow Mod (CSM) (80B) |
| | 123.228226 | Nox | OF-PXC1 | OFP | Flow Mod (CSM) (80B) |
| Set up end-to-end path by inserting new flow entries | 123.228554 | Nox | OF-PXC1 | OFP | Flow Mod (CSM) (80B) |
| | 123.231802 | Nox | OF-PXC3 | OFP | Flow Mod (CSM) (80B) |
| | 123.232351 | Nox | OF-PXC3 | OFP | Flow Mod (CSM) (80B) |
| | 123.232702 | Nox | OF-OBS-ER2 | OFP | Flow Mod (CSM) (80B) |

**Figure 14.46**  Wireshark capture for path 1 setup.

We also evaluated the dynamic lightpath restoration, as summarized in Table 14.8. The control plane processing latency from the generation of a failure alarm to the insertion of new flow entries in every node along the restoration path ranged from 21 to 241 ms, and the overall latency ranged from 269 to 487 ms. Similar with the path setup/release latency, the major contributor to the restoration latency was also the OpenFlow message propagation time. When the nodes are geographically separated from the Nox, this latency increases significantly. Figure 14.47 shows the Wireshark capture of the message procedure for path 5 restoration. It can be seen that an alarm indication is inserted in the Packet In message. If such Packet In message was received by the Nox, the Nox automatically created the restoration path.

Obviously, scalability is a major issue of an OpenFlow-based UCP because, in a centralized architecture, the Nox has to perform both path computation and provisioning. To test the performance of an OpenFlow UCP in the case of high load, 1000 new flows at Client-3 and Client-5 were generated simultaneously, with a flow interval time of 0, 1, and 5 s. Figure 14.48 shows the CPU use of the Nox. It can be seen that, when the interval time of flows was 0 s, the Nox was very busy and highly loaded. However, when the time interval was slightly increased to 1 s, the CPU load was sharply alleviated. However, in any case, considering the scalability issue of a centralized UCP, the introduction of a powerful PCE into the OpenFlow-based UCP is one of the promising solutions to offload the Nox and thus enhance the network scalability, as we introduced in previous sections.

**Table 14.8**   Summary of Dynamic End-to-End Restoration

| NO. | WORKING PATH | FAILED LINK | LOS DETECTION | RESTORATION PATH | CPLR (ms) | OLR |
|-----|-------------|-------------|---------------|------------------|-----------|-----|
| 1 | Path 3 | OF-PXC5–OF-PXC6 | OF-TPND3 | 05→06→01→04→02→07→08 | ~21 | ~269 |
| 2 | Path 2 | OF-PXC1–OF-PXC2 | OF-TPND2 | 78→79→73→71→77→03→01→06→05 | ~83 | ~332 |
| 3 | Path 5 | OF-PXC5–OF-PXC6 | OF-TPND1 | 32→31→04→03→77→71→73→79→78 | ~241 | ~487 |

*Note:* CPLR, control plane latency during restoration; OLR, overall latency during restoration.

| 200.350027 | OF-TPND1 | NOX | OFP+TCP | Packet In (AM) (BufI |
| 200.351261 | NOX | OF-R4 | OFP | Flow Mod (CSM) (80B) |
| 200.351591 | NOX | OF-ROADM1 | OFP | Flow Mod (CSM) (80B) |
| ... | | ... | | ... |
| 200.356906 | NOX | OF-PXC4 | OFP | Flow Mod (CSM) (80B) |
| 200.357447 | NOX | OF-PXC3 | OFP | Flow Mod (CSM) (80B) |
| 200.357816 | NOX | OF-PXC1 | OFP | Flow Mod (CSM) (80B) |
| 200.358170 | NOX | OF-TPND1 | OFP | Flow Mod (CSM) (80B) |
| 200.358515 | NOX | OF-R1 | OFP | Flow Mod (CSM) (80B) |

| 0080 | 00 00 50 00 00 20 4e 92  00 00 41 6c 61 72 6d 20 | ..P.. N. ..arm |
| 0090 | 44 65 76 69 63 65 20 44  57 34 32 30 30 20 70 6f | Device D W4200 po |
| 00a0 | 72 74 20 30 2f 33 2f 31  20 6d 65 73 73 61 67 65 | rt 0/3/1 message |
| 00b0 | 20 63 72 5f 6c 6f 73 | cr_los |

**Figure 14.47**   Wireshark capture for path 5 restoration.

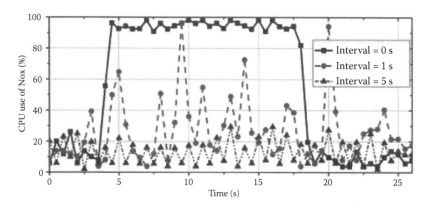

**Figure 14.48**   CPU use of the Nox.

*OpenFlow for Spectrum-Sliced Elastic Optical Path Networks*

The spectrum-sliced elastic optical path network, which is also known as the EON, or the flexible grid optical network has been recently proposed to more efficiently use network spectrum resources [63]. In an EON, optical spectrum ranges are adaptively allocated to an optical path according to the client (e.g., IP) traffic demand, modulation format, and path attributes (e.g., physical length or optical impairments) [64]. For dynamic and intelligent end-to-end optical path provisioning and IP traffic offloading in an EON, a control plane is a key enabling technique.

Therefore, several studies have started to design a GMPLS-based control plane for an EON [65–69]. Despite massive progress, note that such studies mainly focused on the control of the optical layer. To the best of our knowledge, the specific GMPLS-based unified control for both IP and EON layers has not been addressed yet. More importantly, although more mature and intelligent, a GMPLS-based control plane may not be an ideal solution for the deployment in a real operational scenario because of its distributed nature and high complexity, especially for a unified control functionality in IP and optical multilayer networks, as previously mentioned.

In light of this, for the first time, our previous work in Ref. [14] presented experimentally an OpenFlow-based control plane, referred to as OpenSlice, to achieve a dynamically optical path provisioning and IP traffic offloading in an EON. In this section, we propose a more transparent and detailed description of OpenSlice and the main points of this study.

*OpenSlice: Architecture and Protocols*

*Network Architecture* Figures 14.49 and 14.50 show the proposed network architecture. To connect IP routers to an EON with IP offloading capability, a multiflow optical TPND (MOTP) has been proposed [70] and demonstrated [71] recently. In an MOTP, a flow classifier at the transmitter side is deployed to identify the packets of an incoming IP flow according to their destination addresses, virtual local area network tags, etc., and to split such flow into several

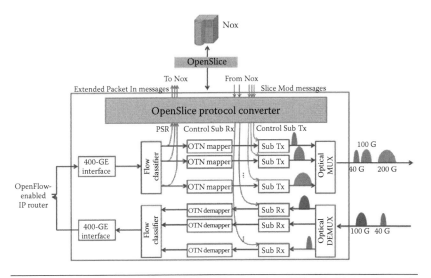

**Figure 14.49** Multiflow OTP [70,71] with OpenSlice extensions. OTN, optical transport network; GE, gigabit ethernet.

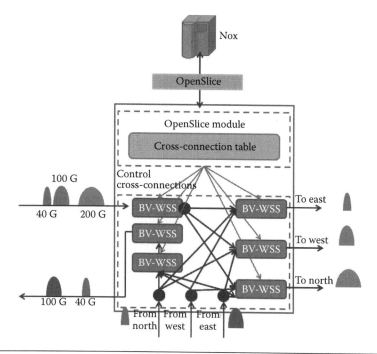

**Figure 14.50**   Bandwidth variable WXC [63] with OpenSlice extensions.

subflows [70]. Before mapping each subflow to an appropriate optical transport unit, the flow classifier generates a path setup request (PSR) for each subflow, containing not only the source/destination addresses, but also the bit rate of each subflow. This requires that the flow classifier is able to monitor or detect the bit rate for a flow, and the approach for this detection is beyond the scope of this chapter.

The EON layer is configured with the bandwidth variable WXC (BV-WXC) [63], which are implemented by using BV wavelength selective switches (BV-WSS) [63]. Both the MOTP and BV-WXC are extended with the OpenSlice functionality, which are referred to as OpenSlice-enabled MOTP (OF-MOTP) and OpenSlice-enabled BV-WXC (OF-BV-WXC), as shown in Figures 14.49 and 14.50, respectively. A centralized controller is introduced to control all the IP routers through the standard OpenFlow protocol and control all the OF-MOTPs and OF-BV-WXC through the OpenSlice protocol. The controller implemented in this chapter is based on Nox. For simplicity, we will still use Nox to represent our controller, but note that it is an extended controller from the original Nox enabled to support the OpenSlice protocol.

Specifically, in an OF-MOTP, an OpenSlice protocol converter is deployed, which is able to convert a PSR into an extended OpenFlow Packet In message (as detailed next) for further processing by the Nox. In addition, it can convert a *Slice Mod* message (as detailed next) into a vendor-specific command (e.g. TL1) to control each Tx/Rx pair in an MOTP for a suitable central frequency (CF), slot width (SW), and modulation format.

In an OF-BV-WXC, an OpenSlice module with a cross-connection table (CT) is introduced. The CT maintains all the cross-connection information within an OF-BV-WXC, including input/output ports, CF, SW, and modulation format (as shown in Figure 14.51). Note that the calculation of CF and SW presented in Figure 14.51 is based on IETF work-in-progress drafts [72]. Conceptually, the CT is similar with the flow table in the standard OpenFlow terminology. The Slice Mod message can add/delete a cross-connection entry into the CT and, thus, control the OF-BV-WXC, allocating a cross-connection with the spectrum bandwidth to create an appropriately sized optical path.

*OpenFlow Protocol Extensions* The key extensions to the OpenFlow protocol to support an EON are briefly summarized as follows: (1) The Feature Reply message is extended to report the new features of an EON (e.g., flexi-grid switching capability, available spectrum ranges, etc.) to the Nox controller. (2) The OpenFlow Packet In message is extended to carry the bit rate of each incoming subflow. (3) The Nox is extended to perform a routing and spectrum assignment (RSA) algorithm, allocating suitable frequency slots and the selected modulation format according to the source/destination addresses and the

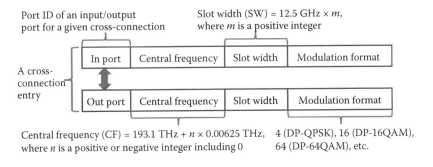

**Figure 14.51** CT entry.

bit rate of the flow, which are obtained from the extended Packet In messages. (4) A new message referred to as Slice Mod is introduced, which is based on the OpenFlow Flow Mod message. This new message carries the RSA results from the Nox, including actions (i.e., add a new cross-connection, delete a matching cross-connection, and modify an existing cross-connect), input and output ports, CF, SW, and modulation format.

*Procedure for Path Provisioning in the EON*   First, a handshake procedure between the Nox and each OpenFlow-enabled NE, for example, an OF-BV-WXC, is required for the Nox to know the feature/capability of each NE, as the procedure shown in Figure 14.52a. Once a new NE is introduced to the network, Hello messages are changed between the Nox and the new NE. Then, the Nox sends a Feature Request message to the NE, and the NE replies with a Feature Reply message that specifies the features and capabilities supported by this NE. Here, we extended the Feature Reply message for each OF-BV-WXC to report its new features/capabilities to the Nox, including its datapath ID, port number, supported switching type, available spectrum ranges for each port, etc., to the Nox controller. In addition, each NE also uses this extended Feature Reply message to report its peering connectivity, including the port ID and the datapath ID of its neighboring NEs. To guarantee the liveness of a connection between an NE and the Nox, Echo Request and Echo Reply messages are used, which can be sent from either the NE or the Nox. Figure 14.52b shows the procedure of path provisioning for one subflow from the ingress OF-MOTP. When the Nox receives a Packet In message from

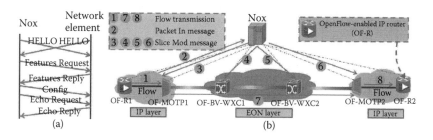

**Figure 14.52**   (a) Handshake procedure between Nox and each OpenFlow-enabled NE; (b) Procedure for path provisioning in the optical domain (The Flow Mod messages between the Nox and IP routers are not depicted in this figure.).

the ingress OF-MOTP, it performs the RSA computation according to the source and destination addresses and the bit rate information of this subflow, and then configures each OF-MOTP and OF-BV-WXC along the computed path by using the aforementioned Slice Mod messages.

*Experimental Setup, Results, and Discussions*   To evaluate the overall feasibility and efficiency of OpenSlice, we set up a testbed, as shown in Figure 14.53, which models the Japan core network topology. The testbed is deployed only with the OpenSlice-based control plane, and the data plane is emulated. All the nodes are connected to a dedicated Nox controller, which is located at node 12. The value close to each node indicates the message propagation latency from the Nox to each node. The DWDM links are characterized by 128 individual slots of 6.25 GHz each. The network topology and resource information is statically configured in the Nox and is dynamically updated by the Nox, notably when cross-connections are set up or released. In this experiment, we consider three different bit rates—100, 200, and 400 Gb/s (112, 224, and 448 Gb/s including overhead)—and three modulation formats—Dual-Polarization 64-ary Quadrature Amplitude Modulation (DP-64QAM), DP-16QAM, and Dual-Polarization Quadrature Phase-Shift Keying (DP-QPSK). The RSA is based on the algorithm presented in Ref. [73], where a route list with a necessary slot number and modulation format is precomputed. The modulation format is selected based on the path distance [64]. An efficient format such as DP-64QAM is selected for short paths, and a more robust one such as DP-QPSK is selected for long paths. We also assume

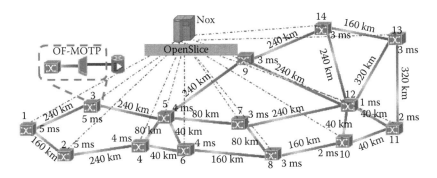

**Figure 14.53**   Experimental setup.

that an OF-MOTP is attached to each OF-BV-WXC (Figure 14.53), and the Tx/Rx in the OF-MOTP is based on Nyquist-wavelength division multiplexing (WDM) [73] and coherent detection. The use of aggressive optical prefiltering with a spectrum shape approaching that of a Nyquist filter, together with a square spectrum, minimizes the required bandwidth to a value equal to the channel baud rate. For example, the required bandwidth for a 400-Gb/s flow with the modulation format DP-QPSK is 112 GHz (i.e., 448/4) [68]. Therefore, the required SW is 9 × 12.5 GHz.

Figure 14.54 shows the Wireshark capture of an extended Feature Reply message during the handshake phase. It can be seen that, when a Feature Request message is received, the OF-BV-WXC automatically replied with a Feature Reply message. The processing latency between Feature Request/Reply messages was approximately 1.3 ms, as shown in Figure 14.54. By using the Feature Reply message, each OF-BV-WXC reported its feature to the Nox controller, including

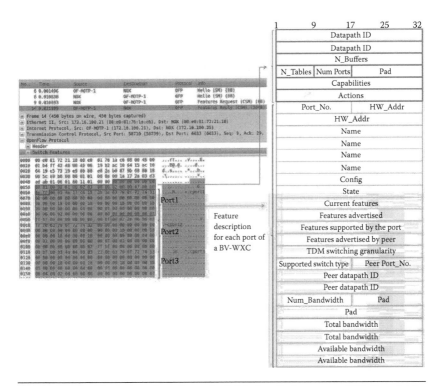

**Figure 14.54** Wireshark capture of an extended Feature Reply message. TDM, time division multiplexing; HW Addr, hardware address.

its datapath ID, port number, supported switch type, neighbor information, available spectrum ranges for each port, etc. Note that the packet format for the extended Feature Reply message presented in Figure 14.54 is based on the OpenFlow circuit switch addendum version 0.3 [29]. The only difference is that we newly defined bit map information to represent the flexible grid switching capability and resource availabilities. More details and the meaning of each field can be referred to in reference [29].

In the experiment, we set up six paths with different hop counts, as shown in Table 14.9. Figure 14.55 shows the Wireshark capture of OpenSlice messages for creating path 1, including the message sequence (Figure 14.55a), the extended Packet In message (Figure 14.55b), and the Slice Mod message (Figure 14.55c). It can be seen that the flow bit rate is encapsulated in the extended Packet In message, and the information for a new cross-connection entry is carried within the Slice Mod message. Table 14.9 also shows the OpenSlice message average latencies, obtained by repeating the experiment 100 times. In our tested scenario, the OpenSlice message latency for creating a path with 1 to 5 hops is approximately 32 to 36 ms. The path release procedure is also evaluated by setting the Action type (Figure 14.55c) in the Slice Mod message to delete a matching cross-connection. The results show that the latency for releasing a path with 1 to 5 hops is approximately 14 to 17 ms.

We also compare the performance of OpenSlice with the GMPLS-based control plane in terms of path provisioning latency for an EON. The results related to GMPLS are measured on the ADRENALINE testbed [39,42] with the same network topology as shown in Figure 14.53, and the GMPLS extensions for an EON are based on references [67,69]. The experimental results are depicted in Figure 14.56. The results show that, for creating a path with more than three hops, the OpenSlice outperforms the GMPLS-based control plane in terms of the average path provisioning latency. It is because, in GMPLS, PATH/RESV messages are processed hop by hop. With longer paths (increased hop count), the GMPLS signaling latency increases proportionally. On the other hand, the path provisioning latency in an EON is less sensitive to the hop count because the centralized controllers control all the nodes almost simultaneously.

**Table 14.9**  Experimental Results

| NO. | SRC. | DEST. | BIT RATE (Gb/s) | ROUTE | MODULATION FORMAT | CF | SW (GHz) | SETUP TIME (ms) | RELEASE TIME (ms) |
|-----|------|-------|-----------------|-------|-------------------|----|----------|-----------------|-------------------|
| 1 | 1 | 2 | 400 | 1-2 | DP-64QAM | 0 × 6A00003D | 3 × 12.5 | ~32.1 | ~14.1 |
| 2 | 1 | 2 | 100 | 1-2 | DP-64QAM | 0 × 6A000039 | 1 × 12.5 | ~32.1 | ~14.1 |
| 3 | 5 | 12 | 100 | 5-7-12 | DP-16QAM | 0 × 6A00003E | 2 × 12.5 | ~33.1 | ~14.4 |
| 4 | 4 | 14 | 200 | 4-5-9-14 | DP-QPSK | 0 × 6A00003B | 5 × 12.5 | ~33.9 | ~15.3 |
| 5 | 3 | 11 | 200 | 3-5-7-12-11 | DP-QPSK | 0 × 6A000037 | 5 × 12.5 | ~34.7 | ~16.0 |
| 6 | 1 | 13 | 400 | 1-3-5-9-14-13 | DP-QPSK | 0 × 6A000029 | 9 × 12.5 | ~36.0 | ~17.2 |

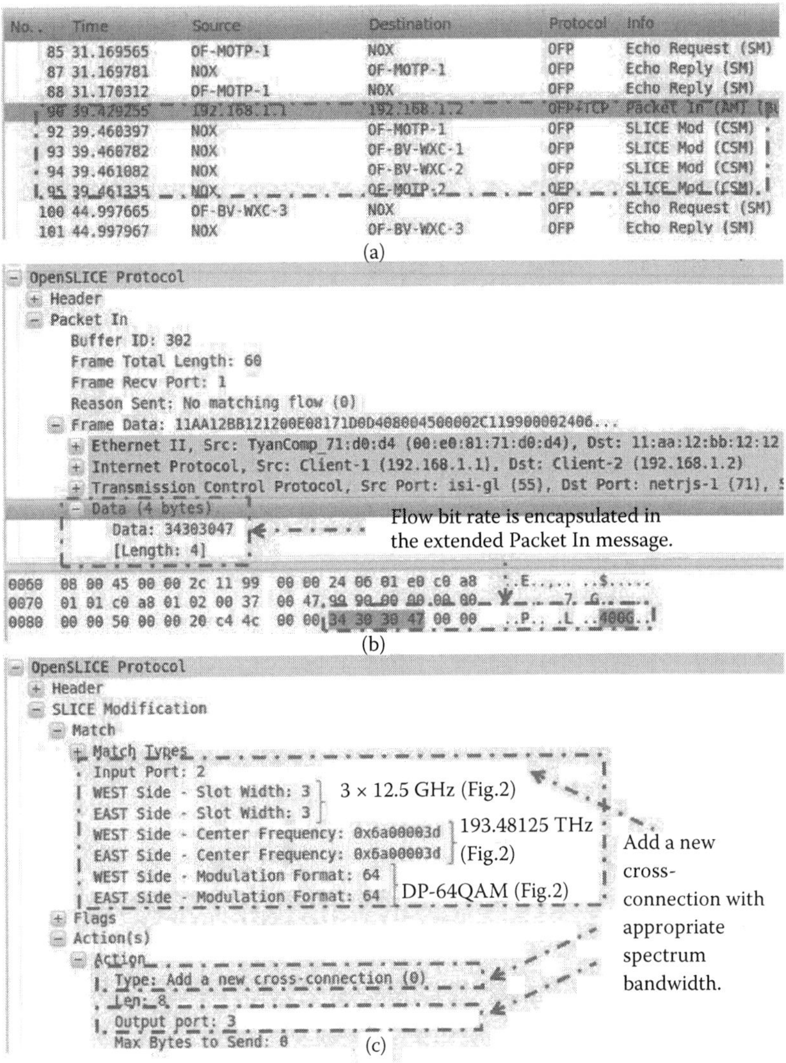

| No. | Time | Source | Destination | Protocol | Info |
|---|---|---|---|---|---|
| 85 | 31.169565 | OF-MOTP-1 | NOX | OFP | Echo Request (SM) |
| 87 | 31.169781 | NOX | OF-MOTP-1 | OFP | Echo Reply (SM) |
| 88 | 31.170312 | OF-MOTP-1 | NOX | OFP | Echo Reply (SM) |
| 90 | 39.425255 | 192.168.1.1 | 192.168.1.2 | OFP+TCP | Packet In (AM) |
| 92 | 39.460397 | NOX | OF-MOTP-1 | OFP | SLICE Mod (CSM) |
| 93 | 39.460782 | NOX | OF-BV-WXC-1 | OFP | SLICE Mod (CSM) |
| 94 | 39.461082 | NOX | OF-BV-WXC-2 | OFP | SLICE Mod (CSM) |
| 95 | 39.461335 | NOX | OF-MOTP-2 | OFP | SLICE Mod (CSM) |
| 100 | 44.997665 | OF-BV-WXC-3 | NOX | OFP | Echo Request (SM) |
| 101 | 44.997967 | NOX | OF-BV-WXC-3 | OFP | Echo Reply (SM) |

(a)

```
OpenSLICE Protocol
  Header
  Packet In
    Buffer ID: 302
    Frame Total Length: 60
    Frame Recv Port: 1
    Reason Sent: No matching flow (0)
    Frame Data: 11AA12BB121200E08171D0D408004500002C119900002406...
      Ethernet II, Src: TyanComp 71:d0:d4 (00:e0:81:71:d0:d4), Dst: 11:aa:12:bb:12:12
      Internet Protocol, Src: Client-1 (192.168.1.1), Dst: Client-2 (192.168.1.2)
      Transmission Control Protocol, Src Port: isi-gl (55), Dst Port: netrjs-1 (71),
      Data (4 bytes)
        Data: 34303047          Flow bit rate is encapsulated in
        [Length: 4]             the extended Packet In message.
```

```
0060  88 00 45 00 00 2c 11 99   00 00 24 06 01 e0 c0 a8   ..E..,.. ..$.....
0070  01 01 c0 a8 01 02 00 37   00 47.99 90.00 00.00.00   .......7. .G.....
0080  00 00 50 00 00 20 c4 4c   00 00 34 30 30 47 00 00   ..P.. .L ..100G.
```

(b)

```
OpenSLICE Protocol
  Header
  SLICE Modification
    Match
      Match Types
      Input Port: 2
      WEST Side - Slot Width: 3    ] 3 × 12.5 GHz (Fig.2)
      EAST Side - Slot Width: 3    ]
      WEST Side - Center Frequency: 0x6a00003d  ] 193.48125 THz
      EAST Side - Center Frequency: 0x6a00003d  ] (Fig.2)
      WEST Side - Modulation Format: 64   ] DP-64QAM (Fig.2)
      EAST Side - Modulation Format: 64   ]
    Flags
    Action(s)
      Action
        Type: Add a new cross-connection (0)
        Len: 8
        Output port: 3
        Max Bytes to Send: 0          (c)
```

Add a new cross-connection with appropriate spectrum bandwidth.

**Figure 14.55** Wireshark capture of the OpenSlice messages: (a) procedure for path 1 provisioning; (b) an extended Packet In message; (c) a Slice Mod message.

In this section, we present OpenSlice, an OpenFlow-based control plane for spectrum-sliced elastic optical path networks. We verified experimentally its overall feasibility for dynamic end-to-end path provisioning and IP traffic offloading through OpenFlow-based protocol extensions and seamless interworking operations. We also quantitatively evaluated its performance in terms of path provisioning latency and compared it with the GMPLS-based control plane. The results

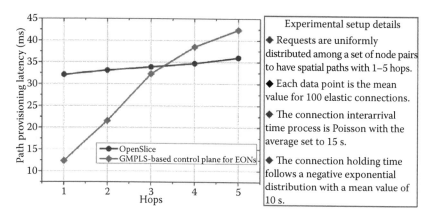

**Figure 14.56** Comparison of OpenSlice and GMPLS for average path provisioning latency in an EON.

indicate that, in our tested scenario, the OpenSlice outperforms the GMPLS-based control plane when creating an elastic optical path with more than three hops.

Future works and open issues for OpenSlice include actual testbed demonstration with intelligent interworking between the OpenSlice and MOTPs, mitigation of the scalability issue for the OpenSlice architecture, performance monitoring, OpenSlice-based control for multidomain EON, and multivendor interoperability tests/field trials. So far, some preliminary studies have been carried out to address these issues [15–17,21,24].

## Conclusion

In today's commercial IP/DWDM multilayer networks, different layers are separately operated without dynamic interaction, which leads to a low network efficiency, a high operational expense, and a long processing latency for path provisioning and restoration. With a decade of development and standardization efforts, traditional MPLS was extended to GMPLS to support multiple types of switching capabilities, especially the wavelength/fiber switching in the metro/backbone optical network, and accordingly, three different MPLS/GMPLS interconnection models (i.e., peer, overlay, and border peer) have been proposed for various multilayer network scenarios as a UCP. However, to the best of our knowledge, there are still no commercial deployments of these models to date because the GMPLS UCP is too complicated. On the other hand, SDN and OpenFlow, which allow operators to control the network using software running on a network operating system (e.g., Nox) within an external controller, provide the maximum flexibility for the operator to control an optical network and match the carrier's preferences given its flexibility, simplicity, and manageability.

In light of this, in the book chapter, we reviewed recent research activities related to the SDN/OpenFlow deployment for metro/backbone optical switching networks. We first introduced the motivation of this deployment and then detailed the SDN/OpenFlow for traditional WSON. After that, we introduced the comparison of the different control plane techniques for metro/backbone optical networks and the interworking/interoperability between OpenFlow/PCE and OpenFlow/GMPLS. The OpenFlow/PCE-based solution is also extended to support multidomain optical networks. Finally, we presented the SDN/OpenFlow deployment for other optical switching paradigms, such as OBS networks, multilayer multigranularity optical networks, and spectrum-sliced elastic optical path networks. The experimental results indicated that OpenFlow is a promising solution for metro/backbone optical networks as a UCP. We hope that the work presented in this chapter will be beneficial for the industrial deployment of intelligent optical networks with a dynamic UCP and shed light on future researches in this area.

## Acknowledgments

The authors thank Dr. Ramon Casellas, Dr. Ricardo Martínez, Dr. Raül Muñoz, Ricard Vilalta, Dr. Itsuro Morita, Dr. Masatoshi Suzuki, Dongxu Zhang, and Dr. Jian Wu for their strong support.

## References

1. Mannie, E. ed. 2004. Generalized multiprotocol label switching (GMPLS) architecture. *IETF RFC 3945*.
2. Farrel, A. 2010. A unified control plane: Dream or pipedream. In *6th International Conference on IP + Optical Network (IPOP '10)*, Paper K-3.

3. Das, S., G. Parulka, and N. Mckeown. 2012. Why OpenFlow/SDN can succeed where GMPLS failed. In *38th European Conference and Exhibition on Optical Communications (ECOC '12)*, Paper Tu.1.D.1, 3pp.
4. Open Networking Foundation. http://www.opennetworking.org/ (access Jan. 1, 2013).
5. The OpenFlow Switch Consortium. http://www.openflow.org/ (access Jan. 1, 2013).
6. Das, S., G. Parulkar, N. McKeown, P. Singh, D. Getachew, and L. Ong. 2010. Packet and circuit network convergence with OpenFlow. In *Optical Fiber Communication Conference and Exposition and National Fiber Optic Engineers Conference (OFC/NFOEC '10)*, Paper OTuG1, 3pp.
7. Liu, L., T. Tsuritani, I. Morita, H. Guo, and J. Wu. 2011. Experimental validation and performance evaluation of OpenFlow-based wavelength path control in transparent optical networks. *Optics Express* 19(27):26,578–26,593.
8. Gudla, V., S. Das, A. Shastri, G. Parulkar, N. McKeown, L. Kazovsky, and S. Yamashita. 2010. Experimental demonstration of OpenFlow control of packet and circuit switches. In *Optical Fiber Communication Conference and Exposition and National Fiber Optic Engineers Conference (OFC/NFOEC '10)*, Paper OTuG2, 3pp.
9. Liu, L., R. Casellas, T. Tsuritani, I. Morita, R. Martínez, and R. Muñoz. 2012. Interworking between OpenFlow and PCE for dynamic wavelength path control in multidomain WSON. In *Optical Fiber Communication Conference and Exposition and National Fiber Optic Engineers Conference (OFC/NFOEC '12)*, Paper OM3G.2, 3pp.
10. Liu, L., D. Zhang, T. Tsuritani, R. Vilalta, R. Casellas, L. Hong, I. Morita et al. 2012. First field trial of an OpenFlow-based unified control plane for multilayer multigranularity optical networks. In *Optical Fiber Communication Conference and Exposition and National Fiber Optic Engineers Conference (OFC/NFOEC '12)*, Paper PDP5D.2.
11. Liu, L., D. Zhang, T. Tsuritani, R. Vilalta, R. Casellas, L. Hong, I. Morita et al. 2013. Field trial of an OpenFlow-based unified control plane for multilayer multigranularity optical switching networks. *IEEE/OSA J. Lightw. Technol.* 31(4):506–514.
12. Azodolmolky, S., R. Nejabati, E. Escalona, R. Jayakumar, N. Efstathiou, and D. Simeonidou. 2011. Integrated OpenFlow-GMPLS control plane: An overlay model for software-defined packet over optical networks. In *37th European Conference and Exhibition on Optical Communications (ECOC '11)*, Paper Tu.5.K.5, 3pp.
13. Channegowda, M., P. Kostecki, N. Efstathiou, S. Azodolmolky, R. Nejabati, P. Kaczmarek, A. Autenrieth et al. 2012. Experimental evaluation of extended OpenFlow deployment for high-performance optical networks. In *38th European Conference and Exhibition on Optical Communications (ECOC '12)*, Paper Tu.1.D.2, 3pp.
14. Liu, L., R. Muñoz, R. Casellas, T. Tsuritani, R. Martínez, and I. Morita. 2012. OpenSlice: An OpenFlow-based control plane for spectrum-sliced

elastic optical path networks. In *38th European Conference and Exhibition on Optical Communications (ECOC '12)*, Paper Mo.2.D.3, 3pp.

15. Liu, L., H. Choi, T. Tsuritani, I. Morita, R. Casellas, R. Martínez, and R. Muñoz. 2012. First proof-of-concept demonstration of OpenFlow-controlled elastic optical networks employing flexible transmitter/receiver. In *International Conference on Photonics in Switching (PS '12)*, Paper PDP-1, 3pp.

16. Liu, L., R. Casellas, T. Tsuritani, I. Morita, R. Martínez, and R. Muñoz. 2012. Experimental demonstration of an OpenFlow/PCE integrated control plane for IP over translucent WSON with the assistance of a per-request–based dynamic topology server. In *38th European Conference and Exhibition on Optical Communications (ECOC '12)*, Paper Tu.1.D.3, 3pp.

17. Channegowda, M., R. Nejabati, M. Fard, S. Peng, N. Amaya, G. Zervas, D. Simeonidou et al. 2012. First demonstration of an OpenFlow-based software-defined optical network employing packet, fixed, and flexible DWDM grid technologies on an international multidomain testbed. In *38th European Conference and Exhibition on Optical Communications (ECOC '12)*, Paper Th.3.D.2, 3pp.

18. Zhang, D., L. Liu, L. Hong, H. Guo, T. Tsuritani, J. Wu, and I. Morita. 2012. Experimental demonstration of OBS/WSON multilayer optical switched networks with an OpenFlow-based unified control plane. In *16th International Conference on Optical Network Design and Modeling (ONDM '12)*, 6pp.

19. Zhang, D., H. Guo, L. Liu, T. Tsuritani, J. Wu, and I. Morita. 2013. Dynamic wavelength assignment and burst contention mitigation for the LOBS-over-WSON multilayer networks with an OpenFlow-based control plane. *Optical Fiber Communication Conference and Exposition and National Fiber Optic Engineers Conference (OFC/NFOEC '13)*, Paper OTh4B.6, 3pp.

20. Liu, L., T. Tsuritani, and I. Morita. 2012. Experimental demonstration of OpenFlow/GMPLS interworking control plane for IP/DWDM multi-layer optical networks. In *14th International Conference on Transparent Optical Networks (ICTON '12)*, Paper Tu.A2.5.

21. Choi, H. Y., L. Liu, T. Tsuritani, and I. Morita. 2013. Demonstration of BER-adaptive WSON employing flexible transmitter/receiver with an extended OpenFlow-based control plane. *IEEE Photon. Technol. Lett.* 25(2):119–121.

22. Casellas, R., R. Martínez, R. Muñoz, L. Liu, T. Tsuritani, and I. Morita. 2013. An integrated stateful PCE/OpenFlow controller for the control and management of flexi-grid optical networks. In *Optical Fiber Communication Conference and Exposition and National Fiber Optic Engineers Conference (OFC/NFOEC '13)*, Paper OW4G.2, 3pp.

23. Liu, L., H. Y. Choi, R. Casellas, T. Tsuritani, I. Morita, R. Martínez, and R. Muñoz. 2013. Demonstration of a dynamic transparent WSON employing flexible transmitter/receiver controlled by an OpenFlow/stateless PCE integrated control plane. In *Optical Fiber Communication Conference and Exposition and National Fiber Optic Engineers Conference (OFC/NFOEC '13)*, Paper NTu3F.5, 3pp.

24. Paolucci, F., F. Cugini, N. Hussain, F. Fresi, and L. Poti. 2012. OpenFlow-based flexible optical networks with enhanced monitoring functionalities. In *38th European Conference and Exhibition on Optical Communications (ECOC '12)*, Paper Tu.1.D.5, 3pp.

25. Yang, H., Y. Zhao, J. Zhang, S. Wang, W. Gu, J. Han, Y. Lin et al. 2013. Multistratum resources integration for data center application based on multiple OpenFlow controllers cooperation. In *Optical Fiber Communication Conference and Exposition and National Fiber Optic Engineers Conference (OFC/NFOEC '13)*, Paper NTu3F.7, 3pp.

26. Zhang, J., Y. Zhao, H. Yang, Y. Ji, H. Li, Y. Lin, G. Li et al. 2013. First demonstration of enhanced software-defined networking (eSDN) over elastic grid (eGrid) optical networks for data center service migration. In *Optical Fiber Communication Conference and Exposition and National Fiber Optic Engineers Conference (OFC/NFOEC '13)*, Paper PDP5B.1, 3pp.

27. Farrel, A., J.-P. Vasseur, and J. Ash. 2006. A path computation element (PCE)-based architecture. *IETF RFC 4655*, 40pp.

28. Nox: An OpenFlow controller. http://www.noxrepo.org/ (access Jan. 1, 2013).

29. Das, S. Extensions to the OpenFlow protocol in support of circuit switching. http://www.openflow.org/wk/images/8/81/OpenFlow_Circuit_Switch_Specification_v0.3.pdf (access Jan. 1, 2013).

30. Wireshark. http://www.wireshark.org/ (access Jan. 1, 2013).

31. Lee, Y., G. Bernstein, and W. Imajuku. 2011. Framework for GMPLS and path computation element (PCE) control of wavelength switched optical networks (WSONs). *IETF RFC 6163*, 51pp.

32. Vasseur, J. P. and J. L. Le Roux, eds. 2009. Path computation element (PCE) communication protocol (PCEP). *IETF RFC 5440*.

33. Liu, L., T. Tsuritani, and I. Morita. 2012. From GMPLS to PCE/GMPLS to OpenFlow: How much benefit can we get from the technical evolution of control plane in optical networks? In *14th International Conference on Transparent Optical Networks (ICTON '12)*, Paper Mo.C2.2, 4pp.

34. Zhao, Y., J. Zhang, H. Yang, and Y. Yu. 2013. Which is more suitable for the control over large-scale optical networks: GMPLS or OpenFlow? In *Optical Fiber Communication Conference and Exposition and National Fiber Optic Engineers Conference (OFC/NFOEC '13)*, Paper NTu3F.2, 3pp.

35. Liu, L., T. Tsuritani, I. Morita, R. Casellas, R. Martínez, and R. Muñoz. 2012. Control plane techniques for elastic optical networks: GMPLS/PCE vs. OpenFlow. In *International Workshop on Flexible Optical Networks, IEEE Global Communications Conference (GLOBECOM '12)*, pp. 352–357.

36. Giorgetti, A., F. Cugini, F. Paolucci, and P. Castoldi. 2012. OpenFlow and PCE architectures in wavelength switched optical networks. In *16th International Conference on Optical Network Design and Modeling (ONDM '12)*, 6pp.

37. Liu, L., T. Tsuritani, R. Casellas, R. Martínez, R. Muñoz, and M. Tsurusawa. 2010. Experimental demonstration and comparison of distributed and centralized multidomain resilient translucent WSON. In *36th European Conference and Exhibition on Optical Communication (ECOC '10)*, Paper We.7.D.3, 3pp.

38. Han, S., K. Jang, K. Park, and S. Moon. 2010. PacketShader: A GPU-accelerated software router. In *The Annual Conference of the ACM Special Interest Group on Data Communication (SIGCOMM '10)*, pp. 195–206 .

39. Muñoz, R., R. Martinez, and R. Casellas. 2009. Challenges for GMPLS lightpath provisioning in transparent optical networks: Wavelength constraints in routing and signaling. *IEEE Commun. Mag.* 47(8):26–34.

40. Shen, G. and R. S. Tucker. 2007. Translucent optical networks: The way forward. *IEEE Commun. Mag.* 45(2):48–54.

41. Le Roux, J. L., J. P. Vasseur, and Y. Lee, eds. 2009. Encoding of objective functions in the path computation element communication protocol. *IETF RFC 5541.*

42. Martínez, R., R. Casellas, R. Muñoz, and T. Tsuritani. 2010. Experimental translucent-oriented routing for dynamic lightpath provisioning in GMPLS-enabled wavelength switched optical networks. *IEEE/OSA J. Lightw. Technol.* 28(8):1241–1255.

43. Liu, L., T. Tsuritani, R. Casellas, and M. Tsurusawa. 2010. Demonstration of a resilient PCE/GMPLS-controlled translucent optical network. In *15th OptoElectronics and Communications Conference (OECC '10)*, Paper 7A1-2, pp. 88–89.

44. Liu, L., T. Tsuritani, R. Casellas, R. Martínez, R. Muñoz, M. Tsurusawa, and I. Morita. 2011. Experimental assessment of a resilient PCE/GMPLS-controlled translucent wavelength switched optical network. *IEICE Transactions on Communications* E94-B(7):1831–1844.

45. Liu, L., R. Casellas, T. Tsuritani, I. Morita, R. Martínez, and R. Muñoz. 2011. Lab trial of PCE-based OSNR-aware dynamic restoration in multidomain GMPLS-enabled translucent WSON. In *37th European Conference and Exhibition on Optical Communication (ECOC '11)*, Paper Tu.5.K.1, 3pp.

46. Liu, L., R. Casellas, T. Tsuritani, I. Morita, S. Okamoto, R. Martínez, and R. Muñoz. 2011. Field and lab trials of PCE-based OSNR-aware dynamic restoration in multidomain GMPLS-enabled translucent WSON. *Optics Express* 19(27):26,568–26,577.

47. Das, S., Y. Yiakoumis, G. Parulkar, N. McKeown, P. Singh, D. Getachew, and P. D. Desai. 2011. Application-aware aggregation and traffic engineering in a converged packet-circuit network. In *Optical Fiber Communication Conference and Exposition and National Fiber Optic Engineers Conference (OFC/NFOEC '11)*, Paper NThD3, 3pp.

48. Liu, L., T. Tsuritani, I. Morita, H. Guo, and J. Wu. 2011. OpenFlow-based wavelength path control in transparent optical networks: A proof-of-concept demonstration. In *37th European Conference and Exhibition on Optical Communications (ECOC '11)*, Paper Tu.5.K.2, 3pp.

49. Casellas, R., R. Martínez, R. Muñoz, and S. Gunreben. 2009. Enhanced backward recursive path computation for multiarea wavelength switched optical networks under wavelength continuity constraint. *IEEE/OSA J. Opt. Commun. Netw.* 1(2):A180–A193.

50. Casellas, R., R. Martínez, R. Muñoz, L. Liu, T. Tsuritani, I. Morita, and M. Tsurusawa. 2011. Dynamic virtual link mesh topology aggregation in

multidomain translucent WSON with hierarchical PCE. *Optics Express* 19(26):B611–B620.

51. Qiao, C. and M. Yoo. 1999. Optical burst switching (OBS)—A new paradigm for an optical Internet. *J. High Speed Netw.* 8(1):69–84.

52. Qiao, C. 2000. Labeled optical burst switching for IP-over-WDM integration. *IEEE Commun. Mag.* 38(9):104–114.

53. Liu, L., X. Hong, J. Wu, Y. Yin, S. Cai, and J. T. Lin. 2009. Experimental comparison of high-speed transmission control protocols on a traffic-driven labeled optical burst switching network testbed for grid applications. *OSA J. Opt. Netw.* 8(5):491–503.

54. Shiomoto, K., ed. 2008. Framework for MPLS-TE to GMPLS migration. *IETF RFC 5145*, 19pp.

55. Kumaki, K., ed. 2008. Interworking requirements to support operation of MPLS-TE over GMPLS networks. *IETF RFC 5146*.

56. Martínez, R., R. Casellas, and R. Muñoz. 2012. Experimental validation/evaluation of a GMPLS unified control plane in multilayer (MPLS-TP/WSON) networks. In *Optical Fiber Communication Conference and Exposition and National Fiber Optic Engineers Conference (OFC/NFOEC '12)*, Paper NTu2J.1, 3pp.

57. Nishioka, I., L. Liu, S. Yoshida, S. Huang, R. Hayashi, K. Kubo, and T. Tsuritani. 2010. Experimental demonstration of dynamic wavelength path control and highly resilient recovery in heterogeneous optical WDM networks. In *15th OptoElectronics and Communications Conference (OECC '10)*, Paper PD1, 2pp.

58. Huang, S., L. Liu, S. Yoshida, I. Nishioka, R. Hayashi, K. Kubo, and T. Tsuritani. 2011. Multifailure restoration demonstrations with multivendor interoperability in control plane–enabled WSON. In *Optical Fiber Communication Conference and Exposition and National Fiber Optic Engineers Conference (OFC/NFOEC '11)*, Paper NThC6.

59. Liu, L., T. Tsuritani, I. Nishioka, S. Huang, S. Yoshida, K. Kubo, and R. Hayashi. 2012. Experimental demonstration of highly resilient wavelength switched optical networks with a multivendor interoperable GMPLS control plane. *IEEE/OSA J. Lightw. Technol.* 30(5):704–712.

60. Nejabati, R., ed. 2008. Grid optical burst switched networks (GOBS). *Open Grid Forum (OGF) Draft, GFD-I.128.* http://www.ogf.org/documents/GFD.128.pdf.

61. Liu, L., H. Guo, Y. Yin, X. Hu, T. Tsuritani, J. Wu, X. Hong et al. 2008. Demonstration of a self-organized consumer grid architecture. In *34th European Conference and Exhibition on Optical Communication (ECOC '08)*, Paper Tu.1.C.3, 2pp.

62. Liu, L., H. Guo, T. Tsuritani, Y. Yin, J. Wu, X. Hong, J. Lin et al. 2012. Dynamic provisioning of self-organized consumer grid services over integrated OBS/WSON networks. *IEEE/OSA J. Lightw. Technol.* 30(5):734–753.

63. Jinno, M., H. Takara, B. Kozicki, Y. Tsukishima, Y. Sone, and S. Matsuoka. 2009. Spectrum-efficient and scalable elastic optical path network:

Architecture, benefits, and enabling technologies. *IEEE Commun. Mag.* 47(11):66–73.

64. Jinno, M., B. Kozicki, H. Takara, A. Watanabe, Y. Sone, T. Tanaka, and A. Hirano. 2010. Distance-adaptive spectrum resource allocation in spectrum-sliced elastic optical path network. *IEEE Commun. Mag.* 48(8):138–145.

65. Casellas, R., R. Muñoz, J. M. Fàbrega, M. S. Moreolo, R. Martínez, L. Liu, T. Tsuritani et al. 2013. Design and experimental validation of a GMPLS/PCE control plane for elastic CO-OFDM optical networks. *IEEE J. Sel. Areas Commun.* 31(1):49–61.

66. Casellas, R., R. Muñoz, J. M. Fàbrega, M. S. Moreolo, R. Martínez, L. Liu, T. Tsuritani et al. 2012. GMPLS/PCE control of flexi-grid DWDM optical networks using CO-OFDM transmission. *IEEE/OSA J. Opt. Commun. Netw.* 4(11):B1–B10.

67. Muñoz, R., R. Casellas, and R. Martínez. 2011. Dynamic distributed spectrum allocation in GMPLS-controlled elastic optical networks. In *37th European Conference and Exhibition on Optical Communications (ECOC '11)*, Paper Tu.5.K.4, 3pp.

68. Sambo, N., F. Cugini, G. Bottari, P. Iovanna, and P. Castoldi. 2011. Distributed setup in optical networks with flexible grid. In *37th European Conference and Exhibition on Optical Communications (ECOC '11)*, Paper We.10.P1.100, 3pp.

69. Casellas, R., R. Muñoz, J. M. Fàbrega, M. S. Moreolo, R. Martínez, L. Liu, T. Tsuritani et al. 2012. Experimental assessment of a combined PCE-RMA and distributed spectrum allocation mechanism for GMPLS elastic CO-OFDM optical networks. In *Optical Fiber Communication Conference and Exposition and National Fiber Optic Engineers Conference (OFC/NFOEC '12)*, Paper OM3G.1.

70. Jinno, M., Y. Sone, H. Takara, A. Hirano, K. Yonenaga, and S. Kawai. 2011. IP traffic offloading to elastic optical layer using multiflow optical transponder. In *37th European Conference and Exhibition on Optical Communications (ECOC '11)*, Paper Mo.2.K.2, 3pp.

71. Takara, H., T. Goh, K. Shibahara, K. Yonenaga, S. Kawai, and M. Jinno. 2011. Experimental demonstration of 400-Gb/s multiflow, multirate, multireach optical transmitter for efficient elastic spectral routing. In *37th European Conference and Exhibition on Optical Communications (ECOC '11)*, Paper Tu.5.A.4.

72. Li, Y., F. Zhang, and R. Casellas. 2011. Flexible grid label format in wavelength switched optical network. IETF draft-li-ccamp-flexible-grid-label-00 (work in progress). http://www.tools.ietf.org/html/draft-li-ccamp-flexible-grid-label-00.

73. Gerstel, O., M. Jinno, A. Lord, and S. J. Ben Yoo. 2012. Elastic optical networking: A new dawn for the optical layer? *IEEE Commun. Mag.* 50(2):S12–S20.

# 15

# OPENFLOW/SDN AND OPTICAL NETWORKS

## LYNDON Y. ONG

**Contents**

**Service Provider Optical Networks**

International Telecommunications Union – Telecommunication Stan-dardization Sector (ITU-T) Recommendation G.805 [1], a core standard for service provider network architecture, defines transport network as "the functional resources of the network, which conveys user information between locations." Optical transport networks (OTNs) provide the underlying connectivity in service provider networks, allowing information to be conveyed between central office locations across metro areas, long distances, and undersea networks.

Optical fiber links typically support the wavelength of multiple optical channels (OCh) multiplexed on a dense wavelength division multiplexing (DWDM) transmission system. DWDM systems carry 80 to 100 wavelengths on a fiber pair, where each wavelength carries a single high-rate signal, such as 100-Gb/s Ethernet, or may carry a multiplex of lower rate signals. For example, a 100-Gb/s channel may be composed of 10 component signals, each of which is a 10-Gb/s Ethernet signal. Services provided by optical networks include the transport of packetized IP and Ethernet traffic and private line services such as an Ethernet private line or private wavelength services.

Optical transport can be used as point-to-point links connecting large packet switches, mesh or ring topology networks incorporating photonic switches or add/drop multiplexers (optical-optical-optical or O-O-O), or mesh networks incorporating electronic time division multiplexing (TDM) switching systems (optical-electronic-optical or O-E-O), depending on the services being offered.

Optical wavelength services offered by service providers can vary in bandwidth from the capacity of a wavelength to a small fraction of that amount. Currently, several carriers offer 40-G services as the high end of the spectrum. The low end is typically 50 to 155 Mbps using synchronous optical network (SONET) or synchronous digital hierarchy (SDH) standards. Some of the fastest growing optical services today are 1 G and 10 G because of the growth of the Ethernet [2]. Optical network services are typically not highly dynamic; only in a few cases have carriers deployed optical network services that are under some dynamic customer control [3]. The promise of new services and applications that could make use of dynamic optical networking is a chief driver for carrier interest in OpenFlow/software-defined network (SDN).

### Optical Transport Network

The current ITU-T standards define an optical transport hierarchy (OTH) [4] for optical network multiplexing and switching. Networks built according to these standards are called OTNs. OTN includes two fundamentally different types of optical switching technologies: photonic switching of wavelengths, which are called OCh, without termination of the wavelength; and electrical switching of the digital components of the terminated wavelength, which are called OCh data units (ODUs). OTN is also used as a framing format to provide interoperability, performance monitoring, and management of a signal even without switching. OTN replaces SONET or SDH standards that are still the majority of optical switches and services deployed today, allowing channel rates of 40 Gb/s and higher.

OTN is organized into a series of layers incorporating both photonic and electronic signals.

In Figure 15.1, the layers from OCh downward are photonic, whereas the layers from OTU upward are electronic/digital. Each layer incorporates its own overhead (OH) for performance monitoring, bit-oriented signaling, etc. Forward error correction (FEC) is added before the signal enters the optical domain to improve performance. The architecture for OTN is defined in ITU-T Recommendation G.872 [5], and the format for OTN, in Recommendation G.709-v1 [6]. Several digital container rates are defined in OTN: in its initial version, ODU1 (2.5 Gbps), ODU2 (10 Gbps), and ODU3 (40 Gbps), subsequently extended by ITU-T Recommendation G.709-v3 to introduce new containers for GbE (ODU0) and 100 GbE (ODU4).

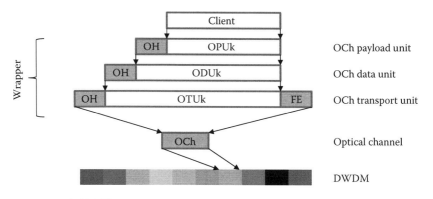

**Figure 15.1**  Layering of electronic and photonic signals (FE, forward error correction).

OTN electrical switches support up to 4 Tb/s of nonblocking switching capacity in a full-height equipment chassis. Multichassis configurations allow switches to scale to 100 Tb/s and beyond if necessary. Digital monitoring points support rapid fault identification and localization. Electrical switching allows for the easy addition and extraction of signals from the transmission system, with a variety of service protection/restoration capabilities using fast electrical reconfiguration.

OTN photonic switches, usually referred to as reconfigurable optical add/drop multiplexers (ROADMs), support the express routing of an OCh through a network node to avoid the additional cost and energy consumption of electrical switching, but with more constraints.

The combination of electrical and photonic switching components is used in today's optical network to support global scalability balanced against cost; electrical switching allows signal regeneration to increase distance while removing optical impairments, but with additional circuitry and energy usage. Embedded management OH information in the digital signal supports performance monitoring and forms the basis for end-to-end service assurance.

Recent trends in optical network technology include the following:

- The use of tunable transmitters and receivers as components that allow more dynamic control over the wavelength to be used to carry the signal over a particular port
- The development of coherent receivers and variable FEC algorithms to increase signal data rates over the basic 12.5-GHz ITU grid to up to 100 Gbps, with improved signal detection ability over fibers of different characteristics

- The introduction of colorless, directionless, and contention-less ROADM designs that (for a price) reduce port-to-port connectivity and wavelength constraints previously encountered with ROADMs [7]
- The introduction of a flexible grid structure for an optical bandwidth that potentially allows variable spectrum allocation for a higher signal bandwidth or longer distances traveled

### Optical Network Management and Intelligent Control Plane

For many years, optical networks were managed using central management systems, which required manual intervention through graphic user interfaces (GUIs) to carry out changes in configuration. The management system provided the database of equipment and components, and each network element (NE) provided a management interface that was used by the management system to control cross-connection.

Beginning approximately 10 years ago, distributed control plane protocols were introduced in the control architecture of optical networks as a way to improve the accuracy of network databases, the speed of provisioning, and the efficiency of recovery. Distributed automatic control of optical networks has resulted in significant reductions in network operations cost; an increase in network availability to 1 defect per million resulting from multiple stages of recovery; and enabled new service offerings, notable of which are customer-controlled optical network services [9].

The distributed control plane is now incorporated into many carrier optical networks because of its ability to automate management functions and support self-healing in response to failures. The control plane has been particularly successful in core networks, where there is a higher degree of connectivity, and in undersea networks, where repairs to failed links is difficult and expensive and the ability to automatically recover from multiple failures is highly valuable. The control plane has mainly been used for the electrical layers of transport networks; the use for most photonic networks is impractical because of the added complexity of path computation for all-optical or photonic links and the greater latency involved in reconfiguring photonic components [10].

In distributed control, control plane messages are exchanged over the signaling control network (SCN), which may consist of in-band signaling links such as the data communication channel (DCC) and general communication channel (GCC) in SONET/SDH and OTN, respectively, and the optical supervisory channel at the OTN optical layer, or use a dedicated out-of-band packet network connecting NEs.

Neighbor discovery is the communication of identity between neighboring NEs. It is enabled by exchanging addresses over the control channel, allowing each device to build up an accurate inventory of link identities and remote link endpoints, and detecting misconnections. Discovery requires the use of an in-band control link or, if an out-of-band control link is used, an additional identifier carried in the OH or the wavelength itself.

The control plane routing protocol is then used to disseminate local link topology and link use to all network elements within the control domain so that each NE builds a complete topology map. This makes it possible for the management system to retrieve the full network topology and status by contacting a single NE. Optical network routing differs from IP routing in that routing is needed only when a connection is initially provisioned, not for every packet. This reduces the criticality of the routing protocol because data will flow correctly on existing connections even if the routing protocol is disrupted.

The topology database is used by the source node of a new connection to compute the optimized path for the connection. Distributed path computation reduces the load on the management system and allows the network to recover rapidly from failures by having recovery paths computed by the affected source nodes.

Finally, new services can be provisioned using a signaling protocol between each of the NEs in the connection path, reducing the requirements on the management system and drastically reducing the latency of connection setup and connection restoration after failure. Control plane signaling protocols set up the connection along a precomputed path end to end, using an explicit route object inserted into the setup message at the source node. As shown in Figure 15.2, the management system plays a reduced role for offline device management, whereas the NEs communicate in real time using signaling to set up the end-to-end path.

Sophisticated planning tools can retrieve the instantaneous topology, status, and occupancy of network links and nodes from an active

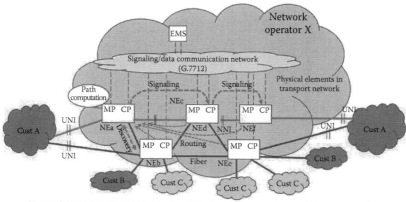

CP, control plane; Cust, customer; DP, data plane; EMS, element management system; MP, management plane; NE, network element; NN, node interface; UNI, user-to-network interface.

**Figure 15.2** Control plane.

network element and perform an analysis of network use. If a more efficient or better performing configuration of the network is computed, the control plane will automatically adjust its routing of new connections once the new configuration has been installed.

### How Can SDN/OpenFlow Improve Optical Network Control?

*Goals of Applying SDN/OpenFlow to Optical Networks*

Although the optical control plane has been successfully deployed in many carrier networks and has reduced the capex and opex of these networks, there are major potential benefits to adopting SDN/OpenFlow for optical network control, as follows:

- A programmatic, abstracted interface for application awareness and greater influence over the network. In the current optical network, the applications and the network are independent, missing potential efficiencies from the coordination of demand and resources. In control plane models such as generalized multiprotocol label switching (GMPLS), the network is expected to react independently to requests from client systems distributed at the network edge, making global coordination of actions more difficult.
- An improved service development and deployment. Because service processing is distributed across network nodes, deployment

of a new service in a distributed control environment may involve upgrading the software at each node in the network. Not only does this introduce deployment timing problems, but it also requires software development coordination across platforms and extensive testing to avoid potential interactions with embedded software and services. The use of OpenFlow/SDN to separate software and hardware would allow service software to be developed on server platforms rather than embedded systems.

- Multiple levels of abstraction. Introduction of the OpenFlow/SDN technology support allow optical network virtualization, presenting a different view of the network depending on application requirements. Different applications may need different levels of information and control over connectivity to specified destinations. The Quantum application programming interface (API), for example, in the OpenStack software suite allows an application to simply request a new layer 2 network connecting peer machine [11]; a similar abstraction would allow the introduction of very high bandwidth point-to-point services with greater customer controllability.

- Cost reduction. Potentially, greater separation of software and hardware using SDN/OpenFlow may reduce the cost of optical network equipment by reducing the processing requirements and software development costs for the network element and centralizing software on common platforms. How great the cost reduction would be is unclear, however, because the cost of optical equipment is primarily in the hardware, the photonics, and the associated electronic components, rather than in the software.

## Potential Applications for SDN/OpenFlow in Optical Networks

### Packet-Optical Interconnection (POI)

An initial application for SDN/OpenFlow is for the control of multilayer packet/optical networks. Packet-optical transport systems (POTS) incorporate both packet switching and optical switching/transmission in one device, simplifying the aggregation of packet transport into optical pipes and allowing for the efficient grooming of traffic into the carrier's optical network [12]. The use of SDN/ OpenFlow for control is relatively easy here because OpenFlow already contains the control semantics for directing the mapping of packet traffic into virtual ports (defined in the OpenFlow 1.2 specification [13]), and these virtual ports can be mapped by a management system or an optical control plane into point-to-point optical paths connecting the POTS systems across the wide area network (WAN). The use of OpenFlow allows the service provider to specify more flexible mapping based on the different components of the packet header and flexible encapsulation and decapsulation actions depending on the desired service, whereas the optical paths provide high bandwidth connectivity with guaranteed bandwidth and performance between sites. This could allow the carrier to introduce new services based on access control or selective class of service.

*Data Center Interconnection and Network Virtualization*

The OpenFlow/SDN technology is already seeing its widest deployment within the data center, where topology and traffic patterns can be controlled and equipment tends to be homogeneous [14]. A logical extension is to apply OpenFlow/SDN to the data center interconnection across the WAN. Inter-DC traffic comprises database distribution and synchronization between geographic sites, virtual machine (VM) migration, and transfer. Much of this traffic will be generated by the need for geographic distribution to meet backup requirements and movement of workload from one cloud to another. Flexibility in workload placement and movement among the pool of provider DCs contributes to a reduction in the total DC resources through the expansion of virtualized asset pools. This will create an important potential source of economies for SPs (Figure 15.3).

The cloud backbone network interconnecting provider DCs poses challenges that the OpenFlow-based SDN is ideally suited to address.

User self-service paradigms, application operational time variations, and a significantly transactional component to many of the traffic types all create an intrinsically and significantly time-variable character to the traffic among DC connection points on the backbone network. Survivability and recovery from disasters or major outages

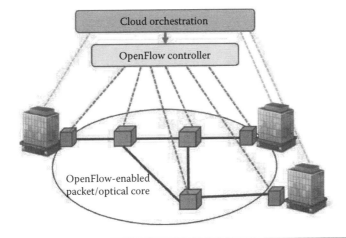

**Figure 15.3** Data center interconnection application.

may add large volumes of unpredictable, transactional traffic on the inter-DC backbone.

Under existing paradigms, both shared packet networks and dedicated connection require adequate bandwidth capacities to support peak traffic loads on all paths and through all aggregation and switching points. This implies a design-for-worst-case planning paradigm, effectively requiring an overdesign to add slack capacity (i.e., peak vs. average or lowest required capacity) to the network.

Under a centralized network control layer that maintains a global view of network resources and controls their allocation in response to evolving traffic demands—the SDN paradigm—the various trends and challenges previously described may be addressed. The cloud orchestrator and network control layer are global; both the demands and the use of network resources may be globally optimized. For example, the network control layer may see that concurrent major data transfers can be accommodated by sending each over a different network path, despite node or link saturation along default routes.

*Private Optical Networks*

A special case application of SDN/OpenFlow for data center interconnection is for smaller private optical networks that a data center operator may deploy to connect their locations. If these networks cover a relatively small geographic area, the simplicity of a centrally run SDN/OpenFlow control plane designed for a small private network may be very attractive to the data center operator, where their requirements for scalability, reliability, etc., are not as extensive as would be for a service provider. This application could be the driver of early, limited deployments of OpenFlow/SDN for optical networks.

### Extending OpenFlow/SDN to Optical Networks

*What Are the Challenges to OpenFlow/SDN for Optical Networks?*

Although significant research has been done on the extension of OpenFlow/SDN to optical networks, there are still some challenges that will need to be overcome before there is real implementation and deployment in service provider networks.

*OpenFlow Extensions for Circuits*   OpenFlow (e.g., OpenFlow v.1.0 and v.1.3) supports the configuration of matching of incoming packets based on port, Ethernet header, IP header, and transmission control protocol (TCP) port values and forwarding rules to modify headers and selectively forward packets to outgoing ports. Basic extensions for circuit control have not yet been integrated into the OpenFlow specification, although research and prototyping has been done at both layers 1 and 0, as will be described below. The complexity of these extensions depends on the level of control, especially on the level of functionality, as follows:

1. The mapping of input to output timeslot and/or wavelength
2. The control of optical transceiver characteristics such as modulation type, power levels, dispersion compensation, etc.
3. The control of the internal functions of the switch, such as adaptation between different TDM layers at layer 1 or wavelength conversion at layer 0

OpenFlow uses a simple Match/Action Table model of the switch that does not easily model internal characteristics, such as port-to-port wavelength connectivity limitations in ROADMs, and assumes that much information about switch constraints is preconfigured in the controller. Similarly, the model focuses on the actions within the switch and not the links between switches, whereas much of the complexity of transport deals with link characteristics.

*Optical Impairments*  At the photonic layer, the handling of optical impairments and characteristics will be critical. This can be divided into two main aspects:

1. Path computation and set up of cross-connection at intermediate nodes. Path computation would be done in the controller and would need to consider the impact of optical impairments on end-to-end signal-to-noise ratio; cross-connection control would need to specify the matching of incoming and outgoing ports, assuming that wavelength continuity is preserved across the switch. If wavelength conversion is possible, then incoming and outgoing wavelength or waveband must also be controlled.

2. Setting of transmitter and receiver for compatibility with each other and matching with the optical end-to-end path. Within a frequency range that has been cleared end to end, there may be a variability of the characteristics of the signal sent by the transmitter and detected at the receiver, such as modulation type, power level, FEC coding, etc., which must be set by the controller; this setting may be done using individual parameters or by groupings of parameter settings that have been standardized for interoperability, such as the application codes standardized in the ITU-T Recommendations for this purpose [15].

*Lack of Visibility into the Data Stream*  In transport networks, the objective is to convey information transparently between endpoints. The information is defined by characteristics such as port, timeslot, and wavelength defining the data stream rather than information carried within the data stream such as packet header fields. As a result, the controller cannot request actions based on an analysis of the data stream, but only actions based on the port, timeslot, and wavelength defining the data stream. This limits the functionality of the controller interface compared with packet networks.

*Technology Specific Discovery*  Discovery in OTNs cannot be done using the Packet_In/Packet_Out method used for packet network discovery because there is no visibility into the data stream. Instead, the controller must be able to take advantage of technology-specific discovery mechanisms such as setting of the trail trace or other header information or, alternatively, exchange of discovery information via the DCC/GCC/OSC control channels of the optical path.

*Service Provider Policy and Interoperability* Service provider networks are typically made up of diverse equipment from multiple vendors and are used to provide service to many different customers. Deployment of SDN/OpenFlow in the service provider network will need to be able to support such an environment by:

- Supporting multiple domains of differing equipment types and vendors, where a single controller instance may not be sufficient for control purposes. Multiple controllers will need to be coordinated in a multidomain network, but there is no standard controller-to-controller interface. Another approach to coordination may be the use of hierarchy where a parent controller coordinates the actions of multiple child controllers; however, this will introduce further requirements for controller-to-controller interoperability and testing.
- Supporting service provider needs for injecting policy and authorization over the control of network resources. OpenFlow/SDN already has the concept of a FlowVisor [16]. A FlowVisor is a mechanism for partitioning control so that a particular client controller only sees and controls a subset of a controlled network; however, the policy aspects of configuring what resources are controlled by which client will need further specification. In general, security is an area of OpenFlow/SDN that is still a work in progress.

*Reliability and Scalability* Service provider networks cover large geographic areas where both the number of network elements and the geographic distance can be a challenge for SDN/OpenFlow control structures. At the speed of light in fiber (~300,000 km/s), control messages require an approximately 30-ms round trip to cross 5000 km, a significant amount of time relative to 50-ms standards for transport network actions such as protection switching in case of failure.

Some functions will clearly need to be controlled locally to meet standard time frames for recovery actions. The number of network elements and the requirements for very high availability of service provider networks (aiming at 0.99999 availability) will require that controllers have active backups with state synchronization and fast failover time, features that are still under development for controller implementations.

*Research on OpenFlow/SDN for Optical Networks*

*Ciena/Stanford Prototyping*   As a proof of concept, Ciena and Stanford University cooperated in 2009 on a prototype OpenFlow-enabled packet and circuit network, using the Ciena CoreDirector (CD) CI SONET/SDH switch and a Stanford-developed OpenFlow controller/ application that set up, modified, and tore down L1/L2 flows on demand and dynamically responded to network congestion. The network and application was demonstrated at the SuperComputing 2009 Conference as the first implementation of integrated packet and circuit control based on OpenFlow [17] (Figure 15.4).

At the start of the demonstration, connectivity between the CD switches and the OpenFlow controller is established over an out-of-band Ethernet control network. The controller was initially configured

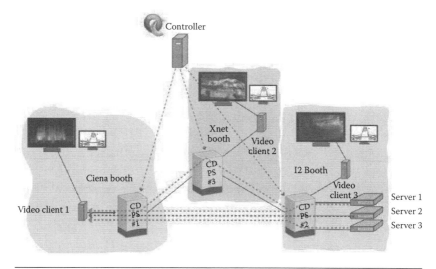

**Figure 15.4**   Ciena/Stanford prototype demonstration.

with the identities of the switches and used communication over the OpenFlow interface to build a switch/topology database. The controller then preprovisioned a SONET/SDH Virtual Concatenation Group (VCG) between the CDs, capable of carrying Ethernet private line connections. After this initial startup phase, a video client would make a request for a video from a remote streaming video server. The request is redirected to the OpenFlow controller using the Packet_In command, and the controller responds by directing CDs 1 and 2 to create an internal VLAN corresponding to the client port (in CD 1) and the video server port (in CD 2), and map the VLAN into the VCG virtual ports, thereby enabling the packet flow to be transported over the VCG. All subsequent packets (in both directions) for this client-server pair are matched at the CD Ethernet port using the existing flow definitions and are directly forwarded in hardware. At the client side, the packets received from the VCG are switched to the client port based on the VLAN tag, which is then stripped off before the packets are forwarded to the client PCs, where the video is displayed on the screen. Using OpenFlow, it was possible to display both circuit and packet states in real time.

The extensions required for OpenFlow included the following:

- OpenFlow specifications (especially OpenFlow 1.0 [18], which was the basis for the prototype) already support control over matching and actions for the input and output ports and packet header information fields below:

| Physical | −Input port |
|----------|-------------|
| Ethernet | −VLAN ID |
|          | −Source address |
|          | −Destination address |
|          | −EtherType |
| IP       | −Source address |
|          | −Destination address |
|          | −Protocol number |
| TCP      | −Source port |
|          | −Destination port |

- In addition to the support of the input port and the L2 to L4 packet header information for matching of incoming packets and configuration of these on egress, extensions were made to allow L1 circuit parameters to be matched and configured and the creation of VCG that, in SONET/SDH, can be hitlessly enlarged or reduced in bandwidth depending on the following traffic requirements:

| | |
|---|---|
| Physical | —Input port/fiber |
| Wavelength | —Input wavelength |
| Electronic group | —VCG ID |
| Electronic TDM | —Starting timeslot |
| | —Signal type |

- Lastly, extensions were made to pass L1 topology information from the switch to the controller (especially discovered peer's switch and port IDs), allowing the use of technology-specific discovery mechanisms in the switch because the Packet_Out function of OpenFlow could not be used over an L1 interface.

The extensions made to the OpenFlow protocol were documented in a publically available software package called pac.c [19] and have been used in subsequent research projects on OpenFlow-based circuit control.

*OpenFlow in Europe—Linking Infrastructure and Applications (OFELIA) Project* Another major research project on the application of OpenFlow/SDN to optical networks is the OFELIA project. The goal of the OFELIA project is to create an experimental facility that allows researchers to control the network that they are using for carrying experimental data precisely and dynamically using OpenFlow-based control. The OFELIA work focuses on the virtualization of the optical network using standard interfaces, including both OpenFlow and GMPLS.

OFELIA is a large research effort, with an European Commission (EC) budget of €4,450,000, a 3-year life span (2010–2013), and a network of five federated island domains, including networks in the United Kingdom, Switzerland, Italy, and Germany [20].

One of the initial studies by the University of Essex [21] looked at the pairing of OpenFlow and GMPLS through the OpenFlow control of the packet mapping at access points and the use of the GMPLS user network interface (UNI) user-network interface to request paths across the optical network, as shown in Figure 15.5.

In this case, the OpenFlow protocol itself is unchanged, and another interface (GMPLS UNI) is used for the control of the optical network. In a further experiment, extensions were defined to the OpenFlow interface that built on the pac.c work, adding flexible labels rather than strict SONET/SDH timeslots and control of adaptation as well. In this model, the OpenFlow protocol acts on a virtual header with circuit characteristics [22].

| CCT ID | in port | out port | label in (e.g. encoding, ST, G-PID) | label out (e.g. encoding, ST, G-PID) | adaptation actions |
|--------|---------|----------|-------------------------------------|--------------------------------------|--------------------|

This format allows great flexibility because the label can correspond to the timeslot in the electronic domain or the wavelength in the optical domain; furthermore, it is possible to specify the type of adaptation to be used between layers. Further studies of the full integration of the OpenFlow control of packet and optical versus combinations of OpenFlow and distributed optical control suggest that the added flexibility of full integration may come with some additional cost when it involves the redesigning of the optical

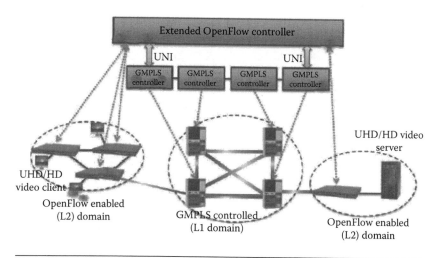

**Figure 15.5** OFELIA study using GMPLS UNI (UHD/HD, ultra high definition/high definition).

control plane, and may not be the least costly solution, although architecturally consistent.

## How Will OpenFlow/SDN Be Deployed in Carrier Optical Networks?

*OpenFlow/SDN Overlay with Transport Tunnels for POI*

As discussed above, the OpenFlow interface already supports the capacity for the control of packet forwarding at L2 and above and the concept of virtualized ports for the ingress and egress that can be the endpoint of a configured packet or circuit tunnel. This can be applied easily to packet/optical interconnection, where OpenFlow is used to control packet mapping into optical circuits, and the optical circuits are controlled separately through a management system, a distributed control plane, or other methods such as stateful path computation element (PCE) (described in more detail under Standards). In this deployment, OpenFlow would be used for one (packet) layer to specify the matching of incoming packets and forwarding them into a virtual port. OpenFlow is only needed for the subset of network elements supporting packet interfaces for customer edge services, whereas the core of the network (optical paths) are set up and controlled by an internal mechanism. Traffic through the tunnels passes through intermediate switches without visibility to the controller, and traffic engineering through the core is done independently.

Recent modeling of the POI control through OpenFlow has taken a different direction than the initial research, which focused on the treatment of wavelength, timeslot, etc., as additional match fields. Instead, the current modeling uses the concept of logical ports introduced into more recent versions of OpenFlow. In this model, the physical port on the switch may have multiple associated logical ports, and each logical port may have characteristics, such as wavelength or timeslot, which are modifiable by the controller (alternatively, the model may include a separate logical port for each wavelength or timeslot and use the match table to configure the mapping of a packet flow to a particular logical port). A generalized model for port characteristics in transport networks would include bit error rate, alarms, and other information available from the digital framing and, possibly, associate link characteristics with either the logical port or the associated physical port (Figure 15.6).

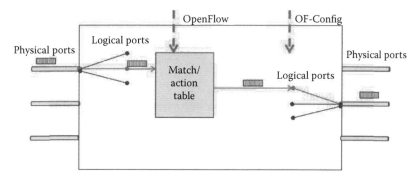

**Figure 15.6**   OpenFlow POI model.

*OpenFlow Overlay Over the Control Plane (with Abstraction)*

With extensions for the control of circuit mapping and packet forwarding, analogous to what was explored in pac.c and OFELIA, OpenFlow can be used directly for control over optical switching elements. To speed up the deployment of such capability in a carrier environment, which typically consists of multiple domains of deployed equipment, it would be more cost effective to deploy OpenFlow/SDN as an overlay rather than introduce an OpenFlow/SDN agent on every network element and add connectivity from each network element to a controller. Moreover, scalability and reliability requirements would be greater. Instead, an overlay deployment of OpenFlow/SDN would use OpenFlow interfaces only to an element management system (EMS) or a subset of network elements, where OpenFlow can be used by the application as a programmatic interface for the control of paths across the network, but the actual instantiation and direct control of resources is managed separately.

The intermediate system would mediate between the requests sent by the client controller and the actual control of network devices, providing an abstract model to the client controller that appears as a real network that is dedicated to that controller, where in fact the real network is being shared by multiple client controllers. This simplifies the function of the individual controller and provides policy control over the resources that are allocated to each application. Performance issues may be a concern here because the mediation function adds latency to control traffic going both from the client controller to the device and from the device to the client controller (e.g., event notifications) (Figure 15.7).

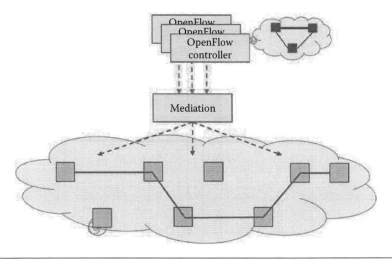

**Figure 15.7**   OpenFlow overlay with abstraction.

*Direct OpenFlow Control of Packet/Optical*

Although the overlay deployment addresses the programmability of optical networks and the improved separation of service and equipment software for faster service deployment and innovation, it does not affect the cost of network equipment, as overlay implies that the OpenFlow/SDN is deployed as an extra interface in addition to whatever legacy control structure is in place. If the goal is to achieve savings by the complete separation of software and hardware, this entails a fully centralized OpenFlow/SDN structure where the software on the network equipment is limited to an agent and to whatever is necessary to control local actions (e.g., protection) and element management. This is also the greatest change from the existing method of network operation, making it the most difficult alternative to implement, test, deploy, and manage for the service provider.

In a fully centralized OpenFlow/SDN control architecture, every network element has an interface to the controller, and the controller impacts all switching actions, including multiple layers if the network element supports multiple layers of switching technology. The controller must then be aware of all supported and unsupported actions, including any limitations on adaptation from one layer to another or connectivity between an incoming port and an outgoing port. To allow data to flow across the network, the controller must talk to each

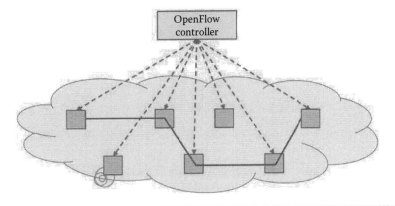

**Figure 15.8**  OpenFlow direct/centralized control.

network element along the datapath and configure the matching and forwarding actions that allows data to flow from the input port to the desired output port connecting to the downstream node (Figure 15.8).

The issue with a fully centralized control is as much related to the implementation and deployment as to the technology: carriers will need to verify that the solution is scalable and reliable, and even then, deployment will be gated by having to modify existing operations systems and procedures.

*Standards and Interoperability*

Going forward, there are several activities that have been established in the industry to develop common standards for the OpenFlow/SDN application to optical networks. The main activities are in the Open Networking Foundation (ONF) and the Internet Engineering Task Force (IETF).

*Open Networking Foundation*    ONF is a nonprofit consortium dedicated to the development and standardization of SDN interfaces, particularly, the OpenFlow protocol and the OpenFlow Configuration and Management (OF-Config) protocol. The mission of OFN is to commercialize and promote SDN and the underlying technologies as a disruptive approach to networking. Its activities are directed by a board made up of members of the user and service provider community. The ONF, by virtue of its control over OpenFlow, is the primary body for the development of SDN standards [23].

ONF has two groups, in particular, that are relevant to the creation of SDN for transport networks: the Extensibility Working Group (WG) and the New Transport Discussion Group. The Extensibility WG of ONF develops the detailed extensions to the OpenFlow specification that are deemed necessary to improve or extend the functionality of OpenFlow. Any technical extensions to the OpenFlow protocol must be passed by the Extensibility WG.

The New Transport Discussion Group was established in 2012 within ONF as a forum for discussing the application of SDN/OpenFlow to transport networks, including both optical and wireless networks in its original scope. The group is currently focusing on optical networks, with the objective of defining the use cases for SDN/OpenFlow over optical networks, the requirements for service provider and private network applications, and the recommendations for extensions to the OpenFlow protocols.

*IETF Standards*   Early efforts to incorporate SDN concepts into Internet standards led to the scheduling of IETF Birds of Feather (BoF) sessions on software-driven networking [24] and cross-stratum optimization (CSO) [25]. The former focused on the control plane architecture that would incorporate a centralized controller that interfaced with network elements to control connectivity at an abstract level. The latter focused on the potential use cases and benefits that could result from the coordination of applications and the optical network, the strata of its name. Both efforts were eventually judged to be not sufficiently of interest to IETF and were terminated with no follow-up standards effort.

Efforts in IETF have now focused on a project called Interface to Routing System (I2RS) [26] that takes a different direction that is less disruptive to the current routing environment. This approach focuses on the creation of a new programmatic interface into the router, which will support greater visibility into the routing information base and greater ability for the application to control forwarding decisions. Although, in concept, this will address the goal of programmability, it does not address service deployment or cost-reduction goals.

*PCE and Stateful PCE*    Because of the limitations of existing distributed routing mechanisms, work on the use of PCEs [27] for complex path computation began in 2006. PCE separates the path computation function from the other functions in a network element, allowing path computation to be migrated to a centralized server. PCE is a broadly applicable technology, which can potentially solve the problems with multidomain routing in optical networks and help with problems of complex path computation for all-optical networks.

The basic entities in a PCE system are the path computation client (PCC) and the path computation entity (PCE). The PCC uses the PCE protocol (PCEP) [28] to request a path from a source to a destination, along with other information, such as transport requirements, constraints, and the type of optimization. The PCE can be implemented on a network element or on a dedicated server and does path computation with its local database.

What ties PCE with SDN/OpenFlow is the concept of stateful PCE [29], where the PCE not only computes the path for a connection, but also controls the state of that connection, including causing the connection to be modified, rerouted, or terminated. Recently, it has been proposed that the PCE be able to initiate new connections as well, in which case, it takes on the functions of a central controller as in the OpenFlow, with the exceptions that:

- Its control of matching and mapping actions is limited to the context of a single layer connection (cannot operate over a flexible set of header values as in OpenFlow); and
- It relies on a distributed signaling protocol to carry out the actions it specifies, for example, to modify a connection, it directs the source node to send out the necessary signaling messages rather than interact with each network element in the path.

In Figure 15.9, the PCE provides an API for communication with client applications and has network topology stored internally. When requested, it both computes the desired network path and initiates the connect setup process at the source NE.

A variety of potential extensions for PCE-driven network action, coordinated by applications, has been captured in a recent IETF draft in this area [30].

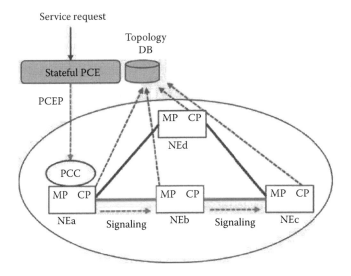

**Figure 15.9** PCE with stateful PCE. DB, database; MP, management plane; CP, control plane; PCC, path computation client.

*Optical Internetworking Forum (OIF)* One other group that may be involved is OIF. OIF is a group of carriers, system vendors, and component vendors that focuses on interoperability and agreements for the deployment of optical networks. OIF is led by a strong Carrier WG that provides carrier requirements for optical technology, both at the system and component levels, and includes major service providers such as AT&T, Verizon, Deutsche Telekom, France Telecom, China Telecom, and NTT. An example of OIF's work on requirements is the work that was done on the definition of a framework for service provider long haul 100-G transmission [31]. For SDN/OpenFlow, OIF has initiated work to identify use cases and carrier requirements for transport SDN through its Carrier WG. OIF could play a significant role in establishing the requirements for carrier-grade OpenFlow/SDN and driving deployment into service provider networks.

Conclusions for the Future

OpenFlow/SDN will need several modifications to be suitable for optical network control, both extensions to the protocol itself to control circuits (with no data stream visibility) rather than packets and progress in implementations to support service provider scalability, security, and reliability requirements. Initially, packet-oriented OpenFlow/SDN can be used in overlay deployment to control packet mapping and combined with existing methods of optical path provisioning to offer improved packet/optical interconnection services.

An extensive body of research and prototyping does exist for the application of OpenFlow/SDN to optical networks at layers 1 and 0, establishing basic functional requirements that would need to be addressed for circuit control. These may be the basis for limited scale deployment in private optical networks, where the requirements on implementation are less stringent. Deployment in the overlay mode where the OpenFlow/SDN interface serves as a service interface and is mediated by a transport network FlowVisor can be an early method of deploying capabilities in a service provider environment to support service innovation.

Eventually, the use of direct OpenFlow/SDN control of optical networks may emerge as a control architecture within domains of optical equipment as high availability controller implementations become available. Further work on the interconnection of controllers is needed for OpenFlow/SDN to scale to carrier environments that typically consist of multiple domains. How rapidly such deployment occurs depends on the ability of implementers to develop hardened, scalable versions of OpenFlow/SDN controllers and how real the advantages will be for cost and service development and deployment.

# References

1. ITU-T Recommendation G.805. 2000. Generic functional architecture of transport networks.
2. Berthold, J. and L. Ong. 2012. Next-generation optical network architecture and multidomain issues. *Proceedings of the IEEE* 100(5):1130–1139.
3. Beckett, S. and M. Lazer. 2006. Optical mesh service. *OIF Workshop on ASON/GMPLS Implementations in Carrier Networks.* http://www.oiforum.com/public/documents/061016-AT&T.pdf.
4. ITU-T Recommendation G.709-200912. 2009. Interface for the optical transport network (OTN).
5. ITU-T Recommendation G.872-200111. 2001. Architecture of optical transport networks.
6. ITU-T Recommendation G.709-200303. 2003. Interface for the optical transport network (OTN).
7. Roorda, P. 2008. Evolution to colorless and directionless ROADM architectures. *OFC/NFOEC 2008*, Paper NWE2.pdf.

8. AT&T presentation on optical mesh service. 2006. *OIF Workshop on ASON/GMPLS Implementations in Carrier Networks.* http://www.oiforum.com/public/documents/061016-AT&T.pdf.

9. Ong, L., E. Roch, S. Shew, and A. Smith. 2012. New technologies and directions for the optical control plane. *Journal of Lightwave Technologies* 30(4):537–546.

10. *Quantum API v.1.0 Guide.* 2012. http://www.docs.openstack.org/api/openstack-network/1.0/quantum-api-guide-1.0.pdf.

11. Perrin, S. 2009. The core packet-optical transport evolution. *Heavy Reading* 7(12).

12. OpenFlow Switch Specification v.1.2.0 (Wire Protocol 0×03). 2011.

13. OpenFlow-enabled cloud backbone networks create global provider data centers. 2012. *ONF Solution Brief.*

14. Martinelli, G., G. Galimberti, L. Ong, D. Ceccarelli, and C. Margaria. 2012. WSON optical interface class. draft-martinelli-wson-interface-class-03.txt (work in progress).

15. Sherwood, R., G. Gibb, K. Yap, G. Appenzeller, M. Casado, N. McKeown, and G. Parulkar. 2009. FlowVisor: A network virtualization layer. OPENFLOW-TR-2009-1. http://www.openflow.org/downloads/technical reports/openflow-tr-2009-1-flowvisor.pdf.

16. Das, S., G. Parulkar, P. Singh, D. Getachew, L. Ong, and N. McKeown. 2010. Packet and circuit network convergence with OpenFlow. *Optical Fiber Conference (OFC/NFOEC '10).* http://klamath.stanford.edu/~nickm/papers/Openflow-OFC10_invited.pdf.

17. OpenFlow Switch Specification v.1.0.0 (Wire Protocol 0×01). 2009.

18. Addendum to OpenFlow Switch Specification v.1.0 (Draft v.0.3). http://www.openflow.org/wk/index.php/PAC.C#Experimental_Extensions_to_OpenFlow.

19. Autenreith, A., P. Kaczmarek, P. Kostecki, J.-P. Elbers, S. Azodolmolky, M. Channegowda, R. Nejabati, and D. Simeonidou. 2012. Introducing OFELIA and OpenFlow extensions to support optical networking equipment. *OFC/NFOEC,* http://www.fp7-ofelia.eu/.../2012-03-08-ONTC-Autenrieth-OFELIA-final.pdf.

20. Azodolmolky, S., R. Nejabati, E. Escalona, R. Jayakumar, N. Efstathiou, and D. Simeonidou. 2010. Integrated OpenFlow–GMPLS control plane: An overlay model for software-defined packet over optical networks. *ECOC '10,* http://www.fp7-ofelia.eu/assets/Publications-and-Presentations/ECOCOpenFlowV3.pdf.

21. Sherazipour, M. and M. Tatipamula. 2011. Design considerations for OpenFlow extensions toward multidomain, multilayer, and optical networks. *ECOC '11,* http://www.fp7-ofelia.eu/assets/ECOC2011-OFELIA-Workshop-Presentations/9DesignConsiderationsforOpenFlow ExtensionsTowardsMulti-DomainMulti-LayerandOpticalNetworks.pdf.

22. Open Networking. http://www.opennetworking.org.

23. Nadeau, T. and P. Pan. 2011. Framework for software-driven networks. draft-nadeau-sdnframework-01 (work in progress).

24. Lee, Y., G. Bernstein, N. So, T. Kim, K. Shiomoto, and O. Dios. 2011. Research proposal for cross-stratum optimization (CSO) between data centers and networks. draft-lee-cross-stratum-optimization-datacenter-00.txt (work in progress).
25. Atlas, A., T. Nadeau, and D. Ward. 2012. Interface to the routing system framework. draft-ward-irs-framework-00.txt (work in progress).
26. Farrel, A., J. Vasseur, and J. Ash. 2006. IETF RFC 4655: A path computation element-based architecture.
27. Vasseur, J. (ed.). 2009. IETF RFC 5440: Path computation element (PCE) communication protocol (PCEP).
28. King, D. and A. Farrel. 2011. Application of the path computation element architecture to the determination of a sequence of domains in MPLS and GMPLS. draft-king-pce-hierarchy-fwk-06.txt (work in progress).
29. King, D. and A. Farrel. 2012. A PCE-based architecture for application-based network operations. draft-farrkingel-pce-abno-architecture-00.txt (work in progress).
30. OIF-FD-100G-DWDM-01.0. 2009. 100-G ultra long haul DWDM framework document.

# 16

# SECURITY ISSUES IN SDN/OPENFLOW

## NAGARAJ HEGDE AND FEI HU

**Contents**

## Introduction

Software-defined networking (SDN) is a new approach to networking. It was invented by Nicira Networks based on their earlier work at Stanford University, University of California at Berkeley, Princeton University, and CMU. The goal of SDN is to provide an open, user-controlled management of the forwarding hardware in a network. SDN exploits the ability to split the data plane from the control plane in routers and switches. The control plane is open and controlled centrally with SDN while having the commands and logic sent back down to the data planes of the hardware (routers or switches). This paradigm provides a view of the entire network and helps make changes centrally without a device-centric configuration on each hardware. The OpenFlow (OF) standard and other open protocols help manage the control planes and allow for precise changes to networks or devices. SDN works by creating virtual networks that are independent of physical networks. To achieve this, it separates the control plane from the data plane and allows the user to control the flow of traffic in the network. Figure 16.1 illustrates the difference between the traditional network and the OF-based SDN.

An OF system typically includes the following three important components:

1. The OF switch. OF provides an open protocol to program the flow table in different switches and routers. An OF switch consists of at least three parts: (1) a flow table, with an action associated with each flow entry; (2) a channel, allowing commands and packets to be sent between a controller and the switch; and (3) the OF protocol, which provides an open and standard controller to communicate with a switch.

2. Controllers. A controller adds and removes flow entries from the flow table on behalf of experiments. A static controller might be a simple application running on a PC to statically establish flows to interconnect a set of test computers for the duration of an experiment.

3. Flow entries. Each flow entry has a simple action associated with it; the three basic ones (that all dedicated OF switches must support) are (1) to forward this flow's packets to a given port, (2) to encapsulate and forward this flow's packets to a controller, and (3) to drop this flow's packets.

**SDN Security Concerns**

Before we discuss the security issues of SDN, let us quickly review the traditional networks' vulnerability and protection as summarized in Table 16.1.

SDN creates some new targets for potential security attacks, such as the SDN controller and the virtual infrastructure. Besides all the traditional network attack targets, SDN has new target points, as follows:

• SDN controller: traditional application, server, and network attacks

Switch

Control plane

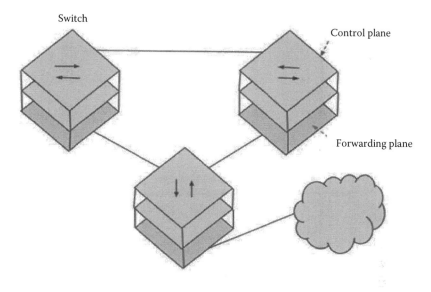

Forwarding plane

Controller

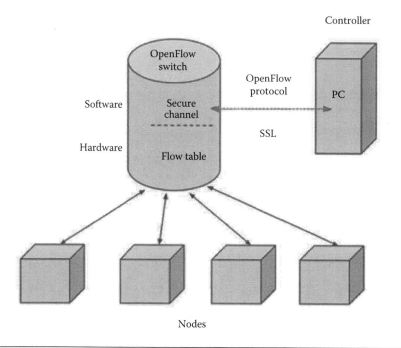

Software

Hardware

OpenFlow
switch

OpenFlow
protocol

PC

Secure
channel

SSL

Flow table

Nodes

**Figure 16.1**   Traditional network architecture (top) and OF architecture (bottom) (SSL, secure socket layer).

**Table 16.1** Traditional Network Vulnerability and Protection

| EXAMPLE ATTACK TARGET LAYER | EXAMPLE ATTACK POINTS | VULNERABILITY EXAMPLES | PROTECTION EXAMPLES |
|---|---|---|---|
| Applications | Network apps, general apps (e.g., database) | – General: cross-site scripting, buffer overflow, SQL injection<br>– Net: DNS cache poisoning | – General: firewall, intrusion, detection, intrusion, prevention<br>– Net: DNSSec, SSL |
| Servers | Transport, OS, hypervisor | – TCP: SYN flood, SYN/ACK scan, spoofed RST, hijack<br>– UDP: smurf DoS attack, spoofed ping-pong | – Encrypt session: SSH, IPSec, intrusion detection/prevention |
| Network | Routers, switches, virtual switches | – IP/routing: MIM routing attack, FIRP attack, IP spoofing, ping flood, ICMP destination unreachable, smurf attack, source attack<br>– FC: target/initiator spoofing<br>– MAC (FCF) spoofing<br>– ARP cache poisoning<br>– Physical link tap | – IP/routing: IP ACL filters, firewall, intrusion detection, OSPF with IPSec, split horizon, infinite hop count detection, override source routing<br>– FC: zoning, FC ACL filters<br>– Ingress/egress MAC ACL filters, VLANs<br>– Physical security<br>– Authentication protocol |

- Virtual infrastructure: traditional application and server attacks on the hypervisor, virtual switch, and virtual machine (VM)
- Network: OF protocol for OF-enabled devices

In the following section, we will describe some important OF/SDN security schemes.

### Enabling Fast Failure Recovery in OF Networks

It is important to have some robust restoration options available in OF networks. In Ref. [1], the addition of a fast restoration mechanism for OF networks is proposed, and its performance is evaluated by comparing the switchover times and the packet loss to the existing restoration options in a current OF implementation. It discusses some mechanisms implemented at the OF and Nox software to recover from a link failure. It also discusses the limitations of these mechanisms in enabling fast recovery in OF networks.

In the case of OF mechanisms, the failure can be recovered if a new correct flow entry is added in OF switches after the failure occurs. The recovery from the failure depends on the time when the OF switch requests the flow entry from the controller. Hence, the recovery from failure depends on the life of the flow entries in the flow table. The life of flow entries is associated with two intervals: idle timeout and hard timeout. Idle timeout indicates the time when the flow entries should be removed because of the lack of activity. It is the time interval of a flow entry with which the OF switch has not received the packet of a particular flow of that entry. Hard timeout implies the maximum life of flow entries, regardless of activity. OF switches remove their flow entries from the flow tables after the expiration of one or both intervals. Recovery from failure is directly proportional to the value of the aforementioned intervals.

In the case of Nox mechanisms, the recovery from failures is possible with a new flow entry only if the controller also knows about the failure. Otherwise, the controller may add an incorrect entry in the OF switches. Thus, recovery depends not only on hard and idle timeouts but also on mechanisms running in the network to detect the failure. Nox implements L2-Learning and routing mechanisms to recover from a failure. They implemented L2-Learning in C++ and Python. The former is called the L2-Learning switch, and the latter is called the L2-Learning pyswitch. They behave differently to recover from a failure.

Fast recovery is only possible if the incorrect flow entries are flushed from all the OF switches and new entries are installed immediately after detecting the link failure in the existing path. This will be possible with the help of a controller, only if the controller (1) knows about the failure, (2) remembers the older path that was established by adding flow entries in OF switches, (3) is able to calculate a new path, and (4) knows all the current flows in the network.

In fast restoration, the controller performs a set of actions to restore the path between affected hosts. The initial task is to check whether its calculated older paths among end hosts are affected or not. If these are affected, then the controller calculates the new path for those end hosts. Other than this, the controller also checks if it has added flow entries in OF switches regarding the older faulty path. If yes, then it deletes the flow entries from all the OF switches regarding the older path and adds flow entries in all the OF switches for the new path.

In Ref. [1], they used Ubuntu v.9.04 for the installation of OpenVSwitch v.1.1.0 and Nox v.0.9.0. More than 11,000 ping packets were sent from Client 0 to Server 1 with an interval of 10 ms. Then, the number of ping replies received by Client 0 was calculated. Hard and idle timeouts for flow entries are set to 20 and 10 s, respectively. The network routing loops can be easily removed by using any loop-free mechanism, for example, by building a spanning tree in the topology. The failure recovery time was investigated in the OF network when the outgoing path is affected by link failure. For fast restoration, we need an efficient scheme to calculate paths. A predetermined mechanism was used in Ref. [1] to calculate the path. The OF architecture allows the implementation of restoration options that are much faster than medium access control (MAC) reconvergence (routing and L2-Learning pyswitch) or the client-initiated recovery with an address resolution protocol (ARP) request (L2-Learning switch). Their fast restoration mechanism can be integrated in any mechanism where a controller is able to detect the failure by some means. In their fast restoration mechanism, the flow is able to switch to another path within 12-ms interval regardless of the time left before the expiration of the flow entries.

### Network Intrusion Detection and Countermeasure Selection (NICE) in Virtual Network Systems

In Ref. [2], NICE in virtual network systems was investigated. The Cloud Security Alliance (CSA) survey shows that, among all security issues, the abusive use of cloud computing is considered as the top security threat. Attackers can exploit vulnerabilities in clouds and use cloud system resources to deploy attacks. The convention schemes that patch known security holes in cloud data centers, where cloud users usually have the privilege to control software installed on their managed VMs, may not work effectively and can violate the service-level agreement (SLA). In Ref. [2], the authors propose NICE to establish a defense-in-depth intrusion detection framework. For better attack detection, NICE incorporates attack graph analytical procedures into the intrusion detection processes. In general, NICE includes two main phases:

1. They deploy a lightweight mirroring-based network intrusion detection agent (NICE-A) on each cloud server to capture and analyze cloud traffic. NICE-A periodically scans the virtual system vulnerabilities within a cloud server to establish a scenario attack graph (SAG). NICE's decision about whether to put a VM in the network inspection state will be based on the severity of the identified vulnerability toward the collaborative attack goals.

2. Deep packet inspection (DPI) is applied when a VM enters the inspection state and/or when virtual network reconfigurations can be deployed to the inspecting VM to make the potential attack behaviors prominent. In Ref. [2], they talk in detail about how to use attack graphs to model security threats and vulnerabilities in a virtual network system and propose a VM protection model based on virtual network reconfiguration approaches to prevent VMs from being exploited.

Different types of models are explained in Ref. [2]: (1) the threat/attack model: It is assumed that an attacker can either be outside or inside of the virtual networking system, and his primary goal is to exploit vulnerable VMs and compromise them as zombies. To improve the resiliency to zombie explorations, the protection model focuses on virtual-network–based attack detection and reconfiguration solutions; (2) the attack graph model: This is a modeling tool that illustrates all possible multistage, multihost attack paths that are crucial to understand threats and then to decide appropriate countermeasures. This is helpful in identifying potential threats, possible attacks, and known vulnerabilities in a cloud system; and (3) the VM protection model: This consists of a VM profiler, a security indexer, and a state monitor. Depending on various factors, such as connectivity, the number of vulnerabilities present, and their impact scores, a security index will be specified for all the VMs in the network.

Regarding the NICE framework, its major components are distributed and lightweight NICE-A software modules in each physical cloud server, a network controller, a VM profiling server, and an attack analyzer. The latter three components are located in a centralized control center connected to the software switches on each cloud server (i.e., virtual switches built on one or multiple Linux software bridges). NICE-A is a software agent implemented in each cloud server connected to the control center through a dedicated and isolated secure channel. OF tunneling or virtual local area network (VLAN) approaches are used to separate them from the normal data packets. The network controller is responsible for deploying attack counter-measures based on decisions made by the attack analyzer.

The NICE system components include the following:

- NICE-A. This is a network-based intrusion detection system (NIDS) agent installed in each cloud server. It scans the traf-fic going through Linux bridges that control all the traffic among VMs and in/out from physical cloud servers.
- VM profiling. To get precise information about the state, ser-vices running, open ports, etc., of VMs in the cloud, we can build profiles for those items. Connectivity with other VMs is the major factor that counts toward a VM profile.
- Attack analyzer. The major functions of the NICE system are performed by the attack analyzer, which includes procedures such as attack graph construction and update, alert correla-tion, and countermeasure selection.
- Network controller. This is a key component to support the programmable networking capability to realize the virtual network reconfigurations based on OF protocols.

In Ref. [2], they also gave an idea about NICE security measure-ment, attack mitigation, and countermeasures. Several counter-measures can be taken to restrict attackers' capabilities. When vulnerabilities are discovered or some VMs are identified as suspicious, it is important to differentiate between compromised and suspicious VMs. The countermeasure serves the purpose of (1) protecting the target VMs from being compromised, and (2) making attack behavior stand prominent so that the attacker's actions can be identified.

The performance of NICE is conducted in two directions in Ref. [2]: the security performance and the system performance. Both have been measured in detail by using virtual environment. The sec-ond one demonstrates the feasibility of NICE and shows that the pro-posed solution can significantly reduce the risk of the cloud system from being exploited and abused by internal and external attackers.

To conclude, NICE used the attack graph model to conduct attack detection and prediction. The proposed solution investigates how to use the programmability of software switches to improve the detection accuracy and defeat the victim exploitation phases of collaborative attacks. NICE only investigates the NIDS approach to counter zombie explorative attacks.

### FRESCO: Modular Composable Security Services for SDNs

In OF, we need to dramatically rethink the relationship between the data and control planes of the network device. From a network security perspective, these networks offer researchers with an unprecedented singular point of control over the network flow routing decisions across the data planes of all OF-enabled network components. An OF security application can implement much more complex logic than simply halting or forwarding a flow.

There are a few challenges that motivate us to design a new security scheme in SDNs: (1) information deficiency challenge: OF controllers do not uniformly capture and store TCP session information, among other key state tracking data. However, this is often required to develop security functionality (e.g., TCP connection status, IP reputation). This is referred to as information deficiency challenge. They incorporated a database module, FRESCO-DB, which simplified the storage and management of session state; (2) security service composition challenge: The proposed framework incorporated a modular and composable design architecture, inspired by the Click router architecture. This fosters a rapid and collaborative development of applications through module composition; and (3) threat response translation challenge: The OF protocol enables the controlling software layer to communicate flow handling instructions to the data plane.

The FRESCO framework consists of an application layer (which provides an interpreter and application programming interfaces (APIs) to support composable application development) and a security enforcement kernel (SEK; which enforces the policy actions from developed security applications). Both components are integrated into Nox, an open-source OF controller.

Nox Python modules were used to implement FRESCO's application layer. This is extended through FRESCO's APIs to provide two key developer functions, as follows:

1. FRESCO development environment (DE) and resource controller (RC). They provide FRESCO application developers with OF switch- and controller-agnostic access to network flow events and statistics. Developers use the FRESCO script language to instantiate and define the interactions between Nox Python security modules, which, in turn, invoke FRESCO internal modules. Those modules are instantiated to form a security application driven by the input specified via the FRESCO scripts. These are accessed via FRESCO's DE database API. These instantiated modules are executed by FRESCO DE as the triggering input events are received.

2. FRESCO SEK. Diverse security policies, such as DROP, REDIRECT, and QUARANTINE, can be enforced by security applications developed in FRESCO scripts to react to network threats by simply setting an action variable. These high-level security policies can help developers focus on implementing security applications that are translated into flow rules for OF-enabled switches by FRESCO DE. FRESCO incorporates an SEK, which is integrated directly into the OF controller on which FRESCO operates.

On the implementation perspective of the FRESCO architecture, Python is used to implement the FRESCO application layer prototype and runs as an OF application on Nox. The prototype operates on Nox v.0.5.0 using the OF v.1.1.0 protocol and is implemented in approximately 3000 lines of Python. FRESCO modules are implemented as independent Python objects, and inputs and parameters of a module are input variables to the Python object. The return values of a Python object are considered as output values of a module.

The FRESCO SEK is implemented as a native C++ extension of the Nox source code in approximately 1160 lines of C++ code. The modified OF command function was used to send OF commands to network switches and to capture flow rules from all OF applications.

To evaluate the components in FRESCO, Mininet was implemented, which provided a rapid prototyping environment for the emulation of OF network switches. Using Mininet, they have emulated one OF network switch, three hosts connected to the switch, and one host to operate their Nox controller. Flow generation has been performed by selecting one or two hosts to initiate TCP or UDP connections.

Despite the success of OF, developing and deploying complex OF security services remains a significant challenge. The proposed FRESCO is presented as a new application development framework specifically designed to address this problem. The FRESCO architecture has been integrated with the Nox OF controller and presents several illustrative security applications written in the FRESCO scripting language [3].

### Revisiting Traffic Anomaly Detection Using SDN

Small office/home office (SOHO) and purely home networks have had an explosive growth over the last decade because of the widespread penetration of broadband Internet in the home market. Moreover, computers in such networks are often vulnerable. SDN offers a natural opportunity to delegate the task of network security to the home network while sparing the home users a natural opportunity to delegate the task of network security to the home network. Moreover, the home user is spared from complex security management tasks. The authors propose a programmable home network router that provides the ideal platform and location in the network for detecting security problems.

In Ref. [4], four prominent anomaly detection algorithms are implemented in the Nox SDN controller. A detailed accuracy evaluation has been performed on real-world traffic data sets collected at three different network deployment points. Efficiency evaluation of the SDN implementations on home and SOHO network data sets shows that, in addition to providing better accuracy, this approach allows line rate anomaly detection.

Here, the implementations of four prominent traffic anomaly detection algorithms in the context of an SDN are briefly described.

1. Threshold random walk with credit-based (TRW-CB) rate limiting algorithm. It detects scanning worm infections on a host by noting that the probability of a connection attempt of being a success should be much higher for a benign host than a malicious one. The algorithm leverages this observation using sequential hypothesis testing (i.e., likelihood ratio test) to classify whether the internal host has a scanning infection. The algorithm maintains a queue of new connection initiations (i.e., TCP SYNs) that are yet to receive a response (i.e., a SYN/ACK) for each internal host.

2. Rate limiting. This uses the observation that, during virus propagation, an infected machine attempts to connect to many different machines in a short span of time. On the other hand, an uninfected machine makes connections at a lower rate and is more likely to repeat connection attempts to recently accessed machines.

3. Maximum entropy detector. This estimates the benign traffic distribution using maximum entropy estimation. Training traffic is divided into 2348 packet classes, and maximum entropy estimation is then used to develop a baseline benign distribution for each class. Packet classes are derived from two dimensions. The first dimension contains four classes (TCP, UDP, TCP SYN, and TCP reset (RST)). In the second dimension, each of these four classes is split into 587 subclasses based on destination port numbers.

4. Network advertisement (NETAD). This operates on rule-based filtered traffic in a modeled subset of common protocols. The filter removes uninteresting traffic based on the premise that the first few packets of a connection request are sufficient for traffic anomaly detection.

In Ref. [4], the data have been collected at benign traffic at three different locations in the network because the aim was to study the accuracy of anomaly detection algorithms in a typical home network, a SOHO network, and an internet service provider (ISP). To collect the attack traffic, denial of service (DoS) (TCP SYN) and fragile (UDP flood) have been launched simultaneously from three end hosts.

An SDN using OF and Nox allows a flexible, highly accurate line rate detection of anomalies inside home and SOHO networks. The key benefit of this approach is that the standardized programmability of SDN allows the algorithms to exist in the context of a broader framework.

### Language-Based Security for SDNs

Analyzing the fundamental problem of how to program SDN in a secure and reliable manner is discussed in Ref. [5]. The solution involves the development of a new programming model that supports the concept of a network slice. The isolation of the traffic of one program from another is achieved with the help of slices. They also isolate one type of traffic from another, within that same program. They have developed a semantics for slices, which illustrates new kinds of formal modular reasoning principles that network programmers can now exploit. It provides definitions of the end-to-end security properties that the slices entail and proves the correctness of a compiler for an idealized core calculus in a slice-based network programming. They have also described their implementation, which is equipped with a translation validation framework that automatically verifies compiled programs using the Z3 theorem prover.

It is challenging today to implement isolation in networks. For this, a large set of devices, including routers, switches, and firewalls, to be manually configured, can be used to block forbidden traffic, but they allow other traffic to traverse the network. Developing and maintaining these configurations will be done using low-level and vendor-specific configuration languages, and this work is tedious for network operators. Simple errors can often lead to serious security breaches. In Ref. [5], it was suggested that using a high-level programming language can make it easy to describe forwarding policies and construct isolated subnetworks while leaving the tedious, error-prone work of generating correct and efficient low-level configurations to a compiler and an SDN. Such an approach seemed to solve the problem that was faced today: networks have traditionally been built out of closed devices that cannot be programmed, except through proprietary and idiosyncratic interfaces. One cannot use Nox as a solution because it lacks mechanisms for isolating the traffic of one module from the traffic of another.

Moreover, we have to see that network programming is insecure and noncompositional. Even the smallest, most trivial modules need to explicitly avoid interfering with other modules whose functionality is completely orthogonal. The programmer determines which traffic will be processed and how it will be forwarded only by analyzing every module in their program. A different approach is described in Ref. [5]. They provided programmers with a high-level, correct-by-construction abstraction that supports the programming-isolated slices of the network, rather than forcing programmers to rely on an external, coarse-grained hypervisor or their own on-off, *ad hoc* techniques for building modular and secure network programs. A slice is defined in terms of the following ingredients: a topology composed of switches, ports, and links, and a collection of predicates on packets, one for each of the outward-facing edge ports in the slice.

Several distinct technical contributions are made [5]. They developed a formal calculus that models a network program as a function from packets to sets of packets to precisely analyze the semantics of slices. The system can execute multiple programs in a single network. The formalized isolation serves as a pair of noninterference conditions: one with respect to a notion of confidentiality and another with respect to integrity. A slice is isolated from another if running them side by side in the same network does not result in a slice leaking packets into the other slice. They defined several intuitive security properties such as isolation and developed an operational condition called separation that implies the isolation property. Finally, they formalized a compilation algorithm and proved that it establishes separation and isolation.

In Ref. [5], a problem that arises in the context of configuring networks is addressed, but how should one express and verify the isolation? This is a fundamental issue. Their solution used technology developed by, and of interest to, the programming languages community: the foundation for their solution is a correct-by-construction programming abstraction. To describe the execution of network programs, they used structured operational semantics. They have used concepts such as confidentiality, integrity, and observational equivalence, drawn from classic work on language-based information-flow security. They used translation validation to obtain assurance.

Moreover, they used familiar proof techniques to validate all of their theorems.

Overall, in Ref. [5], they showed how to define, analyze, implement, verify, and use the slice abstraction to build secure and reliable SDNs. The definition of slices leads to an elegant theory of isolation and proofs of strong end-to-end properties based on observational equivalence, such as confidentiality and integrity. The implementation is highly reliable because they encoded the semantics of their network policies as logical formulae and the use of Z3 theorem prover to validate that their compiler generates outputs that are equivalent to inputs. These encodings also allow them to automatically verify the isolation properties of compiled programs.

In summary, the slice abstraction provides network programmers with a powerful means to seal off portions of their SDN programs from outside interference, such as modules and abstract data types in conventional programming languages. The ability to impose strong boundaries between different program components provides important security benefits and simplifies the construction of programs in settings where security is not an issue. By carving a large program up into slices, a programmer can reason locally about each slice instead of globally when attempting to determine how and where their traffic is forwarded.

### Scalable Fault Management for OF

High reliability is an important requirement because a transport connection aggregates traffic. Automatic recovery mechanisms that are triggered by operations, administration, and maintenance (OAM) tools [6] are required for reliability to reestablish connectivity when a path failure occurs.

Recovery is categorized into restoration and protection. For restoration, detour or alternate paths are computed and configured only after a failure is detected. This method is relatively slow. In contrast, for protection, a backup path is configured parallel to the working path. Hence, a fast switch-over minimizes traffic disruption whenever a failure is detected. Transport applications demand a 50-ms recovery time protection to be supported by any transport network technology. The traffic engineering function can calculate recovery paths in later stages.

In Ref. [6], they proposed to relax the separation of control operations to include connectivity monitoring OAM in the switch to overcome the scalability limitations of centralized connectivity monitoring. The connectivity monitoring must operate in a proactive manner to ensure fast detection of any impairment along the path. The source end of the monitored flow will be the entry point of a transport tunnel. It periodically emits probe packets and interleaves them into the data flow running along that path. The destination end extracts the probe packets from the data flow. If the destination end stops receiving the probe packets for a long enough period (which is referred to as the detection time) or a received packet indicates that the source endpoint detects some errors in the other direction, known as remote defect indication, some node or link in the monitored flow is assumed to have failed, and the destination end node switches over to an alternate path.

They applied the strict integration of the aforementioned functions with the OF switch forwarding to implement an improved fault management efficiently. They proposed to extend the OF v.1.1.0 switch forwarding model with new entities and showed what protocol extensions are essential to provide these novel entities. Nevertheless, they did not intend to provide a comprehensive protocol specification but rather an insight on the necessary switch model and protocol updates.

To generate monitoring messages within the switch, they allow multiple monitored entities to use the same message rate to share a monitoring packet generator. To reduce the amount of processing required in the switch, they separate packet generation from formatting that is filling in identifiers or timers.

Like any other packet, incoming OAM packets enter the switch through ports. First, the monitored tunnel including the OAM packets is identified with a flow table entry. The typical actions associated to that entry are removing the tunnel tags and passing the packet to the next stage, where the payload is further processed. At this phase, demultiplexing is either done as part of popping the tunnel tag or expressed as an additional rule deployed in a later flow table.

The OAM packets can be terminated at a group table entry containing a single action bucket. The content of the OAM packet is processed, and the timers associated to the OAM termination points are updated. The switch notifies the controller on the occurrence of any monitoring and protection events. To support a generic notification message, an OF error message was modified. Two notification messages are defined here: (1) the messages that report any changes to the status of the monitoring entity and (2) the messages that report the protection reactions. This separation is considered to allow mixed scenarios, where controller-driven restoration is combined with switches that perform continuity monitoring.

1. Fault management in MPLS transport profile. In multiprotocol label switching-transport profile (MPLS-TP) protection, switching is performed by the endpoints of the working path and the backup path. The endpoints may notify each other on protection actions using a data plane–associated protocol, such as Protection State Coordination or a mechanism that is part of the control plane, or using the generalized MPLS (GMPLS) Notify message to stay synchronized. The MPLS-TP OAM supports a continuity check packet flow to detect label switch path (LSP) failures. The recommended implementation is based on the Bidirectional Forwarding Detection (BFD) protocol.

2. Adding BFD and protection support to OF. The current implementation is based on an extension of the OF v.1.0 reference switch implementation that supports MPLS forwarding. As a consequence, they rely on configurable virtual ports instead of using group table entries to implement the monitoring packet construction procedure. This means that not only the packet generator, but also the other packet construction steps are implemented with virtual ports and configured through virtual port management commands.

To evaluate their scheme, they measured the failure time in a testbed consisting of two OF label edge routers (LERs) with two LSPs between them (one working and another as a backup). The switches were implemented as modified OF v.1.0 soft switches on Linux, and the controller was a modified version of the Nox open-source controller running on Linux. Two BFD sessions, both running with a packet transmission interval of 10 ms, monitor the working and backup of LSP by giving a maximum detection time of 30 ms. They transmit constant bit rate (CBR) traffic with roughly 800 packets per second from the source across the working LSP. The captured inter-arrival time of the CBR packets before and after recovery at the sink is an approximation of the full protection time.

Central control entity is expected to enhance scalability by means of offloading the controller by placing control functions at the switches. By redefining the role of the link-layer discovery protocol (LLDP)-based centralized monitoring, the aforementioned gain can be achieved. As a first step, LLDP is not used any longer for the continuity check, which relaxes the time constraints from milliseconds to seconds. Afterward, the new link detection and failure declaration are decoupled by using switch monitoring features together with link-down notifications. As a consequence, there is no need to consider accidental packet losses and jitters, which further increases packet sending intervals.

In conclusion, they proposed to slightly relax the separation of control and forwarding operations to overcome the scalability limitations of centralized fault management. To provide a scalable way for data plane connectivity monitoring and protection switching, they argue that OAM is a function that needs to be distributed and deployed close to the data plane. They propose to place a general message generator and processing function on OF switches. They describe how the OF v.1.1.0 protocol should be extended to support the monitoring function. Moreover, they prove that data plane fault recovery within 50 ms can be reached in a scalable way with their proposed extensions, through their experiments.

### A Dynamic Algorithm for Loop Detection in SDNs

The existence of loops, which are cyclical paths through the network's switches, which can cause some packets to never leave the network, will be a potential problem in computer networks. In Ref. [5], a dynamic algorithm that is built on header space analysis is presented, which allows the detection of loops in SDNs like the ones created using OF over a sequence of rule insertions and deletions on the network's switches, and the key ingredient in the algorithm is a dynamic, strongly connected component algorithm. In the article, the network model has been illustrated as a directed graph because it will be easier

**Table 16.2**　Comparison of the Schemes Discussed

| SCHEME SOURCES | [2] | [3] | [4] | [5] | [6] | [7] | [8] |
|---|---|---|---|---|---|---|---|
| Uses OF/Nox (Y/N) | Y | Y | Y | Y | N | Y | Y |
| Introduce new architecture based on OF/Nox | Y | N | Y | N | N | Y | Y |
| Experiments (E) or real cases (R) | E | E | E | E | E | E | E |
| Software (S) or hardware (H) | S | S | S | S | S | S | S |
| Introduces new language for SDN (Y/N) | N | N | N | N | Y | N | N |

to understand. Hence, concepts of header space analysis have been translated into the language of graph theory.

1. Rule graphs and the dynamic loop detection problem. In Ref. [7], they show how to model a network in the same way as a directed graph. Through this, the translation can introduce notions from dynamic graph algorithms to help one compute port-to-port reachability in their network over rule updates in an efficient manner.
2. A dynamic, strongly connected component algorithm. It introduces an algorithm that allows to dynamically track the strongly connected components (SCCs) in a graph over a sequence of edge insertions and deletions. It also shows how to use this dynamic algorithm to solve the dynamic loop detection.

### Discussion

A comparison of all the aforementioned SDN security schemes is presented in Table 16.2.

### Conclusion

In conclusion, SDN is an emerging technology that allows for granular security by giving complete control of the network to the administrator. The controller is the brain of SDN, and without proper security wrapped around the controller, the network becomes completely vulnerable to accidental changes or malicious attacks. There are different approaches that can be used to achieve this task and take the full benefit of this new technology.

# References

1. Chernick, C. M., C. Edington III, M. J. Fanto, and R. Rosenthal. 2005. Guidelines for the Selection and Use of Transport Layer Security (TLS) Implementations. SP 800-52. National Institute of Standards & Technology, Gaithersburg, MD.

2. Sharma, S., D. Staessens, D. Colle, M. Pickavet, and P. Demeester. 2011. Enabling fast failure recovery in OpenFlow networks. *2011 8th International Workshop on the Design of Reliable Communication Networks (DRCN).* Krakow, Poland, pp. 164, 171. http://www.fp7-sparc.eu/assets/publications/10-Recovery-DRCN2011.pdf.

3. Chung, C. J., P. Khatkar, T. Xing, J. Lee, and D. Huang. 2013. NICE: Network intrusion detection and countermeasure selection in virtual network systems. *IEEE Transactions on Dependable and Secure Computing* 99:1. http://www.snac.eas.asu.edu/snac/document/tdsc-vinice-v8.pdf.

4. Shin, S., P. Porras, V. Yegneswaran, M. Fong, G. Gu, and M. Tyson. 2013. FRESCO: Modular composable security services for software-defined networks. *Proceedings of the ISOC Network and Distributed System Security Symposium,* San Diego, CA. http://people.tamu.edu/~seungwon.shin/NDSS_2013.pdf.

5. Mehdi, S. A., J. Khalid, and S. A. Khayam. 2011. Revisiting traffic anomaly detection using software-defined networking. In *Proceedings of the 14th International Conference on Recent Advances in Intrusion Detection (RAID '11),* Robin Sommer, Davide Balzarotti, and Gregor Maier (Eds.). Springer-Verlag, Berlin, Heidelberg, pp. 161–180. http://www.wisnet.seecs.nust.edu.pk/publications/2011/raid2011_paper.pdf.

6. Schlesinger, C. Language-based security for software-defined networks. http://www.cs.princeton.edu/~cschlesi/isolation.pdf.

7. Kempf, J., E. Bellagamba, A. Kern, and D. Jocha. 2012. Scalable fault management for OpenFlow. *2012 IEEE International Conference on Communications (ICC),* pp. 6606–6610. http://www.kth.diva-portal.org/smash/record.jsf?pid=diva2:613237.

8. Kordalewski, D. and R. Robere. 2012. A dynamic algorithm for loop detection in software-defined networks. http://www.cs.toronto.edu/~robere/paper/networkgraph-1214.pdf.

# 17

# Investigation of Anycast Implementation in Software-Defined Networking

## Jingguo Ge, Yulei Wu, Yuepeng E, Junling You, and Chuan Du

### Contents

## Introduction

Anycast is a paradigm of communication for service discovery, which selects the best one of the service providers in an anycast group as a destination in a robust and efficient manner, which has been an important service model adopted in various networks for diverse applications. Anycast technologies are widely used in content delivery networks (CDNs) for the large-scale distribution of content on the Internet and the direction of requests to find the desired content [1]. To avoid the substantial delays because a transmitting node needs to wait for its next-hop relay node to wake up, wireless sensor networks (WSNs) adopted an anycast-based scheme for each node to opportunistically forward packets to the first neighboring node that wakes up among multiple candidate nodes [2]. Mobile *ad hoc* networks (MANETs) also used anycast-like proposal to shorten the transmission paths between the requester and service providers and reduce the amount of request and reply packets [3].

Currently, application-layer anycast and network-layer (or IP) anycast are the two main research directions on anycast communications and have been widely used in many scenarios because of its inherent ability of service discovery. However, the nonawareness of the topology changes and load conditions hampers the deployment of application-layer anycast [4]. In addition, the address translation of application-layer anycast is costly when the network is under heavy load [5]. IP anycast overcomes these issues with simple implementation but needs to modify existing routing protocols with new router configurations to support anycast service, and thus, the anycast server selection process and the packet routing process are accomplished in the switching equipment, resulting in higher efficiency and robustness than that of application-layer anycast [6]. However, the existing network equipment (e.g., routers and switches) act as black boxes, leading to the poor scalability for the deployment of IP anycast. Furthermore, because of its inherent nature, IP anycast makes the address non-aggregatable in the routing table [7].

Software-defined networking (SDN) has been proposed to programmatically control networks by decoupling the control from the data plane, which is a promising technique to lower the barrier for deploying and managing new functionality, applications, and services in the networks. Thus, the preceding issues encountered in the wide deployment of application-layer anycast and IP anycast can be readily solved by this new paradigm of architecture. The main thrusts in SDN are OpenFlow [8] by the Open Networking Foundation (ONF); Protocol Oblivious Forwarding (POF) [9] by Huawei Technologies Co., Ltd.; PEARL [10] by the Institute of Computing Technology, Chinese Academy of Sciences; etc. OpenFlow is currently the most promising and popular realization of SDN and has been commercially produced by CISCO, HP, Juniper, NEC, etc. In OpenFlow, the software running at a logically centralized controller, manages a collection of switches hosting programmable forwarding tables. In an effort to ease the development and deployment of anycast service in the Internet, this chapter makes the following main contributions:

- This chapter presents a new load-aware anycasting based on OpenFlow technology for intradomain environments and develops the Information Gathering Module, Routing Decision Module, Address Resolution Module, and Data Transmission Module to support anycast service and load-aware mechanism at the controller.
- Extensive Mininet experiments are conducted to validate the effectiveness and accuracy of the proposed OpenFlow-based anycast scheme. The results demonstrate that the performance of the developed load-aware anycast scheme outperforms that of existing solutions in terms of anycast request delay and loss probability.
- The developed OpenFlow-based anycast scheme for the intradomain environment is then extended to the solution for interdomain networks of global deployment strategies. The analysis shows that this strategy can be adopted as an evolvable way for the large-scale deployment of load-aware anycast service in the current Internet.

The remainder of this chapter is organized as follows. "Related Work" presents the existing studies on anycast implementation and the issues encountered. The preliminaries of SDN and OpenFlow are shown in "Preliminaries of SDN and OpenFlow." "Implementation of Anycast Service Based on OpenFlow in Intradomain Environments" describes the design of load-aware anycasting based on OpenFlow for a single autonomous system (AS), including the system architecture and, particularly, the design of the OpenFlow controller, with Information Gathering Module, Routing Decision Module, Address Resolution Module, and Data Transmission Module to support anycast service. The extensive Mininet experiments and analysis are conducted in "Numerical Results and Analysis." "Extension of Anycast Implementation in Interdomain Environment" then extends the proposed scheme to the solution for the interdomain of multidomain scenarios to support evolvable deployment of load-aware anycast service in a global environment. This chapter ends with the "Conclusion."

## Related Work

The existing studies on the investigation of anycast service can be classified into two categories: application-layer anycast and IP anycast. For example, Ma et al. [11] targeted the challenges of anycast implementation in the large-scale global environments and managed to solve the problems of scalability to worldwide implementation, anycast query latency minimization, and optimal server selection strategies. The authors in Ref. [12] presented a context-aware anycast multimedia provisioning scheme in the application layer by using the distributed deployed service registry nodes that collect and maintain the server's contexts and content descriptions, and perform the mapping of the anycast address of the client request to the unicast address of the most convenient server based on the contexts of the clients and servers. Bhattacharjee et al. [13] designed a special name structure for application-layer anycast service and adopted a resolver to translate anycast address to IP address. The communication procedure is similar with the domain name system (DNS) resolution. The proposed scheme is simple and easy to be deployed, but it is nonsensitive to the topology and load status changes. The authors in Refs. [14] and [15] used the statistic and stochastic methods in the scheduling of the anycast manager, which improves the performance of application-layer anycast. Garcia-Luna-Aceves and Smith [16] proposed a mechanism to accelerate the process of anycast address translation, which alleviates the performance degradation resulted by address translation, but this scheme cannot solve the problem radically. The authors in Ref. [17] investigated the problem of directing clients to the replicated servers and presented an anycasting paradigm at the application layer by providing a service that uses an anycast resolver to map an anycast domain name and a selection criteria into an IP address.

On the other hand, the main focus of IP anycast is the design of routing strategies and the improvement of protocols to support anycast service in the current IP architecture. For example, Alzoubi et al. [18] presented a practical architecture for load-aware IP anycast based on the traditional route control technology and its usage in CDN environments. The proposed scheme made use of route control mechanisms to consider server and network load to realize load-aware anycast. The authors in Ref. [19] developed a distributed system that can offload the burden of replica selection while providing these services with a sufficiently expressive interface for specifying mapping policies. The prototyping of the system supports the IP anycast and can handle many customer services with diverse policy objectives. Katabi and Wroclawski [20] proposed an infrastructure to achieve global deployment of IP anycast, which was extended by Ballani and Francis in Ref. [6]. Lenders et al. [21] proposed a density-based anycast routing strategy to improve the stability of IP anycast. The authors in Ref. [22] proposed a new anycast communication model based on IPv6 to solve the problems of scalability and communication errors between clients and servers. A proxy-based architecture that provides the support for stateful anycast communications, while retaining the transparency offered by the native anycast, was proposed in Ref. [23] to overcome the routing scalability issues and the lack of stateful communication support in the traditional IP anycast.

To the best of our knowledge, there is little research conducted in the current literature to investigate the performance of anycast service using OpenFlow. Very recently, Othman and Okamura [24] proposed a content anycast solution using OpenFlow to improve the effectiveness of content distribution. However, this study only presented a mapping mechanism between each file and its associated identifier (ID) and redirected the content requests that cannot be handled by one server to another, rather than solving the inherent problems resulting from the traditional anycast service.

## Preliminaries of SDN and OpenFlow

SDN [25] is a promising technique to lower the barrier for deploying and managing new functionality, applications, and services in the networks. OpenFlow [26] was raised in 2008 by the Clean State Team of Stanford University, and it has been the promising and popular technology to realize the SDN because of its ability to support the decoupling of the control plane and the data (forwarding) plane. OpenFlow centralizes the control of the flow table in the switch devices to an external programmable and flexible controller, and provides a secure protocol to facilitate the communication between the controller and switch devices. OpenFlow has been commercially deployed in data center networks and wide area networks such as Google.

A packet is forwarded by the switches based on the entries in the flow table. OpenFlow switches possess a much simpler flow table than ordinary switches. The flow table in the OpenFlow switch consists of many flow entries, each of which includes six parts: Match Fields, Priority, Counters, Instructions, Timeouts, and Cookie, as shown in Figure 17.1. The Match fields is used to match against packets, which consists of ingress port and packet headers; the Priority is adopted for matching the precedence of the flow entry; the Counters is used to update for matching packets; the Instructions is used to modify the action set or pipeline processing; the Timeouts sets the maximum amount of time or idle time before flow is expired by the switch; and the Cookie is the opaque data value chosen by the controller, which can be used by the controller to filter flow statistics, flow modification, and flow deletion.

The standard v1.3 of OpenFlow supports 40 match fields, where 10 match fields are required to be supported by a switch, including Ingress Port, Ethernet Source Address and Destination Address, Ethernet Type, IPv4 or IPv6 Protocol Number, IPv4 Source Address and Destination Address, IPv6 Source Address and Destination Address, transmission control protocol (TCP) Source Port and

| Match fields | Priority | Counters | Instructions | Timeouts | Cookie |
|---|---|---|---|---|---|

**Figure 17.1**   Main components of a flow entry in a flow table. (From OpenFlow switch specification. Available at http://www.opennetworking.org/images/stories/downloads/specification/openflow-spec-v1.3.0.pdf.)

Destination Port, and user datagram protocol (UDP) Source Port and Destination Port [26].

## Implementation of Anycast Service Based on OpenFlow in Intradomain Environments

### The System Architecture

The proposed architecture for an anycast system based on OpenFlow technology consists of anycast servers providing anycast service, the anycast client, OpenFlow-enabled anycast (OFA) switches, and the anycast controller, where anycast servers and anycast clients are directly or indirectly connected by OFA switches. In this section, we consider the case for the intradomain networks of a single AS, where the OFA switches are deployed in the subnetworks of an AS and are interconnected through IP tunnel or dedicated link to form a wide-area layer 2 network (see Figure 17.2). The anycast controller has IP connections with OFA switches.

The anycast controller is the key component in this system, with the aim of assigning and recycling anycast addresses and making the reasonable and appropriate routing decisions. To increase the scalability of supporting more anycast service, we apply a set of addresses for anycast services. The anycast controller can collect the status and information of OFA switches and the connected anycast servers/

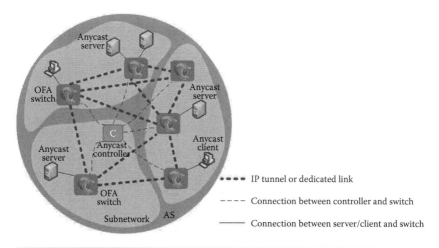

**Figure 17.2**  System architecture of OpenFlow-based anycasting in an AS.

clients, and make the routing decisions for anycast requests according to various performance metrics. The routing decisions are added to the OFA switches as the entries in the flow table. The OFA switches forwards the packets according to the entries in the flow table. In addition, the OFA switches possess the ability to advertise the prefix of anycast addresses.

*The Design of the Anycast Controller*

OpenFlow achieves the higher flexibility in the routing of network flows and the freedom to change the behavior of network traffic by separating the control plan in network devices from the data plane, resulting in the importance of the controller in OpenFlow technologies. In this section, the controller will be designed to support anycast service, as shown in Figure 17.3. Many kinds of OpenFlow controller system have been widely reported, such as POX, NOX, Maestro, Beacon, simple network access control (SNAC), Floodlight, etc. Because of the

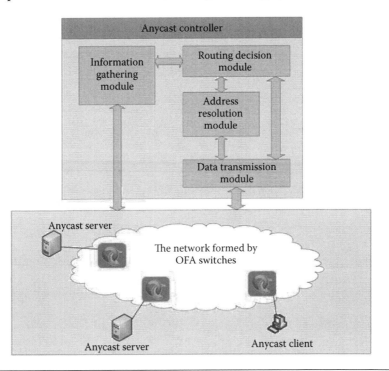

**Figure 17.3**  Design of the anycast controller.

feature of rapid development and prototyping of network control software, in this chapter, we develop the anycast controller based on the POX platform [27] and particularly design the Information Gathering Module, Address Resolution Module, Routing Decision Module, and Data Transmission Module for anycast service. In what follows, the detailed design of these modules will be presented.

*Information Gathering Module*   The anycast controller is responsible for being aware of the topology changes and the status changes of loads at the anycast servers/clients and network links to make appropriate decisions of routing strategies. If the server selection in anycast service depends on the metrics of hop-count from the anycast client who originates the requests to the anycast server, the anycast controller must be aware of the network topology. On the other hand, if the load balance needs to be considered in routing decision making, the load status of each network link and anycast servers and clients should be known by the controller.

The anycast controller possesses the abilities to detect the network topology and load status of anycast servers and clients and network links by communicating with OFA switches based on the OpenFlow protocol [26].

- The OFA switches report the changes of ports to the anycast controller when receiving a Port-Status message of asynchronous type. This can be used by the anycast controller to detect the topology changes of the underlying networks.
- The anycast controller can get the status information of OFA switches by sending a Read-State message of controller-to-switch type. Keeping track of the status statistics of all OFA switches, the anycast controller can be aware of the required information, for example, load status on network links, of the whole network.

For the implementation, we can set a timer in the anycast controller to trigger a request for the required information to OFA switches periodically. According to the response of the OFA switches, the anycast controller can be aware of the topology of the whole network and monitor the load status of each link and port, which is helpful to support routing decision making, stated in the next subsection, under different measurement metrics.

*Routing Decision Module* This module makes routing decisions for packets that fail to match the entries in the flow tables of OFA switches. In the proposed controller, two routing metrics are considered: hop counts and link loads. In what follows, the implementation of the two strategies in the routing decision module will be presented.

*The Routing Strategy Based on Hop Counts* The routing strategy based on the hop counts choose the nearest anycast server to serve the request, where the term *nearest* is calculated by the hop counts from the anycast server and the anycast client who originates the request. To achieve this purpose, it is necessary to calculate the hop counts between any two OFA switches in the network. If the entire network topology is abstracted as a graph, then the problem is transformed into calculating the shortest path between any two points in the graph.

The Floyd-Warshall algorithm has been widely adopted in the literature [28–30] for the fundamental graph problem of all-pairs shortest path, which is then used in this chapter to carry out the shortest path calculation between two OFA switches. In the Floyd-Warshall algorithm, there are two possible shortest paths from Point $A$ to Point $B$ in the graph. One is the path that connects Point $A$ and Point $B$, and the other is the path from Point $A$ to Point $B$ through some other points, say, Point $C$. Let $Distance(A, B)$ represent the shortest path from Point $A$ to Point $B$, and let $P$ denote the set of all points in the graph. Thus, for $\forall C \in P$, $C \neq A$, $C \neq B$,

$$Distance(A, C) + Distance(C, B) > Distance(A, B).$$

Then, once traversing all points in the graph, $Distance(A, B)$ records the shortest path from Point $A$ to Point $B$.

*The Routing Strategy Based on Link Loads* The routing strategy based on link loads choose the best anycast server to serve the request, where the term *best* means that the link between the anycast client and the anycast server has the optimal loads. The link loads can be measured as the number of packets transferred through the link per time unit, which can be obtained by the counters of flow table entries in the OFA switches [26] as specified in "Preliminaries of SDN and OpenFlow," and thus can be captured by the anycast controller through the OpenFlow protocol between the controller and the switches.

In the design of the anycast controller, we also adopt the Floyd-Warshall algorithm [30] to perform routing strategies based on the link loads, where the difference from the routing scheme based on the hop counts is that the links in the network topology are weighted with the loads. Algorithm 1 shows the calculation of the path using the routing strategy based on link loads.

**Algorithm 1: The calculation of the path based on link loads**

**Input**: $Load\_matrix[N][N]$ // representing the link load between node $i$ and node $j$
**Output**: $Path[N][N]$ and $Output\_port[N][N]$ // $Path[i][j]$ denotes the path between node $i$ and node $j$; $Output\_port[i][j]$ represents the output port at node $i$ if the packet destination is node $j$

1.  $N$ is the number of OFA switches;
2.  Initialization $Output\_port[N][N]$;
3.  For $i$ in the range $(1, N)$
4.      For $j$ in the range $(1, N)$
5.          If there exists a link between node $i$ and node $j$
6.              $Output\_port[i][j] = get\_port(link(i, j))$;
7.          Else
8.              $Output\_port[i][j] = NULL$;
9.          EndIf
10.     EndFor
11.  EndFor
12.  $Path = Load\_matrix$;
13.  For $k$ in the range $(1, N)$
14.      For $i$ in the range $(1, N)$
15.          For $j$ in the range $(1, N)$
16.              If $Path[i][j] > Path[i][k] + Path[k][j]$
17.                  $Path[i][j] = Path[i][k] + Path[k][j]$;
18.                  $Output\_port[i][j] = Output\_port[i][k]$;

19.          EndIf
20.       EndFor
21.    EndFor
22. EndFor
23. Return *Path*[*N*][*N*] and *Output_port*[*N*][*N*]

*Address Resolution Module*   The main task of the Address Resolution Module is to generate Address Resolution Protocol (ARP) response packets for anycast requests. Specifically, the anycast controller extracts anycast IP address from the ARP request packets and invokes the Routing Decision Module to select a destination anycast server. The Routing Decision Module then returns the Medium Access Control (MAC) address of the selected anycast server and generates an ARP response packet using the destination MAC address. Finally, the Address Resolution Module invokes the Data Transmission Module to deliver the ARP response to the anycast client.

An anycast client should convert an IP address into a MAC address before communicating with anycast servers in the same AS. In particular, according to the ARP protocol, an IP address can be resolved into only one MAC address in an AS. With an anycast service, there may be more than one anycast servers that share the same anycast address in an AS, and thus, an anycast client who originates an ARP request may receive multiple ARP responses, resulting in ARP resolution collision.

The ARP resolution collision can be resolved by changing the way that ARP requests are sent. In anycast service, only one server is selected to respond to an anycast request, and thus, it does not require all the anycast servers to respond to the ARP request. With the help of the anycast controller, the OFA switches along the path between the anycast client and the given anycast server can be built. The anycast controller can then respond to the ARP request on behalf of the selected anycast server. To avoid the collision, this method ensures that only one ARP response would be generated in the process of ARP resolution.

The entry in the flow table of all OFA switches for the ARP request packet should be added previously to direct this kind of packet to the anycast controller, which will choose an appropriate anycast server with the help of the Routing Decision Module, and then generate an ARP response packet using the selected anycast server's MAC address as the MAC source address of the ARP response packet to send back to the anycast client. Under this circumstance, there is only one ARP response packet sent to the anycast client from multiple anycast servers to accomplish an ARP resolution.

*Data Transmission Module*   The Data Transmission Module is a middleware between OFA switches and the Routing Decision Module and the Address Resolution Module. It receives packets from the OFA switches and transfers the packets into different modules, and vice versa.

The data packets that fail to match the entries in the flow table of an OFA switch will be delivered to the anycast controller. The Data Transmission Module resolves the header of this packet and, according to its protocol type, delivers the packets to the corresponding modules for the further process, as shown in Figure 17.4.

### Numerical Results and Analysis

To evaluate the effectiveness and accuracy of the proposed scheme, we designed the anycast controller using POX and implemented the OFA switches and anycast servers/clients in Mininet [31]. POX is a piece of SDN ecosystem [27], which is a platform used for the rapid development and prototyping of network control software using Python. Mininet is a network emulator that runs a collection of clients/servers, switches, and network links on a Linux kernel. It adopts lightweight virtualization to make a single system over the physical network. The clients/servers in Mininet behaves just like a real machine: one can secure shell (SSH) to it and run arbitrary programs.

In this chapter, we consider the simulation environment with the topology of 100 OFA switches in a 10 × 10 grid structure, and each switch connects with a terminal, where we choose 5% of the terminals as anycast servers, and the others act as anycast clients who originate requests. A POX running on another machine acts as the anycast controller. OFA switches, anycast clients/servers, and anycast controllers are connected by virtual links (VLs) that are generated by Mininet.

Anycast request delay and request loss probability are the two key performance metrics adopted to evaluate the effectiveness and

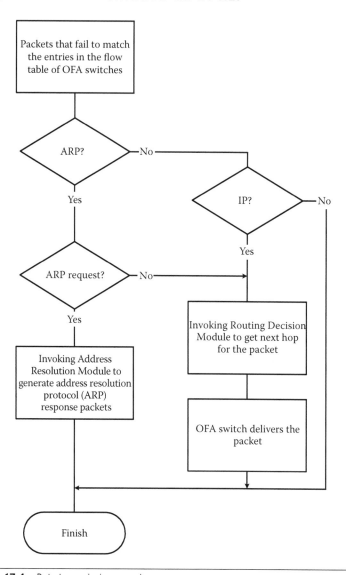

**Figure 17.4** Data transmission procedure.

accuracy of the proposed load-aware anycast scheme. Extensive simulation experiments are conducted under various combinations of the number of OFA switches, the number of anycast clients/servers, the bandwidth of VLs, and the status of background traffic. Each simulation was run until the network reaches the steady state. However, for the sake of specific illustration and without loss of generality, the results are presented based on the parameters shown in Table 17.1.

**Table 17.1**   System Parameters of the Simulation

| SYSTEM PARAMETERS | VALUE |
|---|---|
| Physical machine (PM) | 2.5-GHz Intel Core i5 processor 4-GB 1600-MHz memory |
| Operating system (OS) of PM | Mac OS X 10.7.5 Lion |
| Virtual machine (VM) | Oracle VirtualBox 4.2.4 r81684 |
| VM image | Official Mininet 2.0.0 [32] |
| No. of OFA switches | 100 |
| No. of anycast servers | 5 |
| No. of anycast clients | 95 |
| VL type | Mininet TCLink |
| VL bandwidth | 1 Mbps, 2 Mbps |
| Background traffic | UDP |
| Queue size of OFA switches | 64 packets |
| Queue size of anycast clients | 64 packets |
| Packet size | 64 B |

Figure 17.5 depicts the anycast request delay and request loss probability predicated by the existing hop-based scheme and proposed load-aware scheme against the request generation rate under 1-Mbps bandwidth of the VL. As can be seen from the figure, increasing the request generation rate results in the higher request delay and request loss probability. In addition, as the request generation rate increases, the proposed load-aware anycasting has the lower request delay and loss probability in comparison with those predicted by the existing

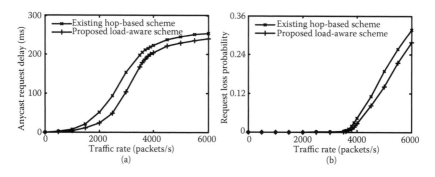

**Figure 17.5**   (a) Anycast request delay and (b) request loss probability predicated by the existing hop-based scheme and proposed load-aware scheme against a traffic rate under 1-Mbps bandwidth without background traffic.

hop-based anycast, especially under a moderate and higher request generation rate.

Figure 17.6 depicts the anycast request delay and request loss probability for both the hop-based anycast scheme and the proposed load-aware anycast scheme against the request generation rate under 2-Mbps bandwidth of VLs and 0.5-Mbps background UDP traffic to mimic the behavior of networks running for a certain period. The results reveal the similar phenomenon as shown in Figure 17.5, although the network has run for some time. The results also emphasize that, as the loads of the network are going moderate, the load-aware anycasting has significant advantage on the request delay and loss probability.

To further evaluate the merit of the proposed load-aware anycast scheme, we consider the case that some VLs are congested in the topology of 100 OFA switches in a 10 × 10 grid structure with 2-Mbps bandwidth of VLs, and each switch connects with a terminal where we choose 2 terminals (i.e., the upper left corner and the bottom right corner shown in Figure 17.7) as anycast servers, and the other 98 terminals act as anycast clients. In particular, we consider the three cases shown in Table 17.2 with VLs 1, 2, and/or 3 added with 2-Mbps background UDP traffic to mimic the congested status of the link. To this end, Figure 17.8 presents the anycast request delay and request loss probability predicted by the existing hop-based anycast scheme and load-aware anycast scheme under the three cases shown in Table 17.2, and the anycast request generation rate is set

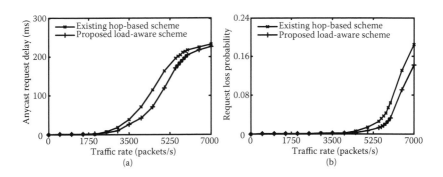

**Figure 17.6** (a) Anycast request delay and (b) request loss probability predicated by the existing hop-based scheme and proposed load-aware scheme against a request generation rate under 2-Mbps bandwidth and 0.5-Mbps background UDP traffic.

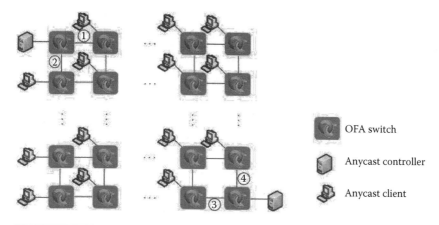

**Figure 17.7**  Network topology in the 10 × 10 grid structure of OFA switches for performance analysis.

**Table 17.2**  Three Cases Considered for Performance Analysis

| CASES | DESCRIPTION |
|---|---|
| Case I | Add 2-Mbps background UDP traffic on VL 1 |
| Case II | Add 2-Mbps background UDP traffic on VLs 1 and 2 |
| Case III | Add 2-Mbps background UDP traffic on VLs 1, 2, and 3 |

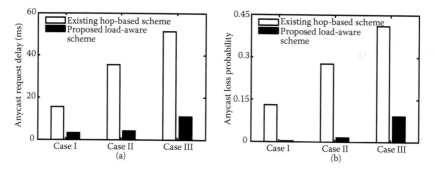

**Figure 17.8**  (a) Anycast request delay and (b) request loss probability predicated by the existing hop-based scheme and proposed load-aware scheme against the three cases shown in Table 17.2.

to be 200 packets per second. From the figure, we can find that the anycast request delay and loss probability predicted by the existing hop-based scheme are greater than those of the proposed load-aware scheme because the hop-based anycasting encountered the congested VLs 1, 2, and/or 3, whereas the load-aware anycast scheme can bypass the congested link and leverage the light-load link to forward packets.

In addition, the performance of the load-aware anycast scheme under Case II does not degrade too much in comparison with that under Case I. This is because at least two alternative links can be shared by the load-aware scheme under these two cases, which alleviates the burden of each VL. In contrast, the performance of the load-aware anycasting under Case III has significant degradation compared with that of Case I and Case II.

### Extension of Anycast Implementation in Interdomain Environment

The anycast has been viewed as a powerful packet addressing and delivery model, and its implementation in the interdomain environment has been widely reported [6,20,33]. In this section, the proposed intradomain anycasting scheme will be extended to the case of interdomain environments with the gradual deployment mode.

In particular, we consider three cases of the configuration and deployment of OFA switches and anycast servers at each AS shown in Figure 17.9: (1) the AS with the deployment of OFA switches and the connected anycast server (see AS 1 and AS 2 in Figure 17.9); (2) the AS with the deployment of OFA switches only (see AS 3 in Figure 17.9); and (3) the AS without the deployment of any OFA switch and anycast server (see AS 4 in Figure 17.9).

The anycast requests originated by an anycast client need to be directed to an OFA switch. In this chapter, we consider the cases that (1) the anycast clients are located at the AS with one or more OFA switches (see AS 3 in Figure 17.9) and (2) the anycast clients are located at the AS without the OFA switch (see AS 4 in Figure 17.9). For both cases, the OFA switches advertise the address prefix for anycast service. Note that only one OFA switch makes the advertisement if multiple switches exist in an AS. For the former case, according to the address prefix advertisement by the OFA switches, the packets with the anycast address as the destination can be directed to the OFA switch in the AS. For the latter case, the anycast requests will be directed to the Border Gateway Protocol (BGP) router in the AS, and the BGP router can then, according to the address prefix advertisement, find the AS that possesses the OFA switches. The anycast requests will then be directed to the BGP router of that AS, and also,

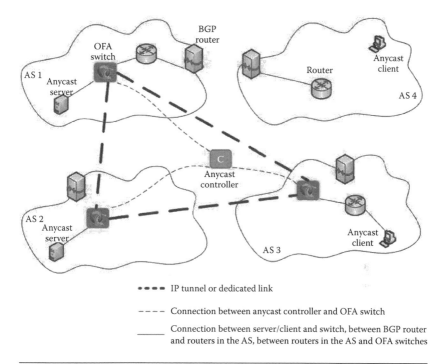

**IP tunnel or dedicated link**

**Connection between anycast controller and OFA switch**

**Connection between server/client and switch, between BGP router and routers in the AS, between routers in the AS and OFA switches**

**Figure 17.9** Implementation of anycast service based on OpenFlow in the interdomain environment.

according to the address prefix advertisement in that AS, the anycast request can be finally directed to the OFA switch.

Once they reach the OFA switch, the anycast requests will be directed to the anycast server according to the entries in the flow table of the OFA switch. If the anycast requests cannot be matched against the entries in the flow table, they will be directed to the anycast controller who can calculate the required routing path based on the design mechanism stated in "The Design of the Anycast Controller" and add the corresponding entries in the flow table of the OFA switches along the path to the anycast server.

### The Merit of the Proposed Anycast Implementation Strategy in the Interdomain Environment

In this section, the proposed scheme will be analyzed in terms of flexibility, feasibility, and scalability. The DNS is a typical example of anycast implementation in interdomain environments. Because the DNS has been widely deployed, its flexibility and scalability has been validated. Therefore, we mainly compare the flexibility between these two schemes.

- Flexibility. Different anycast services should provide diverse strategies on load-aware routing for different scenarios. However, such consideration is quite difficult to be realized for DNS. In contrast, with the proposed interdomain anycast implementation, the strategies for the different load-aware requirements can be readily realized at the anycast controller with a low cost.
- Feasibility. The proposed strategy does not need to change the current network protocols and architecture; however, it needs to deploy OFA switches in the current Internet to gradually support anycast service.
- Scalability. With a new anycast service, the proposed scheme only needs to provide an address in the anycast address space, establish the connection between new anycast servers and the OFA switches, and add the corresponding new entries in the flow tables of OFA switches for this new anycast service by the anycast controller.

Conclusion

This chapter has investigated the anycast implementation in SDN/OpenFlow environments and presented a load-aware anycasting implementation in the OpenFlow networks of intradomain environments. Extensive Mininet experiments have been conducted to validate the effectiveness and accuracy of the proposed load-aware anycast scheme. The results have demonstrated that the performance of the developed anycasting outperforms that of existing solutions in terms of anycast request delay and loss probability. The developed OpenFlow-based anycast scheme for intradomain environments has then been extended to a solution for the interdomain global deployment strategies. The analysis has shown that this strategy can be adopted as an evolvable way for the large-scale deployment of load-aware anycast service for the current Internet.

## Acknowledgments

This work is partially supported by the National Program on Key Basic Research Project (973 Program) under grant no. 2012CB315803, the National Key Technology Research and Development Program of the Ministry of Science and Technology of China under grant no. 2012BAH01B03, the Strategic Priority Research Program of the Chinese Academy of Sciences under grant no. XDA06010201, and the National High-Tech R&D Program of China (863 Program) under grant no. 2013AA013501.

## References

1. Al-Qudah, Z., S. Lee, M. Rabinovich, O. Spatscheck, and J. V. D. Merwe. 2009. Anycast-aware transport for content delivery networks. In *Proceedings of the 18th International Conference on World Wide Web (WWW '09)*, pp. 301–310.
2. Kim, J., X. Lin, N. B. Shroff, and P. Sinha. 2010. Minimizing delay and maximizing lifetime for wireless sensor networks with anycast. *IEEE/ACM Transactions on Networking* 18(2):515–528.
3. Chen, S.-K. and P.-C. Wang. 2011. Design and implementation of an anycast services discovery in mobile *ad hoc* networks. *ACM Transactions on Autonomous and Adaptive Systems* 6(1).
4. Zeng, D., Y. Zhang, Z. Yu, and Y. Qian. 2011. Research and development of anycast communication service. In *Applied Informatics and Communication* 225. Berlin/Heidelberg: Springer, pp. 212–217.
5. Kurian, J. and K. Sarac. 2010. A survey on the design, applications, and enhancements of application-layer overlay networks. *Computing Surveys* 43(1).
6. Ballani, H. and P. Francis. 2005. Toward a global IP anycast service. In *Proceedings of the 2005 Conference on Applications, Technologies, Architectures, and Protocols for Computer Communications (SIGCOMM '05)*, pp. 301–312.
7. Ratnasamy, S. and S. McCanne. 2005. Toward an evolvable Internet architecture. In *Proceedings of the 2005 Conference on Applications, Technologies, Architectures, and Protocols for Computer Communications (SIGCOMM '05)*, pp. 313–324.
8. Vaughan-Nichols, S. J. 2011. OpenFlow: The next generation of the network? *Computer* 44(8):13–15.
9. Protocol oblivious forwarding (POF). http://www.huawei.com/en/about-huawei/newsroom/press-release/hw-258922-sdn.htm.
10. Xie, G., P. He, H. Guan, Z. Li, L. Luo, J. Zhang, Y. Wang et al. 2011. PEARL: A programmable virtual router platform. *IEEE Communications Magazine* 49(7):71–77.

11. Ma, Z.-Y., J. Zhou, and L. Zhang. 2009. A scalable framework for global application anycast. In *Proceedings of the 2nd International Conference on Information and Computing Science (ICIC '09)*, pp. 250–253.

12. Chellouche, S. A., D. Negru, E. Borcoci, and E. Lebars. 2011. Context-aware distributed multimedia provisioning based on anycast model toward future media Internet. In *Proceedings of the 2011 IEEE Consumer Communications and Networking Conference (CCNC '11)*, pp. 880–885.

13. Bhattacharjee, S., M. H. Ammar, E. W. Zegura, V. Shah, and F. Zongming. 1997. Application-layer anycasting. In *Proceedings of the 16th Annual Joint Conference of the IEEE Computer and Communications Societies (INFOCOM '97)*, pp. 1388–1396.

14. Plaice, J., P. Kropf, P. Schulthess, J. Slonim, S. Yu, W. Zhou, F. Huang et al. 2002. An efficient algorithm for application-layer anycasting. In *Distributed Communities on the Web* 2468. Berlin/Heidelberg, Germany: Springer, pp. 74–83.

15. Zhou, Z., G. Xu, C. Deng, J. He, and J. Jiang. 2009. A random selection algorithm implementing load balance for anycast on application layer. In *Proceedings of the 2009 International Symposium on Web Information Systems and Applications (WISA '09)*, pp. 444–448.

16. Garcia-Luna-Aceves, J. J. and B. R. Smith. 2002. System and method for using uniform resource locators to map application layer content names to network layer anycast addresses. U.S. 2002/0010737 A1.

17. Zegura, E. W., M. H. Ammar, F. Zongming, and S. Bhattacharjee. 2000. Application-layer anycasting: A server selection architecture and use in a replicated Web service. *IEEE/ACM Transactions on Networking* 8(4):455–466.

18. Alzoubi, H. A., S. Lee, M. Rabinovich, O. Spatscheck, and J. V. D. Merwe. 2011. A practical architecture for an anycast CDN. *ACM Transactions on the Web* 5(4).

19. Wendell, P., J. W. Jiang, M. J. Freedman, and J. Rexford. 2010. DONAR: Decentralized server selection for cloud services. *ACM SIGCOMM Computer Communication Review* 40(4):231–242.

20. Katabi, D. and J. Wroclawski. 2000. A framework for scalable global IP anycast (GIA). In *Proceedings of the Conference on Applications, Technologies, Architectures, and Protocols for Computer Communication (SIGCOMM '00)*, pp. 3–15.

21. Lenders, V., M. May, and B. Plattner. 2006. Density-based vs. proximity-based anycast routing for mobile networks. In *Proceedings of the 25th IEEE International Conference on Computer Communications (INFOCOM '06)*, pp. 1–13.

22. Wang, X. and H. Qian. 2009. Design and implementation of anycast communication model in IPv6. *International Journal of Network Management* 19(3):175–182.

23. Stevens, T., M. de Leenheer, C. Develder, F. de Turck, B. Dhoedt, and P. Demeester. 2007. ASTAS: Architecture for scalable and transparent anycast services. *Journal of Communications and Networks* 9(4):457–465.

24. Othman, O. M. M. and K. Okamura. 2012. On-demand content anycasting to enhance content server using P2P network. *IEICE Transactions* 95-D(2):514–522.
25. Lantz, B., B. Heller, and N. McKeown. 2010. A network in a laptop: Rapid prototyping for software-defined networks. In *Proceedings of the 9th ACM SIGCOMM Workshop on Hot Topics in Networks (Hotnets-IX '10)*.
26. OpenFlow switch specification. http://www.opennetworking.org/images/stories/downloads/specification/openflow-spec-v1.3.0.pdf.
27. POX. http://www.noxrepo.org/pox/about-pox/.
28. Liu, L. and R. C.-W. Wong. 2011. Finding shortest path on land surface. In *Proceedings of the 2011 ACM SIGMOD International Conference on Management of Data*, pp. 433–444.
29. Penner, M. and V. K. Prasanna. 2006. Cache-friendly implementations of transitive closure. *Journal of Experimental Algorithmics* 11.
30. Ridi, L., J. Torrini, and E. Vicaro. 2012. Developing a scheduler with difference-bound matrices and the Floyd-Warshall algorithm. *IEEE Software* 29(1):76–83.
31. Mininet. http://www.yuba.stanford.edu/foswiki/bin/view/OpenFlow/Mininet.
32. Mininet virtual machine image. http://www.github.com/mininet/mininet.
33. Wu, C.-J., R.-H. Hwang, and J.-M. Ho. 2007. A scalable overlay framework for Internet anycasting service. In *Proceedings of the 2007 ACM Symposium on Applied Computing*, pp. 193–197.

# Index

Page numbers followed by f and t indicate figures and tables, respectively.